# INSECT SEX PHEROMONES

# INSECT SEX PHEROMONES

## MARTIN JACOBSON

Agricultural Environmental Quality Institute
Agricultural Research Service
U.S. Department of Agriculture
Beltsville, Maryland

1972

*ACADEMIC PRESS* New York and London

ACADEMIC PRESS, INC.
111 Fifth Avenue, New York, New York 10003

*United Kingdom Edition published by*
ACADEMIC PRESS, INC. (LONDON) LTD.
24/28 Oval Road, London NW1

LIBRARY OF CONGRESS CATALOG CARD NUMBER: 72-77360

PRINTED IN THE UNITED STATES OF AMERICA

*To*

DEBRA GAIL *and* LINDA HOPE

# CONTENTS

## Chapter VIII    Influence of Age of the Insect on Production of and Response to Sex Pheromones

## Chapter IX    Influence of Time of Day on Sex Pheromone Production and Mating

## Chapter X    Collection, Isolation, and Identification of Sex Pheromones

# PREFACE

The first edition of this reference work, "Insect Sex Attractants," was published in 1965 by Wiley (Interscience). To date it is still the only book of its kind on the subject. However, the surge of interest in the insect sex pheromones as evidenced by the tremendous progress and by the phenomenal additional publications since 1965 has made it imperative to bring the entire subject up-to-date in this revised and expanded version.

In my earlier Preface I stated, "It is my sincere hope that this book will provide an incentive for greater discoveries in this fascinating field." An unprecedented increase in the number of species shown to produce sex pheromones (more than double the number known in 1965), in the number of species whose sex pheromones have been identified (37 now as compared to 3 in 1965), and the extensive literature in the field (about 1400 references as compared to 425 in 1965) indicate that this wish has been fulfilled.

The discovery that many sex pheromones are sexually excitatory rather than attractive has prompted me to substitute the more accurate and encompassing term "pheromones" for the term "attractants" in the title of this edition. Not only have all of the original chapters been extensively enlarged and revised, but a new chapter on test methods and responses has been added. This expanded review of the world literature should be extremely valuable to research and economic entomologists, insect physiologists, chemists, and ecologists.

Photographs and other illustrations are used with the kind permission of the copyright owners, with the source of each indicated. All listed references have been consulted directly. References which have appeared or were noted after the manuscript was prepared have been assigned supplementary numbers, enabling me to include literature which appeared up to a few months prior to publication.

I acknowledge a debt of gratitude to my wife, Nettie, for painstakingly typing the bulk of the manuscript. A special note of thanks is due a number of experts on insect taxonomy at the National Museum of the Smithsonian Institution, Washington, D.C., for checking many of the insect names.

Martin Jacobson

# CHAPTER I

# INTRODUCTION

Insects have managed to persist in hostile surroundings because they have developed extraordinary adaptations or abilities, one of which is a highly specialized sense of smell. Because many insects depend on their sense of smell for survival, they can frequently be attracted to a trap by a chemical for detection purposes, to a toxicant that destroys them, or to a substance that makes them incapable of fertile mating (*599*).

Attractants may be classified as sex, food, or oviposition lures. The type of lure is inferred or deduced from insect behavior, and assignment is frequently uncertain. A chemical is probably a sex attractant if it brings to it an insect, which then assumes a mating position or attempts to mate with the chemical or with an object on which the chemical has been placed. This definition excludes a number of substances, mainly food attractants, which lure only one sex but do not elicit a sexual response. Sex attractants, usually released by a female to lure a male, are important, if not essential, links in the process by which the sexes locate each other for mating. Although odors released by female insects are usually for the purpose of attracting males from a distance, they may also serve to sexually excite the male before copulation and to evoke a courtship response. Sexual odors released by males are primarily for the purpose of sexually exciting the female, making her more receptive to the male's advances (aphrodisiacs). However, species

1

are known in which the males produce distance attractants for the female. These chemical messengers are also called "assembling scents" *(663)* and "sex pheromones," from the Greek *pherein* (to carry) and *horman* (to excite, stimulate) *(647)*. The term pheromone has been attacked by Kirschenblatt *(676)* as being etymologically incorrect, as it gives no clue to its exact derivation. Kirschenblatt *(674, 675)* previously proposed the term "telergones," from the Greek *tele* (afar) and *ergon* (action), to designate all biologically active substances secreted by animals into their environment which influence other organisms ("these substances are products of external secretion and differ principally by their importance from hormones, which display their physiological action within the organism producing them") *(674)*. Micklem *(787)* cited the same objection to the term "pheromone" and suggested that it be changed to "pherormone." In replying to Micklem, Karlson and Lüscher *(649)* gave a new etymological explanation for their term, stating that the ending "mone" is regarded as a proper suffix used in such scientific terms as "hormones," "gamones," and "termones." "Pheromone" is now commonly used and widely accepted to include those substances secreted by an animal to influence the behavior of other animals of the same species *(1268)*. Brown *et al.* *(213)* have recently proposed the term "allomone" to include those chemical substances produced by an organism which evoke a behavioral or physiological reaction in an organism of another species, and the term "kairomone" as a transspecific chemical messenger of benefit to the recipient rather than to the producer.

The use of sex pheromones by organisms other than insects has been conclusively demonstrated in algae *(805, 878)*, nematodes *(484, 485)*, spiders *(532)*, crustaceans *(77, 323, 679, 1011)*, fishes *(1187)*, and mammals such as dogs *(354)*, cattle *(354)*, deer *(810)*, mice *(158, 1129)*, hamsters *(814)*, and primates *(321, 784–786, 1237)*. Indeed, it is possible that sex pheromones play a part in the courtship and reproduction of humans as well *(66, 306, 1177a)*. Excellent reviews of this subject are those by Whitten *(1263)*, Michael and Keverne *(784)*, Bruce *(214, 215)*, and Ralls *(906)*.

General reviews on various aspects of the subject of insect pheromones may be found in references *112, 144–146, 151–154, 229, 244, 247, 305, 375, 375a, 422, 468, 479, 501, 521, 525, 529, 530, 545, 550, 574, 587, 594, 596, 599, 601, 636a, 639, 644, 669, 695, 700, 707, 713, 789, 807, 907, 932, 978, 979, 1092, 1156, 1188, 1254, 1268, 1272,* and *1292*. Brief reviews of insect sex pheromones are found in references *64, 78, 150, 192, 249, 316, 544, 590, 646, 650, 699, 738, 756, 797, 798, 811, 817,* and *850*. The assembling of various moths and butterflies has been reviewed by Poulton *(883)*.

Reviews in references *37, 122, 128, 230, 592, 645, 703, 716, 812, 831a, 900* (140 references), *916, 1112,* and *1131* deal mainly with the chemistry of insect sex pheromones; reference *1079* deals with their specificity (or lack of it), and references *249, 254, 440, 449,* and *1269* are devoted to honeybee pheromones. Sex pheromones among the Lepidoptera are reviewed mainly in references *540, 606,* and *757,* and those among the Coleoptera are discussed in references *78, 223, 774, 997, 1100, 1133a, 1206,* and *1286.*

# CHAPTER II

# OCCURRENCE OF SEX PHEROMONES IN THE FEMALE

As long ago as 1837, von Siebold (*1097*) recognized that a pair of appendages, sometimes colored, opening into the vagina of the females of some insect species may act as an attractant for males. He surmised that the odor emitted by a female insect probably functions to entice the male, while that emitted by a male may be used as a stimulus in copulation (aphrodisiac).

ACARINA

*Amblyomma americanum* (L.), lone star tick
*Amblyomma maculatum* Koch, Gulf Coast tick
*Dermacentor variabilis* (Say), American dog tick
Females of these species produce a pheromone that attracts males of the respective species. Males respond only after reaching a state of maturity initiated by feeding (*138*).
*Panonychus ulmi* (Koch), European red mite
Males tend to aggregate around quiescent female deutonymphs to await the latter's eclosion (*899*).
*Tetranychus urticae* Koch, two-spotted spider mite
Males are attracted strongly to quiescent deutonymphs, remaining until emergence of the adult female, when mating occurs. They are also attracted to ether extracts of the deutonymphs (*307, 308*).

ORTHOPTERA

*Blaberus craniifer* (Burmeister), giant death's head roach
Virgin females produce a volatile sex pheromone which attracts males and elicits antennal waving, alertness, and locomotion toward the females (*107*). It is also interspecifically effective in eliciting male courtship behavior in *Blaberus giganteus* and *Byrsotria fumigata*.

*Blaberus giganteus* L.
Virgin females may produce a volatile attractant for males, releasing courtship behavior (*107*).

*Blatta orientalis* L., oriental cockroach
Females appear to have a nonvolatile sex pheromone present on their body surface which facilitates sex recognition and releases male courtship behavior. The wing-raising display is less complex and variable than that of the male *Periplaneta americana* (*114*).

*Blattella germanica* L., German cockroach
Virgin females produce a substance attractive only to males. It is unattractive to males of *Blatta orientalis* and *Shelfordella tartara* (*1237*).

*Byrsotria fumigata* (Guérin)
Virgin females produce a volatile sex attractant which enables males to perceive them at a considerable distance and elicits alertness, antennal waving, and wing "pumping" (*106, 107, 110, 981*). A number of gynandromorphs are found to produce the female sex pheromone (*116*).

*Leucophaea maderae* (F.), Madeira cockroach
Smyth (*1117*) claims to have collected a volatile material from females that increases the incidence of courtship by males.

*Mantis religiosa* (L.), praying mantis
Caged virgin females can lure large numbers of males from a distance of up to 100 m between 8:30 AM and 1:00 PM (*658*).

*Nauphoeta cinerea* (Olivier), cockroach
The sex pheromone, if it exists, appears to be a nonvolatile substance on the surface of the female. Male display (raising of the wings) is readily evoked, even by unreceptive females (*107*).

*Periplaneta americana* (L.), American cockroach
Females emit an odorous attractant for the male. The substance adheres to paper or other materials with which the females come in contact (*985*). Virgin females, as well as filter papers exposed to them, cause male alertness, antennal movement, searching locomotion, and vigorous wing flutter (*107, 985*).

The wing-raising display, which is released much more readily in groups of males than in single males, is used as the single criterion of response in a bioassay method developed by Wharton *et al.* (*1258, 1259*). However,

Jacobson and Beroza (*600*) have shown that a number of organic compounds, including several that are repellent to males (such as amyl acetate), will elicit wing-raising, thus making it mandatory that an accurate bioassay show a combination of intense excitement, wing-raising, *and attempts to copulate with one another*. The mating urge is so powerful in this insect that males starved for 4 weeks in the laboratory and then given their choice of the female sex pheromone or food, invariably responded to the sex pheromone until they were near death (*764*).

The sex attractant is produced principally by virgin females and sporadically by mated females. Nymphs ordinarily do not produce the attractant, and a newly emerged female produces very little of the substance at first. During this nonproductive phase, the female does not attract the male, and mating does not occur. Maximum production is attained by the second week after eclosion. Carbon dioxide anesthetization and manual manipulation reduce the production of attractant somewhat (*1260*). Attractant synthesis is drastically depressed within 18 hours after copulation, which accounts for the sporadic production of the substance by mated females.

The female sex pheromone is effective in releasing courting behavior in males of other *Periplaneta* species and in males of *Blatta orientalis*, but not in *Eurycotis floridana*, *Leucophaea maderae*, or *Nauphoeta cinerea* (*107*). The sex pheromones of other species of *Periplaneta* also appear to be interspecifically effective within the genus (*114, 980*).

*Periplaneta australasiae* (Fabr.), Australian cockroach

*Periplaneta brunnea* (Burmeister)

Females of these 2 species produce pheromones that attract and sexually excite males. The same behavior is elicited in males by exposure to filter papers over which females have crawled (*107, 114*).

*Periplaneta fuliginosa* (Serville)

Virgin females apparently produce a chemical substance which acts as a releaser of courtship behavior in males, but this has not yet been proved (*107, 114*).

## Hemiptera

*Dysdercus cingulatus* Fabr., red cotton bug

A cardboard box (with pinhole perforations) containing virgin females was placed 15 inches from a group of males. The males were quickly attracted to the box, and vibrated their wings and raised their antennae. A corresponding box without females did not attract males, but a box in which a female had been confined 5 days earlier acted as an attractant (*837*).

*Lygus hesperus* Knight, lygus bug

Virgin females in field traps attracted males. Mating reduces the attractiveness of females for only a few days (*1153*).

*Lygus lineolaris* (Palisot de Beauvois), tarnished plant bug

Traps baited with field-collected females were attractive to males in the field (*1020*).

*Rhodnius prolixus* (Stal.)

A volatile substance produced by mating pairs, but not by females or males separately, sexually stimulates males. In the absence of females, the stimulated males attempt to mate with other males. Feeding is a prerequisite to pheromone production; males and females copulate only after a blood meal, and unfed males will not respond to the pheromone. Males are stimulated only in complete darkness (*83*).

HOMOPTERA

*Aonidiella aurantii* (Maskell), California red scale

Sexually mature virgin females, as well as their crushed bodies, are attractive to males within minutes of the latter's emergence. The pheromone is continuously present in the female, who may release or withhold it (*1018, 1173*). Females become unattractive within 24 hours after insemination (*1175*). Females reared in the laboratory on potatoes are as attractive as those reared on lemons, but they are slower to mature (*930*). Males of both a laboratory strain and a native strain collected from lemons at Corona, California, showed the same degree of copulatory response to a pheromone extract obtained from virgin females of the laboratory strain. Males of the laboratory strain in free flight showed better response to virgin females of the native strain than to those of the laboratory strain, but they were even more responsive to the pheromone extract (*1172*).

*Matsucoccus resinosae* Bean and Godwin, red pine scale

Males in large flight chambers were greatly attracted to virgin females held in screen cages and attempted to copulate with the screen. Ultraviolet light was attractive to males except in the presence of females. Virgin females in petri dishes spun a small amount of silky fluff on the posterior of their bodies. When removed, the fluff was very attractive to males, and they attempted to copulate with it. Copulation was also attempted with filter papers on which females had rested overnight (*350*).

*Myzus persicae* (Sulzer), green peach aphid

Although sex-pheromone production by the females had been suspected, it could not be demonstrated by the use of olfactometers and field traps baited with live males, females, or their extracts. Males sometimes congregate around a copulating pair. It was concluded that females may produce a sex pheromone, but it is not necessary to elicit copulatory behavior in males, and that aggregate response contributes to the congregation of aphids (*1166*).

*Planococcus citri* (Risso), citrus mealybug

Females of this species are wingless, whereas males are winged. Pentane extracts of sexually mature females applied to filter papers were attractive to males but failed to elicit antennal positioning or copulatory attempts, suggesting that some factors operative in normal recognition or sexual behavior were either absent or present in too low a concentration to be effective. Males ignored a female corpse which had been extracted with pentane, but they rapidly aggregated around the corpse after one female equivalent of extract had been applied to it. Extracts stored in a freezer in the absence of nitrogen rapidly lost activity (*477*).

*Schizaphis borealis* Tambs-Lyche, aphid

Organs (pseudorhinaria) situated on the hind tibiae of the oviparous female emit a chemical pheromone attractive to males. The attractant, emitted only during the active copulatory period of the female, is perceived by organs (rhinaria) on the male antennae, even as early as the last 2 larval instars. Males do not emit any substance attractive to other males (*861, 861a*). The pheromone, which is attractive only over a short distance, is not very specific, eliciting responses from males of several other species of the same genus (*861a*).

## DIPTERA

Among Psychodid males, a male that has just mated with a female is very attractive to other males, who attempt to copulate with him (*932*).

*Culiseta inornata* (Williston), mosquito

Copulation usually occurs immediately after female emergence. Field observations and laboratory experimentation have shown that a volatile chemical substance (or substances) is involved in mating behavior. Males will attempt to copulate with dead females, and when exposed to an extract of virgin females they show increased sexual activity characterized by excited flight, searching, and copulatory attempts with other males. Males responded with sexual activity to filter paper strips on which females had rested, and female extracts lured males into traps (*682, 683*).

Neither female flight sounds nor male antennae are necessary for mating, even in darkness, nor is light essential for mating. Males were excited by the vapors of benzene or ether but did not respond sexually, nor were they excited by the observer's breath or air drawn through a container of females (*683*).

*Deinocerites cancer* Theobald, crabhole mosquito

The females are autogenous and are located for mating by means of tactile and chemical responses. Seven-hour-old males are able to copulate. Older males are attracted to females even before the latter are out of the pupal case; copulation occurs on the water as soon as, or even before, the females are free of the pupal case. Adult males must be attracted to female pupae through a chemical emanation. The empty female pupal skin is attractive to males, who attempt to copulate with it for as long as 1 hour after eclosion (500). An emerging female elicits a strong male response up to 15 cm away and males fight for possession (357, 892). Attraction ceases as soon as mating is established.

*Drosophila melanogaster* Meigen, vinegar fly

According to Spieth (1130), Sturtevant found that pairs of this species could be induced to copulate much more readily if the glass vial in which they were placed had just been occupied by another pair of courting individuals. Apparently the courting actions of the first pair resulted in the release of a substance that served as a stimulus for the second pair. However, Ewing and Manning (388, 749) were unable to confirm this. Several days elapse before a mated female is receptive to another male (538).

Shorey and Bartell (1081) have recently shown that sustained courtship behavior, including wing vibration, resulted when a male oriented toward a female. Exposure to the odor of 10 females reduced the average time of initiation of courtship for single males confined with 1 female. Similar increased sexual activity in males resulted from exposure to the odor of fresh sheep's liver (1082).

*Glossina morsitans orientalis* Vanderplanck, tsetse fly

Male flies were not attracted to air passed over virgin females or to extracts of whole female bodies when tested in a laboratory olfactometer. Similar numbers of wild males were found on oxen baited or unbaited with mature virgin females (338).

*Lucilia cuprina* (Wied.), Australian sheep blowfly

Males exposed to an air stream which had passed over females remote from them showed substantial increases in sexual activity (99). Similar behavior resulted from exposure to the odor of banana (aggregation site) (1082).

Volatile, low-molecular-weight fatty acids and their esters, extracted from adult blowflies with methylene chloride, probably contain the pheromones responsible for their sex and ovipositing behavior, since they most closely represent the cuticular secretions of the insect (457). Oviposition depends in part on responses of females to stimulation by a chemical pheromone released by aggregated females (118).

*Musca domestic* L., housefly

Tests conducted by Rogoff (968, 969) using an olfactometer or simulated, treated fly models (pseudoflies) indicated that females produce 1 or more volatile chemical substances which can elicit mating behavior patterns in males. The pheromone is present in the heads, thoraxes, and abdomens of mature females; it is also found in their eggs and in minor amounts in males. It is soluble in benzene and is species specific; extracts of *M. autumnalis* (face fly) or *Stomoxys calcitrans* (stable fly) females do not affect the behavior of male houseflies. Blind males were able to locate pheromone-treated pseudoflies, thus demonstrating a directional response to the pheromone (*970*).

Murvosh (*815, 816*) reported that his laboratory attraction tests showed that the female emits a chemical substance with a low order of attractiveness to both males and females. Virgin males were not attractive to other males, although the virgin males often attempted to mate with other males. Virgin females less than 24 hours old attracted few or no males, but 7-day-old virgin females, alive or freshly killed, consistently attracted a small percentage of virgin males. In general, while the evidence of Murvosh does suggest the presence of some type of a female sex attractant, its activity is undoubtedly of a very low order.

Mayer and Thaggard (*762*) succeeded in extracting from live virgin females, contaminated holding cylinders, and dead and mated females an olfactory constituent with a low order of attraction for males and none for females, as determined in olfactometric tests. These investigators feel that the attraction is probably due to housefly defecation and not to a pheromone per se, since defecation of stable flies is also attractive to houseflies. This was further substantiated by Mayer and James (*761*), who were able to obtain active extracts with methanol or benzene from fecally contaminated, gauze fly-cage covers and from the virgin females themselves. The attractant reported by Murvosh and by Mayer and Thaggard probably has no direct connection with the sexual excitant reported by Rogoff.

*Phytophaga destructor* (Say), hessian fly

Females in small field cages attracted large numbers of males upwind from a distance of 10–15 feet (*285*).

*Stictochironomus crassiforceps* (Kieff.)

Males respond much more readily to females when the latter are alive and active. It is believed that this response is olfactory, with a supplementary tactile stimulus involved (*1162*).

*Tipula paludosa* Meigen, marsh crane fly

Laboratory and field experiments suggest that males detect a mating stimulus only in close proximity (about 1 cm) to a female. The source of the pheromone appears to be the anterior part of the female rather than the abdomen (*1196*).

ISOPTERA

*Paraneotermes simplicicornis* (Banks), desert damp-wood termite

Female alates frequently assume a calling attitude soon after alighting on vertical or horizontal surfaces and males soon respond. The female abdomen is typically raised at a 25-degree angle from the horizontal with the tip slightly downturned and roughly parallel with the surface (*831*).

*Reticulitermes arenincola* (Goellner), termite

*Reticulitermes flavipes* (Kollar), eastern subterranean termite

After flight, females of both species attract males by odor. When the male touches the female, she lowers her abdomen and is followed in tandem. Males also follow the severed tip of the female abdomen, or other males if they were once attracted by a female. The odor is detected by the male's antennae (*380*).

NEUROPTERA

*Agulla adnixa* (Hagen), snakefly

*Agulla astuta* (Banks), snakefly

*Agulla bicolor* (Albarda), snakefly

Males and females, placed together in a container by species, showed excited running and vibrating of the tip of the abdomen in an up-down rhythm or circular stretching of the abdomen. The excitement was also noticed when a male was introduced into a jar from which a female had just been removed (*3*).

SIPHONAPTERA

*Echidnophaga gallinacea* (Westwood), sticktight flea

The female abdominal tip produces an odorous substance attractive to the male, who senses it by means of receptors in his palpi, since removal of the palpi prevents location of the female. Although females must mature for a time (3 days) before they become attractive to males, the latter may copulate immediately after emergence (*1160*).

COLEOPTERA

In his extensive review of female and male scent glands, Richards (*932*) quotes Lengerken to the effect that the degenerate female driid beetles are very attractive to males, which can be "assembled" just as saturniid moths. The male has large branched antennae, whereas those of the female are simple in structure. In numerous beetle species, the male possesses large antennae and uses these organs to find the female.

Evers (*387*) reported that the elytral and cranial organs of male Malachiidae, a family of tiny beetles usually found in the tropics, are involved in
male and female love play, and Matthes (*33, 755*) found that the females
attract the males during the mating season.

Silverstein (*1098*) refers to his work with sex pheromones produced by
female dermestid beetles, *Attagenus megatoma*, to attract the male.

*Agriotes ferrugineipennis* (LeConte), click beetle

The female is attractive to the male. The attractant can be extracted
from female abdomens with ether or 70% ethanol (*733*).

*Amphimallon majalis* (Razoumowsky), European chafer

After numerous laboratory and field tests with live females and extracts
prepared from them with acetone, benzene, gasoline, xylene, boiling water,
and ethyl alcohol, Roelofs *et al.* (*965*) concluded that a chemical attractant
or excitant for males is probably not produced by this insect. The possible
presence of agents masking attraction in the crude extracts was apparently
ruled out by column chromatography of the extracts on Florisil.

*Attagenus megatoma* (Fabr.) [synonym: *A. piceus* (Olivier)], black carpet
    beetle

A volatile substance attractive to males has been obtained from adult
virgin females (*35, 223, 224, 1104*). The attractant is released within 24
hours after emergence, with maximum release occurring about 3–4 days
after emergence (*224*). The location of the production glands is unknown
(*35, 224*). See Chapter X for the isolation and identification of this sex
attractant.

*Blastophagus piniperda* L., bark beetle

Extracts of abdomens from emergent females elicited no response from
females, but showed some indication of male attraction (*832*).

*Callosobruchus maculatus* F., cowpea weevil

A sex pheromone was obtained by passing a stream of air through a flask
containing live virgin females and condensing the stream at low temperature
(*337*).

*Ceruchus piceus* Weber, stag beetle

Recognition of a female odor triggers immediate mounting and copulatory
actions by males. During the peak of emergence it is common to see masses
of struggling males on a female (*753*).

*Costelytra zealandica* (White), grass grub beetle

From field tests conducted in 1967 with caged live females, Kelsey
(*659*) concluded that they can attract adult males from distances up to
200 yards. The idea for the tests was suggested in 1956 when 67 males
were found associated with 1 female in a heavily infested paddock. The
presence of a chemical sex attractant in the adult females was demon-

strated in the laboratory in 1969 by Henzell *et al.* (*542*) using olfactometric tests. Females did not attract females, and males attracted neither females nor males.

*Ctenicera aeripennis destructor* (Brown), click beetle, prairie grain wireworm

Doane (*352*) was the first to suggest the presence of a sex attractant in this species when he reported that males converged strongly on caged virgin females in the field. A single male placed in such a cage immediately sought out and mated with the female, who was hidden in a crack just below the soil surface under the cage. Lilly and McGinnis (*733*) were able to extract the attractant from the abdomens of virgin females with ether or 70% ethanol.

*Ctenicera sylvatica* (Van Dyke), click beetle

The substance attractive to males can be extracted from female abdomens with ether or 70% ethanol (*733*).

*Diabrotica balteata* (LeConte), banded cucumber beetle

A sex attractant produced by females by the time they are 10 days old lures males in the field from as far as 40 feet. Males responding to the lure rise from the plants in which they are resting and approach upwind in a characteristic hovering flight. When air movement is gentle and steady, they locate the source with little difficulty. When a gusty wind is blowing, they frequently lose the scent and wander off in the wrong direction. Unmated females remain attractive to males for as long as 70 days. Most females cease to be attractive after 1 mating and none are attractive after 2 matings. The attractant may be extracted from the female abdomens with ethyl alcohol; extracts of heads and thoraxes and of filter papers on which females have crawled give negative results. Under ideal conditions, a 10-female equivalent of the abdominal extract will elicit a response from males up to 49 feet from the lure, but few respond at this distance. At temperatures below 65°F, both sexes are inactive and there is little response. No consistent, recognizable response from males caged indoors has been detected (*322*).

*Dorcus parallelus* (Say), stag beetle

Recognition of a female odor triggers immediate mounting and copulatory actions by males. During the peak of emergence it is common to see masses of struggling males on a female (*753*).

*Dytiscus marginalis* (L.)

Blunck is reported by Hesse and Doflein (*548*) to have found that females are highly attractive to males.

*Hemicrepidius decoloratus* (Say), wireworm

Emergence of males precedes that of females. Males have been observed congregating around copulating pairs, suggesting that females or copulating pairs produce a sex attractant (*527*).

*Hemicrepidius morio* (LeConte)

Screen cages containing Douglas-fir bark or logs, placed on the ground in the field, attracted large numbers of this species, 99% of which were males. Although the evidence would not permit a definite conclusion to be drawn, Chapman (*291*) theorized that females had emerged in these cages and were releasing an attractant for the males.

*Hylecoetus dermestoides* (L.)

The male detects the female from a distance by means of his maxillary palpi, which are well supplied with nerve stalks (*451*).

*Lasioderma serricorne* (F.), cigarette beetle

Males in laboratory colonies locate females in 30 seconds or less, suggesting the presence of sex pheromones (*1186a*).

*Leptinotarsa decemlineata* Say, Colorado potato beetle

The scent of female beetles attracted males in a wind tunnel (*1264*). The scent is detected by the 2 terminal segments on the male antennae.

*Limonius agonus* (Say), eastern field wireworm

Females are usually found in crevices in the soil, with only the abdomen protruding. They lure the males, who circle on the ground waving their antennae (*130*).

*Limonius californicus* (Mann.), click beetle, sugar-beet wireworm

Males are attracted in large numbers to newly emerged females (*1075*). An extract prepared from 2 dead virgin females with 70% ethyl alcohol caused male excitation, but an extract of fertilized females failed to elicit a response. Microscope slides moistened with the active extract attracted numerous males upwind in an infested field within 10 seconds from as far away as 40 feet. The males moved rapidly toward the slides, crawling and flying excitedly over them and repeatedly extruding their genitalia. In laboratory tests, a positive response was obtained with a female abdominal extract but not with an extract of the heads and thoraxes (*732*). As little as 0.4% of the ether extract of a single female abdomen frequently elicited a maximum sexual response from males in a laboratory olfactometer (*733, 734*). See Chapter X for the isolation and identification of the attractant.

*Limonius canus* LeConte, Pacific Coast wireworm

A sex pheromone extracted from whole virgin females was highly exciting to males sexually when the latter were offered glass rods that had been dipped in the extracts (*833*).

Although the males of *L. californicus* responded only to extracts of females of that species, males of *L. canus* responded to extracts of females of both species (*734*).

*Limonius* sp., wireworm

Traps containing females attracted males in enormous numbers soon after emergence, but the catches dropped off as soon as numerous wild females appeared (*711*).

*Lucanus capreolus* (L.), stag beetle
*Lucanus placidus* Say, stag beetle
*Platycerus virescens* (Fabr.), stag beetle

Recognition of a female odor triggers immediate mounting and copulatory action by males of these species. During the peak of emergence, it is common to see masses of struggling males on a female (*753*).

*Melolontha vulgaris* (Fabr.)
*Pachycarpus cormutus* (Olivier)

Females of these species are wingless and produce a sex attractant that brings males to them (*526, 620*).

*Oulema melanopus* (L.), cereal leaf beetle

Females require a period of diapause to attain sexual maturity; this diapause can be prevented or terminated with hormonal treatments. Males of any age are sexually mature, but they are inactive until placed in confinement with receptive females. When removed from this confinement, they again become inactive. The males are able to differentiate between active receptive females and inactive unreceptive females (*309*).

*Photinus* sp., firefly

Krammerer (*640*) enclosed female fireflies in a cardboard box which could not be penetrated by light signals. He also killed females and crushed them, so that again there was no glow. In both cases, males swarmed around the box and the remains. Sight and smell are apparently active in attraction.

*Phyllophaga lanceolata* (Say), June beetle

Mating begins about daylight, when the adults emerge from the ground. Although males may feed on the same leaf with a female, they are not attracted to her until she extrudes her genitalia. When the genitalia are extruded, males within a radius of 15–20 yards fly toward her. Exposure of a smashed female initiated male flight. Females did not respond to a calling female. A female crushed and thrown on the ground when the sun was obscured by a cloud attracted only a few nearby males, but when the sun reappeared numerous males became active and flew toward the injured female. In the absence of a breeze, males within a radius of 15–20 feet flew toward the point of origin of the attractant; in a breeze, only male beetles within 3–10 feet were attracted from the windward side, whereas those from the leeward side were attracted from a distance of 30–40 feet (*1192*).

*Pleocoma dubitalis dubitalis* Davis, rain beetle

Males attracted by females in the soil dig down to mate with them (*943*).

Numbers of males are attracted to traps in the soil baited with live females (*399*).

*Pleocoma minor* Linsley

The winged males seek out the flightless females in the soil. Zwick and Peifer (*1337*) buried to soil level polystyrene traps containing a live female and confined them under a tight-fitting wire mesh screen baffle. An overlapping plastic container with a top 2 inches in diameter permitted males to drop into the traps. Such traps were very effective in luring males.

*Pleocoma oregonensis* Leach

One or more males may be found entering a burrow occupied by a female. In a number of instances, 7–9 males were found burrowing down to a female, with 2 or more being crushed in the process (*377*).

*Polyphylla decemlineata* (Say), ten-lined June beetle

Males were attracted to virgin females at heights of 8–18 feet on trees, and to crude ether or 70% ethanol extracts of female abdomens (plus heads and thoraxes) at heights of 6–12 feet on wooden stakes. Extracts of female abdomens elicited excited searching in males who landed at the site. It is assumed that the pheromone emitted by females after they have settled in trees, supplemented by visual stimuli from the silhouetted trees, promotes the mass mating flights of males (*734a*).

*Popillia japonica* Newman, Japanese beetle

Smith and Hadley (*1115*) reported in 1926 that numerous males were observed flying low over the ground in search of emerging females. The males would attempt copulation, resulting in a balling up of a large number of males on top of a female. When alighting, the males always approached the female against the wind, apparently attracted by the odor.

It was not until 1968 that the presence of a powerful, volatile sex pheromone from females was confirmed by field studies (*708, 709*). Females, tethered in the open or confined in traps, attracted males in large numbers within seconds after exposure. A single female exposed in 1 trap attracted 380 males in 1 hour, and 9 females in a trap attracted almost 3000 males in the same time period. After the removal of attractive females, glass and metal containers continued to attract males for up to 30 minutes. Exposure of isolated female abdomens and of heads plus thoraxes showed that the attractant is produced in the abdomen, but various solvent extracts of abdomens were not attractive.

Female beetles in confinement, as opposed to the field, do not appear to be attractive to males. Extracts of the abdomens of male and female beetles were not attractive when placed in field traps (*410*).

*Rhopaea magnicornis* (Blackburn)

Large numbers of males were observed trying to dig their way into a field cage containing a virgin female. Cups previously used to house 4- to 15-day-old virgin females elicited in males a graded response of fanning out of the lamellae of the antennal club and of excited walking, culminating in rapid fluttering of the wings and searching behavior. Laboratory-reared females produced the attractant from the third day of adult life and were no longer attractive to males after fertilization. The attractant is emitted from the abdomen; no other portion of the female's body is attractive to males. Field tests with marked males conducted during the insects' active period of 1–2 hours after sunset showed that caged females were attractive to males up to 30 yards away; recoveries of marked males were dependent on wind speed, since males flew at random in unsteady or very low winds, but flew upwind to females in steady breezes of 5–7 miles per hour (*1121*).

*Rhopaea morbillosa* (Blackburn)

*Rhopaea verreauxi* (Blanchard), pasture chafer

Caged virgin females of these species may emit their respective attractants on at least 3 successive nights and, presumably, until a successful mating is achieved. However, in nature, mating probably takes place soon after a female begins producing the attractant (*799, 1121*).

*Sandalus niger* Knock, stag beetle

Masses of struggling males may be seen atop a single female (*753*).

*Sitophilus oryza* L., rice weevil

A sex pheromone was obtained by passing a stream of air through a flask containing live virgin females and condensing the stream at low temperature (*337*).

*Telephorus rufa* (L.)

Two males excited by the same female inserted their penises into the female at the same time (*548*).

*Tenebrio molitor* L., mealworm beetle, yellow mealworm

Shaking a live female in a vial releases a pheromone that lures and excites males. A glass rod, paper, or cotton pressed to a female abdomen becomes wet with a yellow liquid; holding these treated objects in front of and above a male's head makes him highly excited, and he attempts to copulate with or follow the objects (*1204, 1213*). Material obtained from the male abdomen does not excite females or males, but a male may be made attractive to other males by rubbing the tips of his elytra with the squeezed posterior end of a female's abdomen (*512, 1204, 1213*). Both pheromone production and the male response are undetectable in newly emerged adults; both rise to their maximum extent within 1 week after eclosion (*508, 1204*).

*Trogoderma glabrum* (Herbst)

*Trogoderma granarium* Everts, khapra beetle

*Trogoderma grassmani* Beal
*Trogoderma inclusum* LeConte
*Trogoderma simplex* Jayne
*Trogoderma sternale* Jayne
*Trogoderma variabile* Ballion

The virgin females of these species of dermestid beetles produce an attractant for their respective males and, in some cases, for the males of several other species (*35, 223, 224, 823, 949, 1217, 1234, 1325*). A high degree of interspecificity was found for female extracts of these beetles, except that the pheromones of *T. glabrum* and *T. sternale* appeared to be different from those of other species and from each other (*1217*). The pheromone of female *T. granarium* is also attractive to *Callosobruchus maculatus* males and repellent to *Tenebrio molitor, Tribolium castaneum, Dermestes maculatus,* and *Oryzaephilus surinamensis* (*1325*). Extract of female *T. glabrum* is highly attractive to male *T. inclusum,* and extract of female *T. inclusum* is attractive to male *T. glabrum* (*223*).

*Xenorphipis brendeli* LeConte, wood borer

Males were strongly attracted to caged virgin females but not to mated females; approach was from downwind. Antennae of females are serrate while those of males are pectinate; males with their antennae removed are not attracted to virgin females. It is assumed that the receptor organs for the female pheromone are on the elaborate antennae of the male (*1253*).

HYMENOPTERA

*Andrena flavipes* Panzer, solitary bee

Although males seek nubile females in their own nest sites, they are probably attracted by the odor of the nest site rather than of the female (*245*).

*Apanteles medicaginis* Muesebeck, wasp

Unmated females up to at least 5 hours old emit a scent to which males respond by flying against the wind over at least 90 m. The pheromone also elicits wing vibration in males (*303*).

*Apis cerana* (Fab.), eastern honeybee
*Apis dorsata* (Fab.)
*Apis florea* (Fab.)

Ethanol extracts of virgin and mated queens of these species attracted drones of *A. mellifera* in addition to drones of their own species (*253, 1007*). On days of low sexual appetite, the drones showed a significant preference for their own queen (*1007*). Gas chromatography of the extracts (see Chapter X) indicated that the substance responsible in each case was queen substance (9-oxo-*trans*-2-decenoic acid) (*253*), and this was later confirmed (*1017, 1074*).

*Apis mellifera* L., honeybee

A virgin queen attracts males (drones) when she reaches a height of 12 m from the ground; the drones are attracted first by her movement and then her odor (*1006, 1335*). Drones are not attracted to queens below approximately 15 feet above the ground, and drone swarms attracted to tethered queens at greater heights disperse quickly if the queens are lowered too near the ground. Drones swarmed in large numbers around queens tethered at a height of 30–80 feet within 100 yards of an apiary (*438, 1006*).

Honeybee drones fly at a height of 10–30 m and will follow a queen for a short time from 40 m high to ground level. These experiments were run with queens tethered to balloons (*450, 1008*) as well as with queen extracts (*450*) (Fig. II.1).

Gary (*439*) described an apparatus by which tethered virgin queens could be suspended for restricted flight at desired heights up to 11 m, enabling observations on mating. Drones were attracted in large numbers by the scent of queen sex attractants, approaching from windward, flying typically in sparse swarms that assembled and hovered in conical formation below and behind the queens. Drones were consistently observed to mount on top of the queen's abdomen, but successful mating followed rather rarely. The drones clasped the queen's abdomen and everted the genitals into the sting chamber, which had to be open to enable coupling to occur. The entire copulatory act probably occurs in the air in a few seconds (*439*).

Pain and Ruttner (*843*) were able to verify Gary's results (*438*). A virgin queen was tethered around the thorax with a 50-cm length of nylon string attached to a plastic balloon (1.30 m high and 60 cm in diameter) that was regulated from the ground by a 15-m length of nylon cord. Queens attracted males when suspended 6 to 15 m from the ground; those showing signs of fatigue were replaced by fresh queens. A fertile queen appeared to attract and maintain a swarm more rapidly than did a virgin queen.

By elevating virgin queen honeybees to heights of 16–30 m with helium-filled balloons, it was determined that drones are attracted in some areas but not in others. The degree of attraction varied. The data suggested that definite drone congregation areas exist, and it would appear that queens search out these areas and are there pursued by drones (*1334*). This has since been substantiated by Ruttner and Ruttner (*1009, 1010*) and by Gerig (*450*). At a distance of 150–1000 m from the nearest apiary, several regularly visited drone areas were observed. Queens fastened to balloons were followed in a highly spirited manner within the congregation area, but only for a very short distance from that place. Drones followed queens only

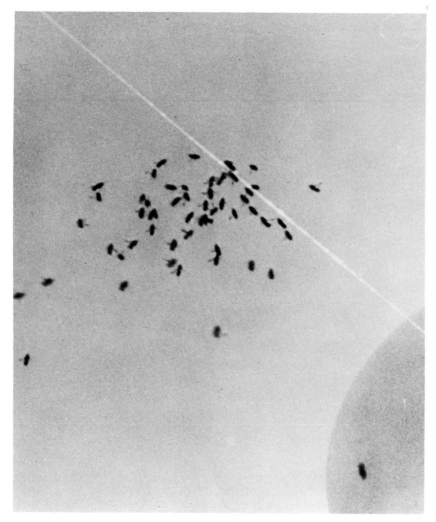

*Fig. II.1.* Drone honeybees following a queen tethered to a balloon aloft. [From Gerig (*450*), Verein Deutschschweizerische Bienenfreunde, Flawil, Switz.]

at heights of at least 10 m. Definite drone assembly places were found to which drones returned regularly; in nature, queens were followed only at these spots.

The queen's mandibular glands were suggested years ago as a possible source of the female odor (*927*). Other investigators (*438, 804*) stated that although the primary source appeared to be the mandibular glands, as shown by testing extracts of various portions of her body, extirpation of these glands does not necessarily render a virgin queen incapable of mating. Velthuis (*1216*) has, in fact, shown that, apart from mandibular gland substances, other queen pheromones are produced by abdominal glands.

Silicic acid fractionation of an ether extract of virgin queen mandibular glands indicated that queen substance and a phospholipid fraction were responsible for the attraction, but reconstitution of the lipid fractions into the lipid complex considerably increases the attraction for drones (*438*). Queen substance and an ether extract of the queen mandibular glands were each impregnated on plastic forms resembling queens, and these were compared with live queens for attractiveness. A fertile queen attracted and maintained a swarm of males, as did old virgin queens, while a control attracted 1 or 2 males for a short period. A swarm of males was attracted and maintained by mandibular extract or by 1.14 mg of its acidic fraction, but not by an ovarian extract of 3 virgin females.

That the queen's mandibular glands are the source of the attractant has been well substantiated by the work of Renner and Baumann (*920*) and Butler (*248*). Butler also showed that queen substance acts as an aphrodisiac in stimulating a drone to mount a queen when her sting chamber is open. Butler (*248*) stated, "The evidence obtained makes it unlikely that any odour from a queen's sting-chamber or anything in the mandibular gland secretion on her body, except 9-oxodecenoic acid, stimulates drones to mount a queen." However, in 1969, Butler (*250, 251a*) obtained results strongly suggesting that another aphrodisiac for the males is produced in gland pockets on the dorsal surface of a queen's abdomen. The glands are most active when the queen is ready to mate, and the substance may be perceived by the drone when he touches the queen's body with his antennae or front legs.

Although queen substance had at first been tentatively identified as the substance responsible for the stabilization of honeybee swarms by queens (*244, 437, 842, 1108*), this phenomenon is actually due to the action of 9-hydroxy-*trans*-2-decenoic acid (*256*).

An odor gland (Nassanoff gland) situated in the abdominal end of various races (*carnica, ligustica,* and *nigra*) of *A. mellifera* protrudes when the bee wishes to attract companions. The pleasant odor, which attracts all races, is dispersed by the simultaneous fanning of the wings. The colony odor, which adheres to the bodies, distinguishes companions of their own colony from foreigners (*919*). Boch and Shearer (*171–174*) reported that the odor appears to be caused by a mixture of geraniol, nerolic acid, and geranic acid, with citral as a minor component (*1073*). However, Butler and Calam (*252*) showed, by isolation from the gland and by field tests, that the cis and trans isomers of citral (neral and geranial, respectively) together make up the most attractive components in the secretion. Citral was more attractive when tested alone (0.39 μg) against a combination of geraniol (100 μg) + geranic acid (100 μg) + nerolic acid (100 μg). Citral (0.77 μg) + geraniol (0.20–0.39 μg) was almost as attractive to workers as the odor of the Nassanoff secretion collected from 10 foraging bees.

When honeybees sting an object they release pheromones that direct the attack of other bees toward it. Boch and Shearer (*175*) isolated from honeybee stings isoamyl acetate, which releases strong alarm behavior, as well as 2-heptanone from the mandibular glands of a foraging bee. Free and Simpson (*426*) report that 2-heptanone is the principal aggression-provoking component in the secretion of the mandibular gland, and that isoamyl acetate is not the only active component of the aggression-provoking secretion at the base of the sting, but that no such component occurs in the venom itself. Butler (*246*) showed that 2-heptanone repels foraging honeybees but that 10-hydroxydecenoic acid, which also occurs in worker mandibular glands, does not.

Butler and Simpson (*259*) in 1965 reported that they could confirm earlier conclusions that, although most of the queen's odor that attracts workers comes from the mandibular glands, some is produced elsewhere, specifically, from the paired gland in the sting chamber (Koschewnikow gland), but not in the subepidermal glands. However, in 1967 and 1968 they concluded that the odor of queen substance together with that of 9-hydroxy-2-decenoic acid, both of which are found in the queen's mandibular glands, enables the workers to find her when swarming (*260, 803a*) and inhibit queen rearing by queenless workers (*255*). The odors of these acids, separately or together, do not attract workers in their hive toward the queen.

*Bombus agrorum* (Fabr.), bumblebee

Males were seen to mount a queen that had been dead for 7 weeks, and one attempted to mate with her. The following year, when she had been dead for 14 months, a male mounted her and tried to copulate (*425*).

*Bombus pratorum* (L.), bumblebee

Young queens suspended by thread from the branches of trees and bushes along a flight path attracted males (*425*).

*Bracon hebetor* (Say), wasp

Males of this insect, a parasite of *Ephestia* larvae, run about excitedly flapping their wings when introduced to females (*487, 1262*). The presence of a camel's hair brush recently used for handling females also excited males (*487*). Stimulation is apparently by odor perceived largely by the antennae (*813*). Some females fail to excite males. Filter papers on which females have been crushed evoke an immediate mating response from males. Several female abdomens elicited attempts by males to copulate, some of which were successful; the anterior half of the abdomen was more attractive than the posterior half (488).

Male *Habrobracon brevicornis* introduced to females of *B. hebetor* are greatly stimulated, and attempt to mount and mate with them. The female may refuse to mate but sometimes mates immediately afterward with a male of her own species (1262).

*Crabro cribrarius* (L.), wasp

An odorous sex attractant for the male is produced by the female's abdomen. Although the active substance was not isolated, dummies rubbed mechanically with freshly caught females attracted males that attempted to copulate with these dummies (696, 697).

*Dasymutilla* spp., velvet ants

Females placed in small cages were very attractive to males (395, 566).

*Diprion similis* (Hartig), introduced pine sawfly

Virgin females in the field attracted exceptionally large numbers of males; one caged female lured more than 11,000 males within 5 days. Males could be attracted approximately 200 feet out of the forest over an open field. Benzene extracts of crushed females or of their abdomens and ethyl ether rinses of glassware in which the females had crawled were highly attractive (310).

Copulation results in a rapid loss of attractancy; mating appears to trigger a mechanism that allows destruction of the attractant in the female. A satisfactory laboratory bioassay method for the attractant has not yet been developed (286). Gynandromorphs were shown to produce sex pheromone to which males typically responded (783a).

*Formica montana* Emery

*Formica pergandei* Emery

Male ants, as they leave their nests in search of reproductive females, fly across wind until they perceive the presence of a female. They then fly directly upwind to the general location of the female and make rapid flitting and crawling movements along plant stems that had come in contact with the female. The most plausible explanation for this attraction is that the males are responding to a pheromone released by the female. However, no males were attracted to crushed parts of a female's body, possibly because other released compounds masked the odor. Also, the pheromone may exist in an inactive form which is activated only on stimulation by the female (643).

*Gorytes campestris* (L.), wasp

An odorous sex attractant for the male is produced by the female's abdomen (*696, 697*).

*Gorytes mystaceus* (L.), wasp

Females are highly attractive to males (*697, 698*).

*Halictus albipes* F., solitary bee

*Halictus calceatus* Scop., solitary bee

The characteristic musk scent of the females of these species, possibly a sex pheromone, is due to the presence of macrocyclic lactones (*32, 700*) (see Chapter X).

*Harpagoxenus sublaevis* Nyl., ant

The wingless females attract males at swarming by assuming a stance with the abdomen elevated, as in calling, during which a sex pheromone is undoubtedly released (*226*).

*Macrocentrus ancylivora* (Rohwer)

Males introduced into a glass vial previously occupied by a female become excited and behave as though they were in the presence of a female. Males are able to respond sexually immediately after emergence (*406*).

*Macrocentrus gifuensis* (Ashmead)

Sexual activity of males is more pronounced when they are not confined in a small space, where the air may become saturated with the female odor. Males may respond sexually immediately after emergence (*845*).

*Macropis labiata* (Fabr.)

Females are highly attractive to males in the field (*697*).

*Megarhyssa atrata* (Fabr.)

Males of this species, a parasite of wood-boring insects, emerge before the females. They are attracted by a female scent, congregating at the point from which the females will emerge. A single individual of a group of 6–10 males on a dead tree was found to have his abdomen inserted through an opening in the wood; cutting into the wood revealed that the male was mating with a female inside (*290*).

*Megarhyssa inquisitor* (Say)

A number of males of this species, predatory on *Malacosoma neustria*, were observed trying to enter into a few openings in a *Malacosoma* pupa. It was found that several female *M. inquisitor* were inside and that the males were lured by the female scent (*1238*).

*Megarhyssa lunator* (L.)

Males congregate on the trunk of a tree awaiting the emergence of females, which are undoubtedly detected by odor (*1, 2*). Males are often observed scraping away the bark to a depth of one-fourth of an inch to find females ready to emerge. Copulation takes place while the female is still in her cell or burrow; she then flies off to oviposit (*430*).

*Neodiprion lecontei* (Fitch), red-headed pine sawfly

Virgin females attract large numbers of males (*133*).

*Neodiprion pratti pratti* (Dyar), Virginia-pine sawfly

Males are strongly attracted to a virgin female, approaching her with the wings spread upward and outward and the antennae expanded. Live females as well as extracts of their abdominal segments lure males to field cages from a distance of 50–100 feet. Females mate only once, becoming unattractive to males within less than 1 minute after copulation. Although males were strongly attracted during 1960 in the wild and a high percentage mated, wild females were not attractive to males in 1963 and only 2.5% mated. No explanation was found for this loss in attractiveness (*170*, *170a*).

*Neodiprion sertifer* (Geoff.), European pine sawfly

Both males and females are ready to mate as soon as they emerge from the cocoon. Males are attracted to females olfactorily, judging from their often sudden and abundant appearance near caged females in the field (*737*).

*Opius alloeus* Muesebeck

Within 30 seconds after a 4-day-old male was introduced into a cage containing a 1-day-old female, he began to walk and fan his wings. Filter paper discs that had been in contact with virgin females, as well as crushed females alone, elicited precopulatory behavior in males. However, ether and water extracts of whole females, of abdomens, or of heads and thoraxes failed to evoke a response (*196*).

*Phaeogenes invisor* Thunberg, wasp

Unmated females up to at least 5 hours old emit a scent to which males respond by flying against the wind over at least 200 m. The pheromone also elicits wing vibration. The shells of *Tortrix* pupae from which females of *P. invisor* had emerged were attractive to the males. An extract of these shells lured males for at least a week (*303*).

*Praon pallitans* (Muesebeck)

A male is ready to mate soon after emergence. He detects the virgin female by odor, becoming highly excited and running about with his wings held vertically above the thorax and his antennae vibrating rapidly. Once contact is made, the male moves his wings rapidly in a vertical position. Once mated, the female loses her attractiveness for males, unless they come in contact with her accidentally (*1023*).

*Pristiphora conjugata* (Dahlb.), sawfly

A male that had just finished copulating was introduced into a vessel containing only males. He immediately became the subject of much attention from his companions, who attempted to mate with him (*343*).

*Pristiphora geniculata* (Hartig), mountain-ash sawfly

Copulation occurs soon after emergence. Males appear slightly earlier than females and seem to be attracted by a scent emanating from the female. An abdomen of a virgin female crushed in a rearing cage rapidly attracted males, who touched it with their genitalia, and finally began to fight with one another. Sometimes, in the field, when populations are high and temperatures favorable, males assemble on the ground in groups of about a dozen, each with his abdomen pointing to the spot on the ground where a female will emerge; they remain in this position vibrating their antennae and moving their genitalia until she emerges. When she reaches the surface of the soil, the males attempt to mate with her (*416*).

*Vespa orientalis* L., Oriental hornet

Delta-*n*-hexadecalactone, isolated from an extract of queen heads, acts as a pheromone for the workers and stimulates their construction of queen cells at the end of the season. The compound may also be a sex attractant for the males in the same way that synthetic undecalactone attracts the bumblebee, *Macropis labiata* (*569*).

*Xenomyrmex floridanus* Emery

A sex pheromone released by females during nuptial flights attracts males and releases copulatory behavior. The pheromone is produced in the poison gland. Crushed female abdomens are also attractive to males. Applicator sticks treated with the poison gland secretion and exposed to males in an arena elicited flight toward the sticks (*557*).

LEPIDOPTERA

*Achroia grisella* (Fabr.), lesser wax moth

*Achroia* sp.

Adult virgin females elicit great excitation in males of these species as well as those of *Galleria mellonella;* copulatory attempts have been made by the latter toward *Achroia* females (*101*).

*Acleris glomerana* (Wlsm.), black-headed budworm

Males sexually stimulated by female pheromone perform precopulatory motions, including rapid wing fluttering, anterior curving of the abdomens, and opening and closing of valvae (*331*).

*Acrolepia assectella* Zeller, leek moth

Virgin females, highly attractive to males, have been used to trap males in the field (*902, 904*).

*Acronicta psi* (L.)

Males are attracted to virgin females by odor (*1212*).

*Actias selene* (Hübner)

Marked males orient toward females from a distance of 11 km (*782*). Males have been shown to produce an olfactory sex pheromone that stimulates oviposition in females (*134*).

*Actias villica* (L.), cream-spot tiger moth

Females attract males at dusk. Countless *Parasemia plantaginis* were also attracted by nearby *A. villica* females and attempted to copulate with them (*661*), leading Ford (*418*) to believe that the females of both species utilize the same sexual scent.

*Adoxophyes fasciata* Walsingham, lesser tea tortrix

Females attract males in the field and elicit a typical mating dance and copulatory attempts in the laboratory (*1169a*).

*Adoxophyes orana* Fischer von Röslerstamm, summerfruit tortrix

Response of males to a methylene chloride extract of virgin females was at its maximum when 3-day-old males kept in continuous light were then kept in darkness for 7 hours. The sex pheromone was bioassayed by counting the number of males that exhibited a mating dance after exposure to a methylene chloride extract of whole females. Male response could be controlled artificially by altering the light and dark periods to which the males were exposed (*1167–1169, 1327*). An inbred laboratory stock was shown to produce less sex pheromone than a stock that was not inbred (*789a*).

*Agathymus baueri* (Stallings & Turner)

*Agathymus polingi* (Skinner)

Observations made by Roever (*967*) indicate that recognition of the males and females of both species for one another is bidirectional in that, while the initial recognition response is made by the male, the receptive female furthers this response by emitting a pheromone when the male approaches.

*Aglia tau* L., nailspot

A virgin female placed in a screen box and carried out to the field attracted 125 males in 2 hours (*883*). The odorous substance responsible for this attraction is secreted by the eighth abdominal segment of the female (*372*). Attraction was also demonstrated by laboratory tests with an olfactometer (*1053, 1054*), and electrophysiologically (*1033*).

*Agrotis fimbria* (L.)

Abdominal glands in the female secrete an attractant for males (*1212*).

*Agrotis ipsilon* (Hufnagel), ypsilon dart, black cutworm

With the aid of a simple apparatus (*409*), it was determined that females at least 1 hour old produce an attractant for males in their last 2 abdominal segments. The excised segments lost their attractiveness when placed in a vacuum, but they became attractive within 15 minutes (or 2 hours) after removal from the vacuum. The activity disappeared completely from the cut segments after $1\frac{1}{2}$–3 hours. The attractant could be collected from the segments with a stream of air ("freezing out"), steam distillation, or extraction with various solvents, especially ether (*407, 408*).

*Alabama argillacea* (Hübner), cotton leafworm

Males exposed in the laboratory to the vapors from an extract of the last 2–3 abdominal segments of virgin females indicated their sexual excitement by lifting the antennae, vibrating the wings, extending the claspers, and attempting to copulate with one another (*136*).

*Amorpha populi* L., poplar hawk moth
Females attract males by scent (*1317*).

*Anagasta* (*Ephestia*) *kühniella* (Zeller), Mediterranean flour moth
Pheromonal attraction of the female for the male was demonstrated almost 4 decades ago by Richards and Thomson (*933*) and by Dickins (*349*). They also showed that a noncalling female placed in a glass-top container with males of *Plodia interpunctella* and *Cadra* (*Ephestia*) *cautella* elicited mating attempts on the part of these males; this nonspecificity has been confirmed by Nakajima (*823, 1292*). Male *A. kühniella* whose eyes had been painted with India ink were placed in a container with a virgin female; the males became excited. Females 50 hours old were more attractive to males than newly emerged females. A male responded more rapidly to a female that had been calling for some time in a box prior to his entrance than to a female added to the box at the same time or slightly before him; this was probably due to accumulation of the scent within the box (*349, 933*). The calling position consists of raising the abdomen and extruding the scent glands (*1193, 1314*). Males respond very strongly to females of this species and of *P. interpunctella* with a long-lasting, characteristic dance and wing flutter (*1053, 1054*).

*Anarsia lineatella* Zeller, peach twig borer
Caged virgin females in the field attracted native males (*70*).

*Ancylis comptana fragariae* (Walsh & Riley), strawberry leaf roller
Males were not stimulated by an ether extract of female abdominal segments in the laboratory olfactometer or in the field, although there is little doubt that a sex pheromone is involved in mating (*962*).

*Antheraea eucalypti* Scott, emperor gum moth
Antennal olfactory receptors enable the male to detect, over considerable distances, traces of a volatile pheromone emitted by the female (*799*).

*Antheraea pernyi* (Guérin-Méneville)
The presence of a sex attractant in the female was demonstrated by Schneider (*1028*), using an electrophysiological technique.

*Antheraea* (*Telea*) *polyphemus* (Cramer), polyphemus moth
Rau and Rau (*911*) found that males marked on the wings with oil paint and released from the second or third story of a city building were able to return to virgin females without difficulty. Few or no males found the females unless they were liberated into the wind blowing from the direction of the females.

Although virgin females caged outdoors attract males from afar, matings do not occur under laboratory conditions because red-oak foliage is necessary to stimulate pheromone production by the female (*934, 935, 937, 938*) (see Chapter V).

*Aphomia gularis* (Zeller)

Males exposed to females become highly excited, showing a typical circling dance. Although the female exhibits very small odor glands, those of the male (situated on the wings) are quite large (*101*).

*Archips argyrospilus* (Walker), fruit-tree leaf roller

*Archips mortuanus* Kearfott

Ether extracts of the last few abdominal segments of these species were highly specific in both the laboratory and the field, attracting only males of the same species (*209, 955–957, 962*).

*Archips griseus* Robinson

*Archips semiferanus* Walker

Live females or ether extracts of female abdomens were attractive only to their own males (*209, 957*).

*Arctia caia* L., tiger moth

Kettlewell (*663*) and Evans (*384*) showed that males are attracted to virgin females by odor. Evans found that 2 virgin females in a muslin cage attracted several males of this species overnight, as well as one male *Phragmatobia fuliginosa*. The nearest locality for *P. fuliginosa* was 2 miles away, where it was not common.

*Argynnis adippe* (L.)

*Argynnis euphrosyne* (L.), pearl-bordered fritillary

*Argynnis latonia* (L.)

*Argynnis paphia* (L.), emperor's cloak

Females of these species are highly attractive to their males (*1212*). However, according to Magnus (*743*), the male finds the female by sight and is then excited by an odor which she releases.

The female attractant-producing glands consist of modified cells between the seventh and eighth abdominal rings. Drops of the liquid obtained from the sacculi laterales and placed on filter paper lure the males, which fan the wings and distribute the attractant particles (*545*).

*Argyrotaenia juglandana* (Fernald)

*Argyrotaenia quadrifasciana* Fernald

*Argyrotaenia quercifoliana* Fitch

Live females or their crude extracts in field traps lured males of the respective species (*957, 962*).

*Argyrotaenia velutinana* (Walker), red-banded leaf roller

A sex attractant for males is produced in glands located dorsally between the eighth and ninth abdominal segments of the female. It is effective from distances up to 3 miles (*950, 951, 961, 962*).

*Autographa biloba* (Stephens)

Methylene chloride extracts of female abdomens attracted males of this species as well as of *Pseudoplusia includens* and *Rachiplusia ou* (*137*).

*Autographa californica* (Speyer), alfalfa looper

Ether extracts of female abdomens, as well as thin-layer chromatographic fractions thereof, were attractive to males in laboratory tests (*441, 442, 1096*). The extracts were also attractive to males of *Trichoplusia ni* (*1093*).

*Automeris* spp.

Extracts of female abdominal glands were attractive or sexually exciting to males when tested by an electrophysiological method (*1033*).

*Biston betularia* L., peppered moth

Virgin females were attractive to males of this species as well as of *Xanthorhoe montanata* and *X. spadicearia*, causing Kettlewell (*664, 665*) to conclude that the 3 species have similar assembling scents.

*Bombyx mori* L., silkworm moth

A newly emerged female placed in a container with males causes great excitement, and copulation quickly occurs. A pupa containing a female soon to emerge is also attractive (*744, 882*). Filter paper wetted with the liquid from the female's sacculi is highly attractive to a male when held close to his antennae (*427*).

The female protrudes a paired scent organ from the hindmost abdominal segment and the male walks nervously about, finds the female, and orients himself for copulation. The protruded female glands are withdrawn into the body immediately after being touched by a male. Excised scent glands, but not the mutilated female, are highly attractive and elicit copulatory attempts. Males will also mate with headless females (*548, 657*). Extract of the female brought close to the male evokes vigorous wing flutter ("schwirrtanz") and attempts to locate the "female" (228).

A female assumes the calling position shortly after emergence, but even mated females will assume this position. These females will then attract males, as will a piece of paper previously rubbed on the female body. The effective distance of the attractant is 3–5 cm, although previously mated males will respond from a greater distance. Turpentine, eucalyptus, and clove oils are not attractive to males, but they do not detract from the female's attractiveness when placed near her. Males kept isolated from females may show sexual excitement, characterized by circus movements, rapid wing vibration, and bending of the abdominal extremity toward the head (*301*).

Males are attracted by unseen females or excised female abdomens. If the male's antennae are removed after he has been excited by the scent, he continues to search for the female, but copulation occurs only if his abdomen touches her (*1065*).

*Brahmaea* spp.

Extracts of female abdomens were attractive to males when bioassayed electrophysiologically (*1033*).

*Bryotopha similis* Stainton

*Bryotopha* sp.

These 2 species are morphologically similar and exhibit identical 4-dot wing patterns, but there is no cross-attraction between them. The grayish *B. similis* males are attracted to *cis*-9-tetradecen-1-ol acetate, identified as the natural sex pheromone of the red-banded leaf roller moth, whereas males of the yellowish, unidentified species are attracted to the trans isomer. These compounds are said to be the natural sex pheromones of *B. similis* and the yellowish species, respectively (*956*).

*Bucculatrix thurberiella* Busck, cotton leaf perforator

Males exposed to virgin females in laboratory olfactometers performed a typical mating dance consisting of antennal movement and wing vibration (*918*).

*Cacoecia murinana* (Hübner)

On a quadratic wooden frame (35 × 25 cm), Franz (*424*) fastened a pane of wire glass covered with glue on both sides. In the center was placed a small cylindrical cage whose front and rear walls consisted of nettle dust and whose side walls were provided with small openings to permit the scent of contained virgin females to escape. These traps were hung in June and July at the top, middle, or bottom of fir trees and attracted large numbers of males from the day of emergence to the eleventh day; no females were attracted to the traps. The largest number of males was attracted to cages hung near the treetop and the smallest numbers to those hung at the bottom. It was difficult to determine whether the catches increased when increasing numbers of females were placed in a trap.

*Cadra (Ephestia) cautella* Walker, almond moth

Virgin females attract not only males of this species but also those of *Plodia interpunctella* and *Anagasta kühniella* (*63, 337, 349, 705, 775, 823, 1165, 1292*). Male *P. interpunctella* were highly responsive to the extract of female almond moths, but male almond moths were much less responsive to female *P. interpunctella* pheromone (*337, 431, 823*). The response to a calling female consists of rapid running, wing flutter, antennal vibration, and antennal curving (*349, 775*). A noncalling female rarely attracts any male. Excised female abdomens were as effective as the calling females in eliciting a response (*775*).

*Caligula japonica* (Butler)

Females are sexually attractive to males (*1212*).

*Callimorpha dominula* (L.), scarlet tiger moth

The female is attractive to the male (*663*). A female imprisoned in a muslin cage in a garden during daylight hours also attracted male *Phragmatobia fuliginosa* (*662*), leading Ford (*419*) to assume that the scent produced by female *C. dominula* is the same as that produced by the female *P. fuliginosa*.

*Callimorpha dominula persona* (Hübner)

Freshly emerged females of *C. dominula* and *C. dominula persona* attracted equal numbers of *C. dominula persona* males released nearby (*860, 1136, 1212*).

*Callosamia promethea* Drury, promethea moth, spice-bush silk moth

Rau and Rau (*911*) reported that a single female, which has a strong odor perceptible to humans, attracted 40 male moths when placed near an open window. Large numbers of males were observed to fly through heavy rain to reach a room in which females were confined. The attractive distance of a female is reported to vary from a few yards to a maximum of 3 miles.

According to Soule (*1123*), males are so excited by the odor of the females that they fasten their claspers on any part of a female body or even on each other. Females are also attractive to male cynthia moths (*Samia cynthia*), which then mate with the female promethea moths; of the eggs laid after these matings, all may hatch to give insects of both the "promethea" and "cynthia" form.

*Celaena haworthii* (Curtis), Haworth's minor

Hordes of males seen flying excitedly around a tuft of grass in the evening were probably attracted by a female pupa about to hatch, according to Barrett (*93*).

*Chaerocampa elpenor* (L.)

Federley (*398*) reported several examples of cross-attraction and cross-mating between species of the sphingids. Male *C. elpenor* lured to field cages by their own females preferred to mate with *Metopsilus porcellus* females present in the same cages. In captivity, male *Deilephola galii*, *D. euphorbiae*, and *Hyloicus pinastri* are attracted to and easily mated with female *C. elpenor*. However, in more cases than not, if a male *C. elpenor* mates with a female *M. porcellus*, he is unable to withdraw his penis and, although the sperm has been introduced, the mating is unsuccessful.

*Chilo plejadellus* Zincken, rice stalk borer

Females produce a sex pheromone for males (*505, 506a*).

*Choristoneura fumiferana* (Clem.), eastern spruce budworm

Calling virgin females extrude abdominal sex pheromone glands before sunset to attract males (*402, 1014, 1015*). This can be demonstrated by blowing air which has passed over such a female into a screen cage containing males; they respond by twirling around with raised wings, opening and closing the claspers, and attempting to copulate (*1014*).

Males of *C. fumiferana* and *C. pinus*, the jack-pine budworm, will attempt to mate with dead females of either species (*1116*), although live males respond only to sex pheromones of their own species in the laboratory and field, according to Sanders (*1015*). Rinses of glassware that had previously contained live females sexually stimulated males of *C. fumiferana, C. occidentalis,* and *C. biennis.* Males of *C. orae* and *C. viridis* did not respond (*1016*).

*Choristoneura occidentalis* Freeman, western spruce budworm
*Choristoneura viridis* Freeman, green spruce budworm

Males responded sexually to air passed over live virgin females in the laboratory (*331*). Rinses of glassware that had previously contained female *C. viridis* evoked a sexual response from male *C. pinus.* A crushed female *C. viridis* abdomen evokes a sexual response in *C. viridis* males (*1016*).

*Choristoneura orae* Freeman, coastal budworm
*Choristoneura pinus* Freeman, jack-pine budworm

Males responded sexually in laboratory bioassays when exposed to rinses of glassware that had previously contained live females (*1016*). Females start calling after sunset (*1015*).

*Choristoneura rosaceana* (Harris), oblique-banded leaf roller

Extracts of virgin females placed in traps lured males in the field (*962*).

*Chrysopeleia ostryaella* Chambers, leaf minor

Field observations showed that a single female could attract many males, and pairs which had just commenced to mate could be detected by the fluttering of the supernumerary males about them. The female ceased to be attractive 1–3 minutes after copulation commenced (*735*).

*Colocasia coryli* (L.)

Virgin females will attract males (*1212*).

*Colotois pennaria* (L.)

Males are attracted to unmated females (*663*).

*Cossus robiniae* (Pck.)

One female lured 70 males within a few hours (*1060*).

*Crambus mutabilis* Clemens
*Crambus teterrellus* (Zincken), bluegrass webworm
*Crambus trisectus* (Walker), sod webworm

Live female moths were effective baits for luring males to traps. Banerjee (*86*) reported that virgin female *C. trisectus* are very attractive to males about 2–3 hours following emergence and remain attractive for 2 days.

Females in field cages were always visited by males of the same species (*84, 86*). Ten virgin females of *C. trisectus* in the field attracted a total of 527 males, and 7 virgin females of *C. teterrellus* attracted a total of 203 males during their entire lives (*85*).

*Cryptophlebia* (*Argyroploce*) *leucotreta* Meyr, false codling moth

The female produces a sex attractant for the male (*912*). It has recently been identified (see Chapter X).

*Cucullia argentea* (Hufnagel), silver monk

*Cucullia verbasci* (L.), brown monk

Females produce a sex pheromone to attract males of the same species (*1212*).

*Danaus gilippus berenice* (Cramer), Florida queen butterfly

Intensive studies conducted by Myers and Brower (*819*) lead to the conclusion that the female may produce an odor important to the male in courtship, since painting the male antennae reduces mating success.

*Dasychira fascelina* (L.)

Males are attracted to females in the field (*1212*) and in a laboratory olfactometer (*1054*).

*Dasychira horsfieldi* (Saunders)

Females contained in a glass jar covered with a thin sheet of paper attracted no males until several pinholes were made in the paper, after which males began to arrive quickly (*478*).

*Dasychira pudibunda* (L.), pale tussock moth

Females at emergence are attractive to males (*373, 1212*). The attractant is an odorous material secreted by the eighth abdominal segment of the virgin female (*372*).

*Dendrolimus pini* (L.)

Adult females, who are capable of mating immediately upon emergence, possess special organs whose odor lures the males (*373*).

*Diatraea saccharalis* (F.), sugarcane borer

Sticky traps baited with virgin females, their abdomens, or extracts of the abdomens have been used to attract and trap males in a cane field (*857, 858*). Females emit the attractant soon after emergence, are most attractive during the first 3 days of life, and lose their attractiveness after mating. Sexual activity is mostly confined to the period between 1:00 AM and 4:00 AM. (*857*).

*Dioryctria abietella* (Denis & Schiffermüller)

Virgin females produce an attractant for males that may be extracted from the female abdomens with ether (*396, 397*).

*Diparopsis castanea* (Hmps.), red bollworm

Males are strongly attracted to caged virgin females placed in cotton fields in Central Africa (*36, 279, 280, 1210*). Mated females are not attractive.

*Diparopsis watersi* (Roths.)

Males are strongly attracted to virgin females in the field (*199*).

*Endromis versicolora* (L.), Kentish morning glory

Males are strongly attracted to virgin females by odor (*663*).

*Ephestia elutella* (Hübner), tobacco moth

Adults are ready for mating almost as soon as the wings are dry. Virgin females ready for mating begin calling, sitting with their wings folded and the apical half of the abdomen bent over the back between the wings (Fig. II.2). The apical abdominal segments are alternately extended and retracted, exposing the intersegmental membranes. There is little doubt that during this process a scent attractive to the males is emitted. The segmental membranes, especially in the neighborhood of the orifice of the ductus bursae, have an appearance strongly suggesting the presence of secretory tissues. Males become very excited in the presence of these calling females; a pillbox from which a female had recently been removed had the

*Fig. II.2.* Typical calling pose of female Lepidoptera (*Vitula edmandsae*). [From Weatherston and Percy (*1250*), Entomological Society of Canada, Ottawa.]

same effect. The male begins fluttering around the female, who at first appears to take little notice or even runs away. Eventually she comes to a standstill and mating occurs (*933*). Females also elicit excitation and evoke copulatory attempts in males of *Anagasta kühniella* and *Plodia interpunctella* (*101, 349*).

*Epiphyas postvittana* (Walker), light-brown apple moth

The female produces and releases a pheromone causing males to orient to the source over some distance. In the laboratory, affected males elevated their antennae, became active, took flight, oriented upwind, and finally mated after performing a dance in her vicinity. Males attained their peak responsiveness to the pheromone by the second night after emergence, and the level was maintained until at least the tenth night (*95, 97, 98*).

*Erannis aurantiaria* (Esper)

*Erannis defoliaria* (Clerck)

Live females placed in sticky traps in Norwegian forests attracted numerous males of the respective species during the night. Traps baited with females 1 night, then emptied, did not attract males the next night. Traps with male and female *E. aurantiaria* caught no males (*1211*).

*Estigmene acrea* (Drury), salt-marsh caterpillar

Methylene chloride extracts of the ninth abdominal segment of virgin females attract and sexually excite males (*740*).

*Eumeta crameri* (Westw.)

Campbell (*278*) reported that a female placed in a closed tin box and put into his pocket attracted a male to his waist while walking in the field at dusk; the male hovered around Campbell's pocket. A muslin bag that contained several females was tied to branches in the garden and attracted several dozen males.

*Euproctis chryssorrhoea* (L.), brown tail moth, gold tail moth

Males responded excitedly in the laboratory to the presence of females (*1054*).

*Eupterotida fabia* (Cram.)

*Eupterotida indulata* (Blanch.)

Males of both species were attracted to their respective females (*478*).

*Eurycyttarus confederata* Grote & Robinson, bagworm

*Eurycyttarus edwardsi* L., bagworm

An emerging male of *E. edwardsi* assumed the mating attitude on a female pupal case 48 hours after the female had emerged and had been preserved in alcohol. A wild male of *E. confederata* attracted to breeding cages attempted to mate with a larval case containing a dead larva and other cases nearby containing live females (*626*).

*Euxoa ochrogaster* (Guenée), red-backed cutworm

Males responded sexually in laboratory olfactometers to a pheromone released by females. The response included wing vibration, extension of the claspers, and copulatory attempts (1155).

*Feltia subterranea* (Fabr.), granulate cutworm

The female produces a sex pheromone to attract or excite males (624).

*Galleria mellonella* (L.), greater wax moth

Females elicit excitation in males of this species as well as of *Achroia grisella* and *Plodia interpunctella*. Such excited *A. grisella* males will attempt to copulate with the female *G. mellonella* (101).

*Grapholitha funebrana* Tr., plum fruit moth

Males marked with a $^{32}$P isotope and released in the field were attracted to traps containing virgin females placed at points of concentric rings of different radii around the site of release. The number of males captured showed no proportional decrease by distance, even at a distance of 9 m. Attraction was intense during the first 3 days after hatching, then it rapidly decreased but was still existent on the eighth day (day of death) (1018).

*Grapholitha molesta* (Busck), Oriental fruit moth

A chemical sex pheromone produced by virgin females in the abdominal tips attracts and sexually excites adult males. Affected males twirl with fanning wings, pausing with the abdominal tips curved upward and opening and closing of the claspers; they appear to expose their scent pencils as if directing a male odor (361, 446, 447, 962). The female ceases to be attractive once she has mated. Males may become so excited when near pairs attempting to mate that they lock their own genitalia firmly together (361, 446).

*Grapholitha sinana* Feld., hemp moth

Virgin females attracted their own males as well as those of *G. compositella* (821).

*Gypsonoma haimbachiana* (Kearfott), cottonwood twig borer

Females produce a pheromone that is highly attractive to males and unattractive to females. Attractiveness decreases markedly after the female is about 1 day old (854).

*Harrisina brillians* (B. & McD.), western grape leaf skeletonizer

Males were lured in fairly large numbers to traps baited with extracts of virgin females (92).

*Hedia nubiferana* Haworth

Ether extracts of virgin female abdominal tips attracted males of this species as well as of *Laspeyresia pomonella* and *Argyrotaenia velutinana* in laboratory olfactometer tests, but only *H. nubiferana* were lured to field traps (962).

*Heliothis phloxiphaga* Grote & Robinson

Females secrete and emit a sex pheromone to lure and excite males (*624*).

*Heliothis virescens* (F.), tobacco budworm

Traps containing live virgin females, ether extracts of such females, or hexane extracts of the last 2 abdominal segments attracted and caught males in a greenhouse; equal numbers of males were caught by 4 and 8 female equivalents (*445*). The presence of a sex pheromone produced by the female has been reported by several other investigators (*139, 441, 469, 1088, 1093, 1096*).

*Heliothis zea* (Boddie), bollworm, corn earworm, tomato fruitworm

The presence of a sex pheromone produced by females was suggested by Graham and co-workers (*469*) and substantiated by Berger *et al.* (*139*) and Shorey *et al.* (*1088, 1093, 1096*). When the female extends her ovipositor she exposes glandular-appearing structures at its base; males attempting to mate with her brush this area with their antennae (*139*).

*Hemileuca maia* (Drury), buck moth

Virgin females placed in screen cages in the field rapidly attracted numerous males flying upwind who then attempted to copulate with the females. A mated female was no longer attractive to males (*368*).

*Hepialus humuli* L., ghost moth

The male attracts the female visually and she then releases a sex pheromone effective at close range (*1241*).

*Heterusia cingala* (Moore)

Males are attracted to females by a chemical scent (*478*).

*Holomelina aurantiaca* complex, tiger moth

Females of this species, as well as members of the aurantiaca complex (*H. immaculata* (Reakirt), *H. lamae* (Freeman), and *H. rubicundaria* (Hübner), *H. ferruginosa* (Walker), *H. fragilis* (Strecker), *H. laeta* (Boisduval), and *H. nigricans* (Reakirt) attract males by means of a pheromone produced in glands situated between the eighth and ninth abdominal segments. The female calls at about 2-second intervals by raising its wings slightly and protracting and retracting the ovipositor and gland (*953*). The pheromone has been identified (see Chapter X).

*Homoeosoma electellum* (Hulst), sunflower moth

Caged virgin females placed in traps lured males when placed within a sunflower field but not outside the field. Live males attracted neither males nor females (*1176*).

*Hyalophora calleta*

*Hyalophora euryalus* (Boisduval), ceanothus silk moth

An electrophysiological method demonstrated that females of both

species produce a sex attractant or stimulant for their respective males
(*1028, 1033*).

*Hypocrita jacobaeae* (L.)

*Hypogymna morio* (L.)

Virgin females attract males by means of a sex pheromone (*1212*).

*Kotochalia junodi* (Heylaerts), bagworm

Virgin females produce a pheromone attractive to males (*192*). Females
are reported to have been visited by males that could only have come from
some miles distant (*626*).

*Lasiocampa quercus* L., oak eggar moth

Several males were attracted in the field to a bag that had contained a
female a week before (*503*). A single female, of which the veteran naturalist,
Fabre, had seen no specimen during 20 years of collecting in his locality,
attracted 60 males (*389*). A box that had contained a virgin female 24 hours
earlier attracted numbers of males in the field (*71, 883*). Males were also
attracted to females in the laboratory (*1212*), and responded to female
extract in electrophysiological experiments (*1033*).

A virgin female was besieged by larger numbers of male *Zygaena
filipendulae* than by males of its own species; male *Z. trifolii* were not
attracted (*418, 1161*).

Kettlewell (*666*) reported that male *L. quercus* f. *callunae* released
three-fourths of a mile downwind from a calling female reached her,
whereas those released upwind required much longer to find the female.

*Lasiocampa trifolii* (Schiff.), grass eggar moth

Males are attracted to females in the field (*663*).

*Laspeyresia (Carpocapsa) pomonella* (L.), codling moth

Proverbs (*890*) reported that males were attracted to cages of live virgin
females hanging in an apple orchard, and this was confirmed by Butt and
Hathaway (*262*) using extracts of female abdomens. These investigators
also showed attraction to males in laboratory and field cages. Males in
small cages exposed to the air from medicine droppers that had contained
female extract solutions responded with a circling dance and attempted to
copulate with other males, both dead and alive, empty pupal cases, and
pieces of corrugated cardboard that had been in contact with the extract.
Males also responded to small hardwood applicator sticks dipped in active
extract.

Barnes *et al.* (*91*) were able to localize the site of the pheromone in
abdominal glands, extracts of which elicited intense sexual excitement in
males. Extracts of the female tips exposed in the orchard attracted only
males, which flew upwind to reach the pheromone source, alighted, and
made clasper responses. Roelofs and Feng (*962*) showed that extracts of

female abdominal tips attracted only males of this species when exposed to a number of tortricid species in the laboratory and field. Howell and Thorp (*560b*) found that mated females had almost no attractancy to males, whereas virgin females were very attractive.

*Lobesia botrana* Schiff., grape vine moth

Males are attracted at least 20–25 m by virgin females in the field (*288*, *462*). Virgin females squashed between the fingers also lure males (*463*).

*Lymantria ampla* (Walker)

Male moths flocked to a room in which a female had been kept even after she had been mated and then killed several days before. Males continued to arrive for 10–14 days following her death, fluttering around the feeding cage in which she had been confined (*478*).

*Mahasena graminivora* (Hampson), bagworm

The female, following complete development, pushes the tip of her abdomen slightly out of the pupal case and emits a strong characteristic odor which pervades the surrounding air. The male flies about seeking the female; after settling on the pupal case, he inserts his abdomen between the wall of the case and the ventral surface of the female to mate (*1071*).

*Malacosoma disstria* Hübner, forest tent caterpillar

Male moths were attracted to field cages containing virgin female moths. Females adopted a calling position with wings folded downward and the terminal portion of the abdomen turned slightly downward. Dead virgin females and empty cages did not attract males. Extracts of female abdominal tips evoked a characteristic sexual response in males in laboratory olfactometers (*1154*).

*Malacosoma neustria* (L.), lackey moth

The female attracts the male by odor (*21*, *363*).

*Manduca (Protoparce) sexta* (Johannson), tobacco hornworm

Virgin females placed in gauze-covered cages can lure males, and mating sometimes occurs between midnight and 1:00 AM. The female does not appear to be attractive to the male until she lowers her abdomen in the calling position; mating occurs a short time afterward (*15*, *16*).

An active extract can be obtained with ethyl ether, benzene, or acetone but not with ethyl alcohol. Filter papers wetted with the active extract were placed in small cages exposed to free-flying males in a large walk-in cage; attraction was shown because the small cage was visited by males between 10:30 PM and 3:00 AM. So-called attractive females chosen for clipping were determined by exposing them in small cages suspended from the roof of the large cage (*17*).

*Metopsis porcellus* (L.)

See *Chaerocampa elpenor* (page 33).

*Micropteryx* spp.

Unpaired males of species of *Micropteryx* often pay more attention to a copulating pair than to solitary females (*932*).

*Nemoria viridata* L.

Poulton (*884*) observed a male flying upwind for about 210 yards to reach an attractive female.

*Nudaurelia cytherea* F., emperor pine moth

The female produces a sex attractant for the male which also elicits a characteristic response in male *N. capensis*, according to electroantennograms (*1241*). An extract of 2 virgin females attracted hundreds of males to baited traps (*192*).

*Orgyia anartoides* (Walker), austral vapourer moth

The female, being wingless through evolution, is dependent upon her sex attractant for finding a mate (*799*).

*Orgyia antiqua* (L.), vapourer moth, rusty tussock moth

The larvalike female with her threadlike antennae sits on her pupal case and is sought out by the males (*1054*). A newly emerged female held in a gauze-covered container lured numerous males during a period of 2–3 hours. Minute drops of liquid may be seen on the surface of the abdominal attractant-producing glands. This liquid absorbed on blotting paper and held before a freshly emerged male causes him to behave exactly as if he were in a female's presence; he flutters his wings and attempts to copulate with the paper (*427*).

*Orgyia ericae* (Germ.)

The dorsal odor glands of the female are very highly developed. She does not leave the pupal case; she merely extends the tip of her abdomen through an opening in the sac to await the arrival of an attracted male (*1212*).

*Orgyia gonostigma* (Fabr.)

The female emits a chemical attractant for the male (*1212*).

*Orgyia leucostigma* J. E. Smith, white-marked tussock moth

Soon after eclosion the wingless virgin female adopts a calling position and begins a rhythmic protraction and retraction of abdominal segments 8, 9, and 10; the male responds by raising the antennae, vibrating the wings, and flying toward the source of the stimulus. Females less than 2 days old were observed to call continuously for several hours (*856*).

*Ostrinia nubilalis* (Hübner), European corn borer

Observations made in the laboratory by infrared photography led to the belief that the female stimulates the male sexually by releasing a sex pheromone (*1128*). This belief became fact when such a sex pheromone, which causes strong sexual excitement in males, was isolated from extracts of female moths (*56, 687, 690*).

*Pachythelia villosella* Ochs.

Newly eclosed females remain within the pupal case and attract males to them (*1013*).

*Panaxia dominula* L.

A caged female was attractive to males of this species as well as those of *Phragmatobia fuliginosa* (*660*).

*Pandemis limitata* (Robinson), three-lined leaf roller

Live females and crude ether extracts of their abdominal tips were attractive to males (*957, 962*).

*Paralobesia viteana* (Clemens), [synonym: *Clysia ambiguella* (Hübner)], grape berry moth

Experiments conducted in the laboratory showed that 2-day-old females excited males immediately, as shown by the vibration of the male's wings and dancing around the female until copulation occurred. Such excitation occurred only in the evenings. Males placed in containers that had previously held a female became highly excited and made searching movements. Females, whose odor is undetectable by humans, mate only once and are then unattractive to males. There is no cross-attraction between *P. viteana* and *Lobesia botrana* (*461*).

Virgin females are known to draw males in the field from at least 25 m distant; the numbers attracted depend on the age of the females and on the weather. Females 1 day old were more attractive than those 2 days old (*462*).

Crude ether extracts of the virgin female abdominal tips attracted males of this species as well as of *Laspeyresia pomonella* and *Argyrotaenia velutinana* in the laboratory (*962*).

*Parasemia plantaginis* L., wood tiger moth

Although females attract males during daylight hours (*661, 666*) and *Actias villica* assemble at dusk (*661*), countless male *P. plantaginis* attracted by females of this species will attempt to mate with *A. villica* females nearby. Female *P. plantaginis* also attract male *A. villica* (*661*). This led Ford (*419*) to assume that the females of both species produce the same attractant.

*Pectinophora gossypiella* Saunders, pink bollworm moth

Before mating, the males exhibit a state of excitation (premating dance) including rapid wing vibrations, with intermittent curving of the abdomen upward while stationary or crawling. Males are attracted to a mating pair and begin the premating dance. A pair crushed during copulation is especially attractive to males. Specimens of both sexes 1–6 days old were placed in cages, and pairs that mated were collected in a jar containing methylene chloride and stored at 45°F. The pairs were then homogenized with mortar and pestle, the extract was filtered, and the solvent was removed at 100°–120°F with the aid of a stream of air. Filter paper was impregnated with an aliquot of oily extract diluted with methylene chloride and placed in small cup-type traps (24 × 32 × 24 inches) containing free-flying males. A gentle stream of air was found to be essential to obtain positive male responses (preliminary trials showed that only males were attracted into these traps when placed among a mixed population). The tests were conducted during daylight hours (although it could be shown that twice as many males could be caught in complete darkness), using the equivalent of 9–11 mating pairs in each trap. The attractant extract remained highly effective on the lure retainer for at least 32 days at approximately 85°F (*840, 841*).

The terminal 2–3 segments of 4- to 5-day old females were later found to be the best source of the sex attractant. A crude methylene chloride extract of such segments elicited a positive response from males when tested at one-fiftieth of a female equivalent per milliliter. Active extracts were obtained from all females, regardless of time of day collected, with methylene chloride, acetone, benzene, chloroform, ethanol, or methanol. Males responded readily with their characteristic dance to the vapors expelled from a glass pipette contaminated with the attractant. Males used in bioassays were aged for 4–5 days under continuous light, since they did not respond as readily during daylight hours if the lights were turned off at night (*140*).

*Phalera bucephala* (L.), moonspot

The odor of an unmated female enclosed in a room will attract males, but the odor is not detectable by humans (*1212*).

*Philosamia cynthia ricini* Donovan, eri-silkworm moth

The female emits a sex pheromone from its abdominal tip that attracts males (*1190*).

*Phlogophora meticulosa* (L.), angleshade moth

Males remain at rest until the female exposes her sex gland and releases an attractive pheromone. As this dispenses, males become active and fly upwind toward the female (*161, 419, 617*).

*Phthorimaea operculella* (Zeller), potato moth, potato tuberworm moth

Males caged with newly emerged females became highly excited, exhibiting sexual responses such as clasper extension, wing flutter, and spinning flight. Extracts of the terminal abdominal segments of the virgin female evoke the characteristic response in the laboratory and lure males in the field (*9, 989*).

Laboratory bioassays and field trapping tests showed that the pheromone is primarily an excitant rather than an attractant. It elicits a response from 2-day-old virgin males at a concentration of $0.25 \times 10^{-3}$ female equivalents (*564*).

*Platynota stultana* Walsingham, omnivorous leaf roller

Males attracted by virgin females in both laboratory and field tests showed typical sexual behavioral responses including raised antennae, wing vibration, and rapid walking toward the females (*14a*).

*Platysamia cecropia* (L.), cecropia moth

The male moth is attracted to a female, flying upwind until he comes close to her, then he flutters in an aimless manner or directly toward her and mates immediately. Males released 5 feet from several excised female abdomens flew immediately to them, while dead or dying females and those from which the eggs had been removed were not attractive. Cocoons from which females had recently emerged were similarly unattractive. Although males prefer to go to virgin females, previously mated females are also attractive and males mate with them readily. Males make frantic efforts to mate with a female who is in the act of mating with another male (*392, 758*). Boxed males unable to see caged females will fly to the females within 2 minutes after the entire ovipositor is protruded (*1124*).

*Plodia interpunctella* (Hübner), Indian meal moth

Virgin females ready for mating begin calling and soon are surrounded by highly excited males (*933*). Males introduced into boxes from which calling females had just been removed began to flutter their wings violently and behave as though they were in pursuit of a female. Males never become sexually excited in the presence of a noncalling female. The assumption of the calling position, by stretching the intersegmental membrane of the ventral body wall, exposes the mouths of the glands so that the scent is emitted. Calling by the female is not continuous; it is characterized by a dorsally bent abdomen that projects between the wings (*205, 829*). Males show immediate excitement when placed in a dish that contained a female for 3 or 5 minutes, fleeting or weak excitement in a dish exposed to a female for 1 minute, and no response when placed in a dish exposed to a female for one-half a minute. The following effects were shown by a male introduced into a dish at various times after it had contained a female for 5 minutes: 1 or 2 minutes, great excitement; 5 minutes, weak excitement; 10 minutes, excitement of very short duration; 20 minutes or longer, no response (*720*).

Males respond to females with a long-lasting, characteristic dance and wing fluttering. After whirring for 40 seconds to several minutes, short resting periods are observed before the activity begins again. Males introduced into petri dishes that have contained a female always show the characteristic response, even though a long period may have elapsed since removal of the female (*205, 1053*). Surviving males at 1 and 2 days after treatment with an LD$_{50}$ dose of Bidrin (3-hydroxy-*N*,*N*-dimethyl-*cis*-crotonamide dimethyl phosphate) were as responsive to the female sex pheromone as untreated males.

Males of this species introduced into a container of females of *P. interpunctella, Cadra cautella,* and *Anagasta kühniella* were sexually stimulated by all species and attempted to mate (*63, 349, 823, 1053, 1292*). Males of all species responded to their own and to the other females (*1053*), although Barth (*101*) had reported that female *P. interpunctella* do not excite males of the other species sexually.

*Porthesia similis* (Fuessly)

Females produce a sex pheromone to attract males (*1212*).

*Porthetria (Lymantria) dispar* (L.), gypsy moth

Male gypsy moths are attracted to the females by scent (*351, 417, 672, 673*). Urbahn (*1212*) reported that 17 males flew excitedly into a room adjoining his garden, and followed by other males, they went directly to a pupal rearing box. Careful searching located a newly emerged female in a corner, not readily apparent from outside the box.

The female does not fly but the male is a strong flier. The male finds the female by following her scent through zigzag flight, usually against the wind carrying the scent (Fig. II.3) (*304*). Schedl (*1021*) stated that the females cannot fly prior to laying their eggs but are then able to fly for a short distance.

Jacentkovski (*579*) placed virgin females in traps located in infested woods (*578*) and checked them twice a day for male catches. Of numerous males caught in these traps, the majority were trapped during daylight hours, mainly about noon. The effective distance of the lure was about 100 m. Empty containers that had previously held females remained attractive to males for 2–3 days.

Sexually excited males show a whirling dance, moving the wings first with small then with large amplitude; they then attempt to copulate with the females. These dances are quite different from flight movements. Almost all males marked with oil paints and released from 4 directions flew into the wind (*1054*).

When virgin females actively release sex attractant they assume a typical calling position and begin a rhythmic protraction and partial retraction of the last abdominal segments or ovipositor. Following mating, females do not call and avoid males attempting to copulate. In the presence of the attractant, sight apparently assists the male in locating the female (*351*).

*Porthetria dispar japonica* (Motsch), gypsy moth

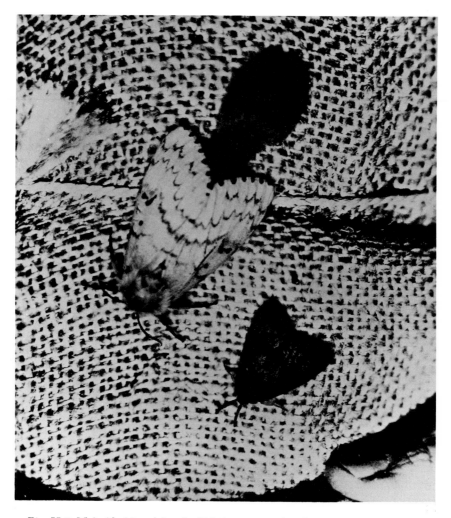

*Fig. II.3.* Male (dark) and female (light) gypsy moths after mating. [By permission of the U.S. Department of Agriculture.]

Olfactometer tests (*1053*) with this insect as well as with *P. dispar* showed definite chemical attraction of the female for the male (*1054*).

*Porthetria* (*Lymantria*) *monacha* (L.), nun moth

The female, who emerges from the pupal case ready to mate and containing ripe eggs, possesses special organs to lure the male (*373*). A single live female placed in a trap lured males in an infested area (*12*).

Although Komareck (*694*), Jacentkovski (*578*), and Ambros (*18*) reported that a female ceases to be attractive upon oviposition, Nolte (*828*) could not substantiate this and believed it to be true only if the female had first been fertilized by the male. Glass vials containing a female and a piece of cotton wool were kept until oviposition occurred; the female was then removed and the vial was stoppered and made airtight with lacquer. Such vials could be used at least a year later to trap males (*363*).

*Prionoxystus robiniae* (Peck), carpenterworm moth

In 1921, Burke (*221*) reported that male moths were attracted during the afternoon to an outdoor cage containing emerged females and males. This was substantiated in 1966 by a field experiment carried out by Solomon and Morris (*1120*).

*Pseudohazis hera* Harris, hera moth

According to Wilson (*1274*), males fly high (8–10 feet) and fast when seeking females. On sensing a virgin female, they backtrack, dropping or slowly flitting around her. They never fly directly toward her when upwind, and if they are some distance to either side of her they never change their flight pattern. Only when they are downwind does their flight pattern change. If the female is encased to prevent her from becoming fertilized, males seem to lose their sense of direction and attempt to copulate with other males.

*Pseudoplusia includens* (Walker)

Rapid production of sex pheromone occurred within 1 day before or after emergence of the female from the pupa (*1095, 1096*). Males respond sexually not only to extracts of females of their own species but also to extracts of female *Autographa biloba* and *Rachiplusia ou*. Extracts of female *P. includens* were likewise attractive to males of *A. biloba* and *R. ou* (*137*).

*Pterosoma palpina* (L.), snout spinner

Females produce a sex pheromone to attract males (*1212*).

*Ptilophora plumigera* (Schiff.)

Females produce a sex pheromone to attract males (*663*).

*Pygaera curtula* (L.)

*Pygaera pigra* (Hufnagel)

Females of these species produce sex attractants for their respective males (*1212*).

*Pyralis farinalis* (L.), meal moth

Males responded with high sexual excitement to methylene chloride extracts of female bodies (*823, 1292*).

*Pyrrharctia isabella* Abbot and Smith, tiger moth

Females attract males by means of a pheromone produced between the last 2 abdominal segments (*953*).

*Rachiplusia ou* (Guenée)

Rapid production of sex pheromone occurred within 1 day before or after emergence of the female from the pupa (*1095, 1096*). Males respond sexually to extracts of female *R. ou, Pseudoplusia includens,* and *Autographa biloba,* and extracts of female *R. ou* are likewise attractive to males of the other species (*137*).

*Rhyacionia buoliana* (Schiff.), European pine shoot moth

Shortly before or after sunset, on the day of emergence, females crawl to exposed places where they release a scent that strongly attracts flying males, frequently from distances (*881*). The production of sex attractant is necessary to stimulate mating (*329*).

Males exhibit precopulatory movements typical of sexual stimulation, comprising rapid wing flutter, upward curving of the abdomen, and opening and closing of the claspers (*329, 880, 1191*). With a "weaving" pattern of flight, stimulated males "home in" on the attractive source. Such precopulatory movements are exhibited only by males downwind from attractive females or methylene chloride extracts of the eighth and ninth virgin female abdominal tips (*329–331*). Males respond to the pheromone over distances of at least 185–190 m (*332*).

*Rhyacionia frustrana* (Comstock), Nantucket pine tip moth

A total of 24 plywood traps, similar to those used by Coppel *et al.* (*310*), were baited with virgin females, the boards were coated with Tanglefoot, and the traps were suspended about 5 feet above the ground in a stand of loblolly pine at 15-foot intervals; traps were arranged randomly in 2 rows. Males were attracted to these traps, even to those in which mixed sexes had been placed. Unmated females were not attractive after death. Virgin females were most attractive early in their life span (up to 9 days old), and became unattractive with age. No females were caught in any trap during the 10-day exposure, while a total of 200 males were trapped. Females were most attractive during the period from 8:30 to 9:30 PM (*748, 1296*).

*Rhyacionia zozana* (Kearfott), ponderosa pine tip moth

Males become sexually stimulated by female attractant, performing precopulatory motions, including rapid wing fluttering, anterior curving of the abdomen, and opening and closing of the claspers (*331*).

*Rothschildia orizaba* (Westwood), orizaba silk moth

Electrophysiological bioassay showed the female to be attractive to the male (*1028*). Glands of females of the species of *Saturnia* and *Antheraea* likewise elicited electroantennograms from male *R. orizaba* (*1033*).

*Samia cynthia* (Drury), cynthia moth

In 1902, Soule (*1123*) reported that males became so excited by the odor of females that they would fasten their claspers on any part of a female body or even on each other. Females also attracted male promothea moths. Females assume a characteristic calling position with the tip of the abdomen raised (*1113*). Electroantennograms were also obtained by exposing the female gland to male antennae (*1033*).

*Sanninoidea exitiosa* Say, peach tree borer

The moths copulate during the morning of the day they emerge from the pupal sacs, except if the weather is cool. The male, attracted by an odor emanating from the female, comes up rapidly against the wind. A newly emerged female in a trap was placed in an apple tree situated approximately one-third of a mile from the nearest peach orchard; she lured males for 7 days (*74, 198*). When ready to mate, the female lifts the abdomen above the plane of the resting position and protrudes her genitalia; many males are readily attracted (*460*).

Soon after emergence the virgin female fixes herself at rest, elevates the abdomen, projects the ovipositor with the genital organs directed downward with the tufts expanded, and waits for a male (*311, 460, 1113, 1114*). Males are strongly attracted in nature between the hours of 10:00 AM and 1:00 PM (*311, 581*), but female calling could be shifted in the laboratory by exposing females to light cycles occurring both earlier and later than the solar day (*580*). According to Jacklin *et al.* (*581*), caged females in an orchard first assumed the calling position at about 12:05 PM on the day they emerged and at about 11:13 AM on subsequent days. The activity of responding wild males was greatest between 11:00 AM and 1:00 PM. Calling females 2–3 days old were most attractive to males.

*Saturnia carpini* (Schiff.)

Forel (*420*) reported that several females placed in a window in Lausanne brought a swarm of males; this was also reported by Poulton (*883*).

*Saturnia pavonia* (L.), emperor moth, peacock moth

A newly emerged female attracted, from a distance, 127 males between 10:30 AM and 5:00 PM (*1136*). A newly emerged female lured 40 males in 1 evening and numerous others on 8 additional evenings (*389*). The female, who possesses scent sacs, remains inactive all night and until forenoon, then begins calling until evening; males fly mainly during the forenoon and early afternoon (*1212*).

*Saturnia pavonia minor* (L.), lesser peacock moth

A female may attract numerous males (*71*). Many males visited a captive female between noon and 2:00 PM (*391, 883*). However, 2 males with their antennae removed could not locate a female they had previously found with ease (*360*).

*Saturnia pyri* (L.)

A female attracts numerous males (*1212*). A newly emerged female held captive in a wire cage in Fabre's home attracted, in the hours before midnight, numerous males through an open window of another room (*390, 883*). Electroantennograms were obtained by exposing male antennae to female extracts (*1033*).

*Sitotroga cerealella* (Olivier), Angoumois grain moth

A pheromone produced by females is attractive to males, eliciting darting about, rapid wing whirring, and attempts to copulate with nearby objects, including other males. Virgin females are more attractive than mated females of the same age. Attractiveness is greatest when the females are 48–77 hours old, but some are still attractive several hours after death (*201, 667*). An extract of female *Pectinophora gossypiella* moths, a closely related member of the Gelechiidae, caused no response in male *S. cerealella* (*201*).

*Smerinthus ocellata* (L.), eyed hawk moth

A single virgin female in a muslin cage hung outside a window attracted 42 males in 8 days (*1252*). A caged female set next to a cage containing a male *S. ocellata* and a female poplar hawk moth, *Amorpha populi*, stimulated the male to copulate with the female *A. populi*. The insects resulting from this union, called "popeyed" hawk moths, were sterile. A female *S. ocellata* likewise mated with a male *A. populi* in preference to a male of her own species (*1317*).

*Solenobia fumosella* (Hein.)

Males of this species immediately begin to flutter if a newly emerged, calling female *S. triquetrella* is brought into their vicinity, but copulation does not occur. However, if the male is first excited by a female of his own species and, at the last moment, a female *S. triquetrella* is substituted underneath him, copulation is successful (*1058*).

*Solenobia lichenella* (L.)

*Solenobia seileri* (Sauter)

*Solenobia triquetrella* (Hübner)

The members of these 3 species are parthenogenetic and can be crossed specifically. Females do not leave the pupal sac but protrude their abdomen from it and mating occurs with attracted males. Introduction of a pupal case with a calling female into a container holding a resting male causes the male to begin fluttering; he flies to her, the ovipositor is withdrawn into the sac, and mating occurs (*1059*).

*Sparganothis directana* Walker

Live females and their crude extracts were attractive to males (*957*).

*Sphinx ligustri* (L. ), privet hawk moth

Males of many hawk moths are able to detect the virgin females from incredible distances (*808*). Females of this species attract their own males as well as those of *Smerinthus ocellatus* with whom they may copulate (*860*).

*Spilosoma lutea* Hufn., buff ermine moth

Female *S. lutea* held captive in a wire-net insectary in a garden attracted male *Arctia caia* in the morning (*370*). Virgin female *S. lutea* placed in a gauze-covered box in an assembling cage caught 23 male *S. lutea* and 1 male *A. caia* by morning (*1297*).

*Spodoptera* (*Prodenia*) *eridania* (Cramer), southern armyworm

In 1962, Butt (*260a*) showed that the female possesses a sex attractant for the males which is produced in the abdominal tip, and Redfern (*913*) has developed a bioassay method for determining activity.

*Spodoptera exigua* (Hübner), beet armyworm

The sex pheromone produced by the female can be bioassayed by Shorey's method (*442, 1096*).

*Spodoptera frugiperda* (J. E. Smith), fall armyworm

Virgin females produce a pheromone within the last abdominal segment that causes males to become very excited, vibrate their wings, fly to the source of the stimulant, and attempt to copulate with it and with one another. It is primarily a mating stimulant and not a distance attractant (*43, 1062, 1063*). The male response lasts from 1 to 3 minutes but can be reinduced after a rest period of 3–5 hours (*1061*).

*Spodoptera litorallis* (synonym: *Prodenia litura*) Fabr., Egyptian cotton leafworm

The female attracts the male with an odor that appears 1 hour after emergence (*407, 408*), but it is produced in highest yield when the insect is 1–2 days old (*504*). A bioassay technique has been developed (*96*).

*Spodoptera ornithogalli* (Guenée), yellow-striped armyworm

The female emits an attractant for the male; it has been bioassayed by Shorey and his co-workers (*442, 1096*).

*Stilpnotia salicis* (L.), satin moth

Males are attracted and sexually excited by virgin females (*1054*).

*Synanthedon pictipes* (Grote & Robinson), lesser peach tree borer

The female is highly attractive to the male by scent (*198, 454, 460*). Manual squeezing of the female's abdomen causes the protrusion of the terminal 3 segments. They are normally protruded to their utmost when the females are calling, at which time the scent will attract males from as far away as 500 feet. Excised abdomens remain attractive for about 20 minutes (*302, 460*). One-day-old females seem to be most attractive (*1280*).

*Thyridopteryx ephemeraeformis* (Haworth), evergreen bagworm moth

Males are attracted to female pupal sacs to which they cling during mating. Receptive females call from inside these sacs by means of a sex pheromone (*653*). The attracted male introduces his abdomen into the concealed lower opening of the case to mate. The male may be attracted over a mile or more (*626*).

*Tineola biselliella* (Hummel), webbing clothes moth

Males attracted to calling females by odor move about actively, vibrating their wings. Antennaeless males cannot perceive the odor (*1182*). The sexually excited male walks about rapidly with his abdomen extended, vibrating or fluttering his wings continuously. This behavior may be observed as soon as 4 hours after emergence. Females observed in the calling pose protrude and retract the ovipositor and vibrate its tip. Excised abdomens, but not the heads and thoraxes, of 1-day-old females were attractive to males; a few males made genital connection with the severed abdomens. Males are often attracted to copulating pairs. One female abdomen will activate a male from a distance of 1.5 cm, whereas 10 severed abdomens will attract from 7 cm. Males do not respond to male abdomens, but they may court in the presence of other males (*986*).

*Trabala vishnu* (Lef.)

A newly emerged female will attract large numbers of males (*478*).

*Trichoplusia ni* (Hübner), cabbage looper

An attractive substance emitted by the female during or immediately preceding copulation stimulates males to activity (*469, 567*). The pheromone may be extracted from the abdominal segments of virgin females with several solvents (*441, 567, 1093*). Excellent behavioral bioassay methods have been developed for the attractant (*441, 442, 567, 1083, 1093, 1096*). Males of *Autographa californica* attempted to mate with female *T. ni*, but successful coupling did not occur (*1093*).

*Vanessa urticae* (L.)

Females attract and excite males by means of a chemical scent (*1212*).

*Vitula edmandsae* (Packard), bumblebee wax moth

Receptive, virgin females assume a typical calling pose, with the wings resting along the dorsal side and the abdomen bent sharply out from the normal body position. The sex pheromone scent released at this time from the glands in the abdominal tip causes intense excitement, antennal movement, and wing flutter in males (*47, 1250*).

*Zeadiatrea grandiosella* (Dyar), southwestern corn borer

Sticky board traps provided with caged live virgin females lured numbers of males in the field (*334*).

*Zeiraphera diniana* (Guenée), larch bud moth

Traps baited with live virgin females attracted large numbers of males in European alpine larch forests (*953a*).

# CHAPTER III

# SEX PHEROMONES PRODUCED BY MALES

ACARINA

*Amblyomma maculatum* Koch, Gulf Coast tick
Females released on a bovine in an environmentaly controlled stall failed to attach in the absence of males, but did attach close to attached males on other bovines (*455*).

ORTHOPTERA

"Male tergal glands in cockroaches of the family Blattaria serve to maneuver the female into the proper precopulatory position and arrest her movement (while she feeds on the gland secretion, or palpates the male's dorsum) long enough for the male to clasp her genitalia." Roth (*979*) has reviewed in great detail the evolution of the male tergal glands in this family. Where behavioral observations are lacking, the presence of tergal specializations in the adult male *only* is strong evidence that a male sex pheromone is produced to attract the female to the temporary position above the male prior to copulation. However, the absence of visible tergal glands does not mean that the female is *not* attracted to the male's dorsum. Males without visible tergal specializations, but who attract females to

their backs to feed, undoubtedly have pheromone-producing glands opening
somewhere on the tergites.

*Byrsotria fumigata* (Guérin), cockroach

A pheromone produced by the courting male causes the female to
straddle the male's abdomen and begin feeding on his tergum (*981, 987*).

*Ectobius pallidus* (Olivier), cockroach

Receptive females are attracted to the male's back by an odorous
substance produced by the tergal gland, which is exposed by the male
raising his wings. The female then palpates or feeds on the secretion just
prior to mating (*978*).

*Eurycotis floridana* (Walker), Florida wood roach

The mating behavior of this roach is considerably different from the
pattern typical of cockroaches. The female initiates behavior by approaching the wingless male from a short distance (there is no wing-raising
display). The male stands near the female repeatedly vibrating his body
from side to side and extending his abdomen to reveal the intersegmental
membrane between the sixth and seventh tergites (*107, 987*). The male
pheromone is volatile and serves the function of sex recognition and
attraction of females. The female mounts the male and feeds on his tergal
secretion for a considerable time; tactile stimulation of the abdominal
dorsum elicits the male's copulatory thrusts (*113*).

*Leucophaea maderae* (F.), Madeira cockroach

Upon sensing the female, the male becomes restless, waves his antennae,
and may engage in vertical body vibration. He then assumes a typical
wing-raising position in front of the female, who mounts and feeds on the
terga if receptive (*383, 1117*).

*Nauphoeta cinerea* (Olivier)

The male stridulates during courtship of nonreceptive females and raises
his wings and tegmina to expose his tergum. The sound is produced by
displacement of the pronotum rubbing against the costal veins (*523, 524*).
The males produce a sex pheromone, named "seducin," principally in the
abdomen, which attracts the female over a short distance and functions as
an arrestant to keep her in the proper position long enough (about 6
seconds) for connection to be made. Seducin stimulates receptive virgin
females as well as females shortly after parturition (*978, 983.*)


HEMIPTERA

*Lethocerus indicus* (Lepetier & Serville), giant water bug

During sexual excitement the male is readily recognized by his odor; his
abdominal glands secrete a liquid with an odor reminiscent of cinnamon

(*242*, *265*). This substance, produced in 2 white tubules, 4 cm long and 2–3 mm thick, occurs to the extent of 0.02 ml per male and is used in southeast Asia as a spice for greasy foods. The female does not secrete the substance, which is believed to act as an aphrodisiac to make her more receptive to the male.

*Musgraveia* (*Rhoecocoris*) *sulciventris* (Stal.), bronze orange bug

A substance obtained from both sexes of this insect may act as an attractant or aphrodisiac (*844*).

*Nezara viridula* (L.), southern green stink bug

Virgin males produce a pheromone that is highly attractive to virgin females (*792*).

*Oncopeltus fasciatus* (Dallas), large milkweed bug

Adult males at least 10 days old, reared in isolation or with only male container mates, produce a pleasant fruity aroma in their containers. The scent, readily detectable by humans, is not produced by females. The aroma apparently serves as a pheromone for the female (*723*, *724*). All males do not produce the aroma, nor do secreters do so consistently. Aroma production is irregular and unpredictable. The secretion does not occur when an adult female is in the same container (*724*).

*Rhodnius prolixus* (Stal.)

Air passed over live males was attractive to females (*1215*).

## TRICHOPTERA

*Sericostoma personatum* (Spence)

A strong odor of vanilla emitted by scent glands on the maxillary palpi of males is thought to be attractive or excitatory to females. An anatomic description of the glands is given by Cummings (*320*).

## COLEOPTERA

Males of Malachiidae, a family of tiny tropical beetles, entice females first with a tasty nectar and then expose them to an aphrodisiac. The males possess tufts of fine hair growing out of their shells (in some species on the wing covers, in others on the head). These hairs are saturated with a glandular secretion that the females cannot resist. During the mating season, the male searches for a female; when he finds one he offers his tuft of hair, which the female then accepts and nibbles upon. In so doing, her antennae come in contact with microscopic pores in his shell through which the aphrodisiac substance is secreted, thus putting her in a state of wild excitement (*33*, *755*).

*Acanthoscelides obtectus* (Say), dried bean weevil

Males, but not females, produce a sweet, fruity odor, mainly 3–8 days after emergence, that may stimulate the emergence of females or act as a female attractant. Hexane washes of various parts of the male body show that the secretion is associated with the thorax and abdomen, but not with the genitalia, associated glands, hindgut, malpighian tubules, or anus. Since maceration of the insect does not increase the amount of material obtained over that recovered by washing of the whole insect, it is probable that the material is present in the outer cuticular surface or in a gland opening on it. Hexane washes of whole males showed a single peak on gas chromatography, but this may not have been caused by the pheromone itself (*559*). See Chapter X for the identification of this pheromone.

*Anthonomus grandis* Boheman, boll weevil

A substance produced by males attracts females from a distance of 2 to more than 30 feet. Females aggressively seek the males, who become attractive at 2 days of age, and peak response occurs in the laboratory when both sexes are 4–6 days old. Virgin males are twice as attractive and virgin females are 3 times as responsive as mated males or females. Males do not respond to females over distances greater than 1 or 2 inches (*102, 103, 233, 655*). Males sterilized with apholate were about half as attractive to virgin females as untreated males when both were fed on laboratory diet, but were equally attractive when both were fed on fresh cotton squares (*520*). During the peak period of sexual activity, females irradiated with 6388 or 12,775 rad of ⁶⁰Co gamma radiation were as responsive to both irradiated and unirradiated males as were untreated females during the same time period. Treated males were as attractive to treated females as untreated males (*117*). Field tests showed the pheromone to be windborne (*317*). Live males attract approximately equal numbers of males and females in the spring and fall, indicating that the active substances may act as an aggregating pheromone as well as a sex pheromone (*1205*).

*Carphoborus minimus* Fabr.

Males feeding on host conifers are attractive to females, as are methanol or ether extracts of males killed by exposure to a temperature of −20°F (*296*).

*Ips pini* (Say), pine engraver

A male makes an initial attack on the inner bark of jack pine, *Pinus banksiana*, prepares a nuptial chamber, and is soon joined by 1 or more females. The attraction of such infested logs is not dependent on the odor of the extruded boring dust. A pleasant odor detectable close to freshly made entrance holes suggests that the attraction may be due to an odoriferous substance emitted by the males (*31*).

*Ips* spp.

Female *I. plastographus*, *I. confusus*, and *I. calligraphus* responded at a much lower level to frass (a mixture of wood fragments and excrement) produced by males of species other than their own. Female *I. confusus* responded at the same level to frass of male *I. confusus* and *I. montanus* (*1286*).

*Mylabris pustulata* Thunbg., blister beetle

Adult males possess a pair of white tubular glands lying ventrally below the alimentary canal in the region of the first 3 abdominal segments and the metathorax, and opening on the mesosternum. Gently pressing the thorax of a live beetle causes the discharge of a whitish, slightly acidic fluid from the gland opening. Although the function of the glands is not definitely known, the fact that they are only rudimentary in females suggests that their secretion is a sex pheromone that attract females (*120*).

*Necrophorus vespillo* (L.), carrion beetle

Males vibrate, sometimes for hours, the tips of their abdomen, releasing at the same time an odorous attractant for females (*571, 756*).

*Necrophorus* spp., carrion beetle

Males of *N. fossor*, *N. germanicus*, *N. humator*, *N. investigator*, *N. vespillo*, and *N. vespilloides* dig up a dead object that they have previously buried, stand on the object with the head down and body raised, and vibrate the last abdominal segment to attract females to them (*896*).

*Pityogenes chalcographus* L.

Males feeding on the wood of their host plants attract numbers of females (*296*).

*Pityokteines spinidens* Reit.

Males feeding on the wood of their host plants attract numbers of females (*296, 297*).

*Pityokteines curvidens* Germ.

*Pityokteines vorontzovi* Jacob.

*Pityophthorus pityographus* Ratz.

Males of these species attract their respective females through feeding on host plants to produce a pheromone (*296*).

*Tenebrio molitor* L., yellow mealworm

Tschinkel (*1204*) found that ethanol extracts of whole males from all-male colonies caused other males to attempt copulation with a glass test rod in a manner indistinguishable from that toward rods treated with female extract. The production of pheromone per individual male attained a maximum after about 5 days.

According to Happ (*508*), males produce an excitant that attracts females and an antiaphrodisiac which inhibits the response of other males to female scent. In laboratory olfactometer tests, females responding to the male scent often extruded their ovipositors, suggesting that male scent may promote rapid oviposition. Extracts of males were 50 times more potent toward the females than were extracts of females. The male emitted the inhibitory pheromone only after stimulation by female scent, and the pheromone appears to be transmitted to the female during mating.

HYMENOPTERA

*Acanthomyops claviger* (Roger), ant
Males exhibiting preflight behavior were crushed, whereupon the sweet odor of volatile terpenes followed by the fecal odor of an indole compound were detected. Both odors appear to originate in the reservoir of the mandibular glands; the thorax and abdomen have no odor. The mandibular-gland reservoirs are relatively large structures, turgid with volatile liquids prior to the nuptial flight. The substances may serve as sex pheromones *(714)*.
*Anthidium* spp.
*Anthophora acervorum* (L.)
Males of these species produce in their mandibular glands pheromones that draw females to them during swarming *(498)*.
*Bombus* spp., bumblebee
The flight patterns of 11 species of bumblebee males bring them to flowers and bushes that are then visited by queens and possibly other males. During the course of this attraction, copulation occurs. The species are *B. agrorum* Fab., *B. elegans* Seidl., *B. hortorum* L., *B. hypnorum* L., *B. lapidarius* L., *B. mendax* Gerst., *B. pratorum* L., *B. pomorum* Pz., *B. silvarum* L., *B. terrestris* (L.), and *B. variabilis* Schmied. *(497, 498, 1110)*. The materials responsible for this attraction are floral odors, apparently secreted by the mandibular glands. The odors of *B. hortorum* and *B. terrestris* are roselike. Although the males of all species are fragrant, *B. distinguendus* Morawitz and *B. latreillellus* Kialy are especially so. Males of 1 species do not pause at spots frequented by those of the other species, suggesting that each species emits a different scent *(496, 701, 1110)*.

These substances are now recognized to be perfume markers used for maintaining biological isolation between species by acting as species-specific attractant pheromones. They probably also have the ability to enhance the disposition of the males for copulatory activity. They are deposited on different objects during male flight *(141)*.
*Camponotus herculeanus* L., carpenter ant
*Camponotus ligniperda* Lat., carpenter ant
Swarming (mating flight) starts at 5:00–7:00 PM in these species. The

males release a strong-smelling secretion from their mandibular glands that stimulates the females to take off. Although it was first believed that females were also attracted by these male scents, this was disproved by means of a balloon test similar to those conducted with honeybees. Females of *C. herculeanus* are also stimulated by the secretion of *C. ligniperda* males (*556, 558*). Crushed, decapitated males, crushed whole females, jasmine oil, and geraniol failed to stimulate males (*558*).

*Eufriesia* spp., euglossine bees
*Euglossa* spp., euglossine bees
*Eulaema* spp., euglossine bees
*Euplusia* spp., euglossine bees

Males of these genera are attracted to orchid flowers by various fragrances. Females are not attracted to these fragrances, but they are attracted to the territorial displays and odors of males, and copulation occurs at the site. It is possible that males convert the floral secretions into sex attractants (*353*).

*Lasius alienus* (Forster), cornfield ant
*Lasius neoniger* Emery, ant

Males of these species discharge most of their mandibular gland contents during the nuptial flight, almost certainly as an aerial courtship signal. The glands contain 2,6-dimethyl-5-hepten-1-ol, citronellol, geraniol, and a few other simple compounds. Perhaps as an evolutionary solution to the problem of interspecific sexual isolation, the proportions of the components in each species differ considerably (*714, 1269*).

*Psithyrus* spp.

Males produce pheromones in their mandibular glands that attract females to them during swarming (*498*).


DIPTERA

*Ceratitis capitata* (Wiedemann), Mediterranean fruit fly

Mature virgin females are attracted over a short distance and sexually excited by a volatile chemical substance emanating from the erectile anal ampuls of sexually mature males (*400, 401, 610, 822*).

*Cochliomyia hominivorax* (Coquerel), screw-worm fly

A male remaining quiescent for a few minutes after being introduced into a cage containing mature virgin females is often suddenly approached and touched by one or more females. The approach is usually accomplished by a short jump, a series of jumps, or by a short flight. Females will often approach a mating pair and initiate mating overtures, the most prominent feature being extension of the ovipositor. Mated females and virgin males do not appear to be affected by males. A volatile substance has been collected from virgin males which elicits wing fluttering, preening, and exploratory movements in virgin females (*414*).

Unmated females of a Florida strain that had been colonized for 11 years responded vigorously to male pheromone from that strain and from a laboratory-reared Mexican strain, but females from the Mexican strain did not respond to the pheromone from either strain. This behavior may possibly be explained by the selection taking place during mass rearing, since adults of the Florida strain were kept in constant darkness for 11 years and therefore may have needed a sex pheromone for locating the opposite sex (*413, 415*).

*Dacus cacuminatus* (Hering), solanum fruit fly

*Dacus cucurbitae* Coquillett, melon fly

*Dacus dorsalis* Hendel, oriental fruit fly

*Dacus oleae* (Gmelin), olive fruit fly

*Dacus tryoni* (Frogg.), Queensland fruit fly

Failing light increases the activity of *D. cacuminatus* and *D. tryoni;* the male always initiates this activity by commencing to call with rapid wing vibration (stridulation). The calling, clearly heard as rather high, flutelike notes emitted in series, stimulates females who are ready for copulation to approach the male. The females of *D. cacuminatus* were attracted to the calling male of the opposite species only in the absence of their own males (*820*). The male *D. tryoni* also produces a sweet-smelling sex pheromone which is released at the time of stridulation. It is likely that the pheromone acts both as an attractant and an excitant, while the sound produced during stridulation assists in the precise location of the male (*411, 412*). Pheromone release in these species is accompanied in the male by wing movements which probably facilitate the evaporation and circulation of the pheromone by producing air currents. The sex pheromone gland is associated with the rectum of the male (*369a, 1051*).

*Drosophila melanogaster* (Meigen), vinegar fly

Spieth, cited by Mayr (*763*), surmised that the wing flutter shown by males before mating probably serves to direct an air stream, containing a sex scent, toward the courted female to stimulate her and increase her receptivity. A lipid found exclusively in the ejaculatory bulb of adult males has been isolated and identified. Although its physiological function has not yet been established, it may involve some aspect of reproduction since the lipid is transferred to females during mating (*207*).

*Drosophila victoria* (Sturtevant)

Males produce an attractant (or aphrodisiac) for the females (*1132*).

*Musca autumnalis* De Geer, face fly

The presence of males may be a stimulus affecting the activity of the females, attracting them for copulation (736).

*Nezara viridula* (L.), southern green stink bug

Males produce a pheromone that is highly attractive, in both the laboratory and the field, to females and to its tachinid parasite *Trichopoda pennipes* (F.) (792).

*Rioxa pornia* (Walker), island fruit fly

Just prior to mating, the male takes up a stance with his abdomen raised from the substrate and the pleural regions of abdominal segments 3–5 distended; the wings are moved occasionally in long, slow sweeps. The odor produced is easily detectable by humans as an unpleasant smell at distances of 50 cm or more from the fly. Females downwind from such a male become active, walking about and quivering their abdomens, and then walk directly toward the male; this strongly suggests that a volatile sex attractant is involved. The male then produces a mound of white foam, consisting of a mass of air bubbles in a gummy, semitransparent substance, toward which the female moves and upon which she proceeds to feed. This apparently serves to keep the female quiescent, since the male immediately leaps on her and assumes the copulatory position (888).

## Lepidoptera

Much has been written, particularly in the older literature, about the odor glands of male moths and butterflies, describing their anatomy and physiology. A review of the subject (327) appearing in 1885 extends back to the seventeenth century. In 1878, Müller (807) described the occurrence of an odor, perceptible to humans and probably serving to lure or excite the female, in males of 44 species of butterflies. The odoriferous organs are usually located on the wings, but they sometimes occur on the hind legs or abdomens (160).

*Acherontia atropos* (L.)

The male abdomen contains odor organs important in sexual behavior (572).

*Achroia grisella* (Fabr.), lesser wax moth

The male attracts and excites the female. A total of 25 males placed in a gauze-covered beaker set in the corner of a large rectangular dish lured 25 virgin females from the opposite corner. An attempt to lure males with virgin females was unsuccessful. A female placed in a cage with a male immediately approached him; she began to flutter and copulation followed shortly (324, 702).

Dahm *et al.* (325) have shown that females attracted by males begin to search for them and are guided by sound perceived by abdominal tympanic organs.

*Apamea monoglypha* (Hufnagel)

Males possess scent brushes supported on a lever attached to the base of the first apparent abdominal sternite; they lie in a lateral pocket stretching from sternites 3 to 5. The characteristic scent (see Chapter X) released by these brushes probably acts on females as an aphrodisiac to induce mating *(72, 73)*.

*Aphomia gularis* (Zeller)

Wing glands in the male secrete an odorous substance attractive to the female *(100)*.

*Argynnis adippe* (L.)

*Argynnis aglaja* (L.)

*Argynnis paphia* (L.), emperor's cloak

Sex organs situated on the wings of males of these species secrete odorous substances which excite the females sexually *(102, 743)*.

*Bapta temerata* (Denis & Schiff.), clouded silver moth

The male possesses elaborate scent brushes to disseminate a scent that calms the female prior to courtship *(163)*.

*Cadra (Ephestia) cautella* (Walker), almond moth

Males bend their abdomen in a dorsal direction similar to the female calling position; this exposes scent glands that produce an odor that is sexually exciting to females. The absence of these scent tufts in the male may prevent mating *(349)*.

*Caligo arisbe* (Hübner)

Glandular lamina of the fourth and fifth abdominal segments in males produce an odor that is initially pleasantly aromatic but later becomes rancid. This odor probably serves to sexually excite the female *(103)*.

*Colias edusa* (Fabr.)

Wing scales of the male release an odorous substance which is sexually exciting to the female *(572)*.

*Creatonotos gangis* (L.)

Tubelike organs (coremata) supporting the scent-dispersing hairs of the male inflate with air to more than 3 times body size; the scent released calms the female during courtship *(163)*.

*Danaus gilippus berenice* (Cramer), Florida queen butterfly

The male pursues the female, overtakes her in the air, and induces her to alight by rapidly brushing her antennae with 2 scent-disseminating hair-pencils which are extruded from the posterior of his abdomen *(210)*. Males deprived of the 2 hair-pencils are capable of courting females but are incapable of seducing them. Of 2 substances identified from the secretion (see Chapter X), 1 acts as the chemical messenger inducing the females to mate. The only known function of the other compound is to serve as a glue that sticks the scent dust to the female (879).

*Danaus gilippus xanthippus* Felder, Trinidad queen butterfly

Males evert their hair-pencils into their wing pockets several times each day prior to courtship. Since the use of the hair-pencils (which produce the scent) during courtship is prerequisite to mating, it is inferred that interaction of the 2 glands has a physiological function (211).

*Danaus plexippus* (L.), monarch butterfly

Males possess odor glands on the wing folds and at the abdominal tips. The glands present in the posterior abdomen are extrusible, brushlike structures (hair-pencils) that serve for pheromone dissemination during courtship; the scent may also serve to excite the female sexually (572, 776).

*Elymnias undularis* (Dru.)

Males emit a vanilla odor which excites the females just prior to mating (499).

*Ephestia elutella* (Hübner), tobacco moth

Wing glands of the male secrete an odorous substance that increases the female's excitement during copulation. Vibration of the wings disperses the odor (100).

*Erynnis tages* (L.)

Scent glands on the costal margins of the male's wings release an aphrodisiac for females (572).

*Eumenis semele* (L.), velvet butterfly

Special scent organs of the male are displayed to the female by means of a peculiar bowing movement. The odor produced evokes the mating attitude in the female (1178, 1181).

*Euplagia quadripunctaria* Poda, Jersey tiger moth

Males possess scent brushes that disseminate a scent to calm females during courtship (163).

*Euploea phaenareta* (Schall)

Males emit a vanilla odor which excites the females just before mating (499).

*Euploea* sp.

The male abdominal tips and wing scales contain glands that emit an odor that is sexually exciting to females (572).

*Eurytides protesilaus* (L.)

Hairy scent glands, pierced by a single canal, on the male's wings, secrete an odorous substance that may act as an aphrodisiac for females (1252).

*Galleria mellonella* L., greater wax moth

Vöhringer (*1236*) and Barth (*103*) reported that males release a scent perceptible to man that they described as "musklike," and they presumed that the attraction of females to males was related to this scent. The male vibrates his wings and dances around, dispersing the odor. The odor, released from forewing glands present only in males, may be produced as soon as 12 hours after the male emerges; the female approaches with a circling dance, fanning her wings rapidly and undulating her abdomen. Although Röller *et al.* (*974*) detected the odor in males about 2–24 hours old, it was not musklike; it was at its maximum during evening hours. Röller *et al.* (*972, 973*) confirmed the release of a sex attractant by the male wing glands and succeeded in identifying it (see Chapter X).

*Hammaptera frondosata* Guérin

Males possess wing glands that secrete an odorous substance that, in flight, keeps a female from flying away. Also, hair-pencils situated between the eighth and ninth abdominal segments produce a substance whose odor sexually excites females just prior to copulation. Flight occurs in early evening and is not observed after 10:30 PM (*105*).

*Hepialus behrensi* (Stretch.)

Glands on the leg of the male emit a scent attractive to females (*942*).

*Hepialus hectus* (L.)

Glands on the tibia of the male emit a scent attractive to females (*342, 942*). The scent was described as being "pineapplelike" by Deegener (*341*), who also studied the anatomy and physiology of the glands.

*Hipparchia semele* (L.), grayling butterfly

The male is provided with scent scales on his wings. At the climax of courtship, he clasps the female's antennae between his wings, bringing them in contact with the scent scales and causing the female to become receptive to his advances (*1179*).

*Illice fasciata* (Schaus), lichen moth

Males approach females with vibrating wings to spread the aromatic scent from their wing glands, which prevents the females from flying away. A male abdominal scent then excites the female just before copulation (*756*).

*Lethe rohria* (F.)

Males emit an odor that excites the females just before mating (*499*).

*Leucania conigera* (Schiff.), brown-line bright-eye
*Leucania impura* (Hübner), smoky wainscot moth
*Leucania pallens* (L.), common wainscot moth

Males of these species possess scent brushes supported on a lever attached to the base of the first apparent abdominal sternite; they lie in a lateral pocket stretching from sternites 3 to 5. The characteristic scent (see

Chapter X) released by these brushes probably acts on females as an aphrodisiac to induce mating (*72, 73, 163*).

*Lycaena* spp.

The male has scent glands scattered over his wings that release an odor sexually exciting to females (*572*).

*Lycorea ceres ceres* (Cramer)

The males of this Trinidad butterfly possess a pair of elaborate organs (hair-pencils) that can be extruded from the end of the abdomen. These organs play an important role in courtship, probably producing and disseminating a sexual scent that serves to seduce the female (*778, 780*). (Several constituents identified in this secretion are discussed in Chapter X.)

*Mamestra brassicae* (L.)

*Mamestra persicariae* (L.)

Males of these species possess scent brushes whose characteristic scent (see Chapter X) probably acts on females as an aphrodisiac to induce mating (*73*).

*Mycalesia suaveolens* (W. M. & N.)

Males emit an odor that excites the females just before mating (*499*).

*Oncopera alboguttata* Tindale

*Oncopera rufobrunnea* Tindale

*Oncopera tindalei* Common

After eclosion, males take to the air first with the abdomen curved upwards in a characteristic manner. A few minutes later the females take to the air. It is probable that the males produce a pheromone to attract the females (*119*).

*Opsiphanes invirae isagoras* Fruhst.

Males possess glandular lamina on the fourth and fifth abdominal segments that secrete a volatile material with a vanillinlike odor which may serve to excite the female before or during mating. It can be extracted from the glands with chloroform. The odor can be detected by humans after the material has stood for about 2 weeks at 25°–35°C; it becomes weak after 3 weeks, and is no longer detectable after 4 weeks. Crystals found in the secretion probably do not account for the odor, as they can still be seen after the odor is no longer detectable (*103*).

*Otosema odorata* L.

The hind legs of the male possess a honeylike odor that may serve to excite the female sexually (*104*).

*Panlymnas chrysippus* L.

Males possess odor glands on the wing folds. The odorous substance produced by these glands probably serves to excite the female sexually (*572*).

*Pantherodes pardalaria* (Hübner)

Males emit a musklike odor from abdominal hair-pencils and also from wing glands that probably attracts and (or) excites females (*806*).

*Papilio aristolochiae* F.

The male emits an odor to excite the female just prior to mating (*499*).

*Pechipogon barbalis* (Cl.)

Glands situated on the tibia and femur of the male secrete a substance that excites the female sexually (*572*).

*Phassus schamyl* (Chr.)

Males possess odor glands on the tibia which secrete a substance attractive to females (*342*).

*Phlogophora meticulosa* (L.), angleshade moth

The male produces a substance sexually attractive to the female (*419*). The attractive substance (see Chapter X) is produced by abdominal scent brushes, which are everted by the male in the immediate vicinity of the female immediately before copulation. Copulation appears to be attempted only after eversion of the brushes (*72, 73, 161*).

*Phragmatobia fuliginosa* L., ruby tiger moth

The male produces a scent in his scent-brush organs, which is disseminated by the coremata and calms the female during courtship (*163*).

*Pieris napi* L., mustard white

*Pieris rapae* L., imported cabbageworm

In these species, glands scattered on the male's wings secrete an odorous substance which is sexually attractive to their respective females (*572, 1242*).

*Plodia interpunctella* (Hübner), Indian meal moth

A male attracted to a female by her odor dances around her vibrating his wings (*101*). In so doing, he releases and disperses from his wing glands an odorous substance that serves to increase the female's excitement during copulation (*100*). Females placed in a dish previously occupied for 5 minutes by a sexually excited male immediately began moving about excitedly, waving their antennae, and assuming a calling position (*720*).

*Polia nebulosa* (Hufnagel)

Males possess scent brushes whose characteristic scent (see Chapter X) probably acts on females as an aphrodisiac to induce mating (*73*).

*Sanninoidea exitiosa* Say., peach tree borer

Males possess arrow-shaped tufts of hair at the end of the abdomen which probably secrete an aphrodisiac for the female (*1114*).

*Sphinx ligustri* L.

Scent glands at the base of the male's abdomen secrete an aphrodisiac for the female (*572*).

*Spilosoma lubricipeda* (L.), ermine tiger moth
The male produces a scent in his scent-brush, which is disseminated by the coremata and calms the female during courtship (*163*).
*Stichophthalma camadeva* (Westw.)
Males emit an odor that excites the female just before mating (*499*).
*Syrichtus malvae* L.
The costal margins of the male's wings contain scent glands that produce an aphrodisiac for the female (*572*).
*Terias hecabe fimbriata* (Wall.)
The male's wing scales contain glands that emit an odor sexually exciting to females (*572*).
*Tineola biselliella* Hummel, webbing clothes moth
Males attracted by females move about excitedly, vibrating their wings. This probably serves to distribute the male's odor, exciting the female sexually (*1182*).
*Trichoplusia ni* (Hübner), cabbage looper
Behavioral experimentation failed to show that the male produces an attractant or excitant for the female, despite a previous report (*1080*) that adults of either sex assume an attractive attitude for the opposite sex, remaining stationary with the wings spread out horizontally or fanning (*1091*). However, an electrophysiological technique showed that a pheromone secreted by the male's abdominal hair-pencil scales acts as an aphrodisiac on females (*474*).
*Xanthorhoe fluctuata* (L.), garden carpet moth
The male possesses tubelike organs (coremata) supporting the scent-dispersing hairs; the scent acts as a sedative to calm the female during courtship (*163*).
*Xylophasia monoglypha* (Hufn.), dark arch moth
The male produces a substance sexually attractive to the female (*419*).

## MECOPTERA

*Harpobittacus australis* (Klug), scorpion fly
*Harpobittacus nigriceps* (Selys), scorpion fly
Males of both species hunt for the soft-bodied insects on which they feed; females have never been observed hunting, capturing, or killing prey in the field. When the male holds its prey and begins to feed, two reddish-brown vesicles are everted on the abdomen between tergites 6–7 and 7–8. These begin to expand and contract in a slow, rhythmic motion and discharge a sex pheromone with a butyric-musty odor (to humans) that attracts the female to the male's vicinity, moving upwind. As soon as the female is

within reach, the male retracts his vesicles and brings the prey to his mouthparts. The female attempts to get hold of the prey but is prevented by the male, whose abdomen seeks out the tip of the female's abdomen and copulation takes place. At this point the male passes the prey to the female who feeds on it as a nuptial meal while mating is in progress. In *H. australis*, the nuptial meal remains in the male's possession, the same prey being used for several mating actions with different females. In *H. nigriceps*, the female discards the prey just prior to termination of copulation, and a fresh nuptial meal is provided by the male on each mating occasion (*189, 191*).

*Harpobittacus similis* E.-P., scorpion fly

*Harpobittacus tillyardi* E.-P., scorpion fly

Laboratory observations of these species revealed mating habits; the scent of their sex lures are comparable in every respect with those of *H. australis* (*191*). When caged pairs were exposed to the pheromone of a second species, production of pheromone and even mating by the former ceased abruptly (*190*).

# CHAPTER IV

# PHEROMONES PRODUCED BY ONE SEX THAT LURE BOTH SEXES (ASSEMBLING OR AGGREGATING SCENTS)

The following insect species are treated separately in this presentation because 1 sex (in some cases both sexes) produces a substance that under special circumstances causes both sexes to assemble for mating.

ORTHOPTERA

*Acheta domesticus* (L.), European house cricket
Chemical attractants are produced by both sexes and both sexes respond to them. Both sexes in the laboratory also appear to produce a dispersant, which is avoided by both sexes. Each sex responds more negatively to the dispersant of its own sex than to that of the opposite sex. The attractant is rather stable, but the dispersant is ineffective after about 24 hours (*1066*).

*Blattella germanica* L., German cockroach
Young nymphs of both sexes produce a chemical pheromone in their rectal pads that causes them to aggregate. The substance has been found in the feces, in ether washings of the abdomen, and to a lesser extent in ether washings of the rest of the body (*575–577, 1292*).

*Locusta migratoria migratorioides* R. & F., African migratory locust
Mature adults produce a pheromone which causes members of both sexes

to aggregate. The pheromone is also reported to be somewhat attractive to the adults of *Schistocerca gregaria* (*830*). Nolte *et al.* (*827*) have recently shown that all hoppers in both the solitaria and gregaria phases produce a pheromone responsible for aggregation into swarms. It is secreted in the feces of hoppers but not in that of adults and can be extracted with solvents as well as from the air of rearing rooms. Reception is not through the antennae but through the stigmata.

*Schistocerca gregaria* Forskal, desert locust

Males and females in laboratory colonies produce a chemical substance which causes them to aggregate (*41, 453*). In addition, males in laboratory colonies produce volatile nitrogenous substances which accelerate the maturation of young adults (*165*). The aggregating pheromone is present on the surface of all parts of the body but not in the internal organs; it also causes aggregation of ovipositing females (*830*). *Locusta migratoria* also responds to it to some extent.

*Bombus agrorum* (Fabr.), bumblebee
*Bombus derhamellus* (Kirby), bumblebee
*Bombus lapidarius* (L.), bumblebee
*Bombus lucorum* (L.), bumblebee
*Bombus pratorum* (L.), bumblebee

Males of these species produce volatile substances that appear to attract both males and nubile queens, increasing the likelihood of mating. The scents, produced in the head, seem to be species specific because males of 1 species do not visit places marked by males of another (*266*).

## HEMIPTERA

*Cimex lectularius* L., bedbug

Both sexes release a chemical pheromone that causes the sexes to aggregate. Adults from which the antennae have been removed fail to respond to the odor of the pheromone, which is soluble in methanol but not in ether and which volatilizes at 32°C at atmospheric pressure (*730*).

## COLEOPTERA

*Anthonomus grandis* Boheman, boll weevil

Males emit a wind-borne pheromone that lures both males and females to male-infested cotton plants. In field tests, the weevils preferred male weevils over uninfested cotton plants in a ratio of 25:1 (*48*). Approximately equal numbers of males and females are attracted in the field in spring and fall (*1205*). The production of pheromone was not affected by feeding adult males for 6 days on a diet containing 0.1% of busulfan (1,4-butanediol dimethanesulfonate), a chemical sterilant (*680*).

*Blastophagus piniperda* L., bark beetle

The female is the first to attack the host tree, *Pinus sylvestris*, and bores into the phloem of the pine rind. Olfactometer tests with females dug out of the rind, as well as with males in the swarming phase, showed that neither males nor females were attracted. It was concluded that neither sex, collected from phloem just after penetration, affected each other in the swarming phase. However, in field tests, traps containing pine logs infested with new female galleries caught beetles of both sexes (*642, 859*).

*Cardiochiles nigriceps* Vierick

Virgin females of this species, an important parasite of tobacco budworm larvae, were attractive to both sexes when used to bait traps. Traps captured predominantly males when placed at the 1-foot level, whereas they captured mostly females at 3 or 6 feet (*730a*). Males attracted other males, but did not attract females.

*Dendroctonus brevicomis* LeConte, western pine beetle

Sexually mature, unmated females feeding on fresh Douglas-fir phloem produce a sex pheromone which is excreted with the frass. Both males and females respond to this pheromone over long distances (*45, 1226, 1230*). The attractiveness of the trees and logs infested is highest during the initial phase of the attack but ceases as feeding gains momentum. Billets and phloem strips infested manually with females and the frass collected were rarely attractive in field bioassay, in contrast to billets cut from freshly attacked trees (*1226*). Although the attractant may be produced over a long period of time, production is terminated when mating occurs (*1225*).

Pheromone biosynthesis does not appear to be specific to species, since *D. frontalis* (*924, 1225, 1226*), *Gnathotrichus sulcatus* (*1001*), and *G. retusus* (*1001*) also respond to the pheromone in *D. brevicomis* borings.

*Dendroctonus frontalis* Zimmerman, southern pine beetle

Sexually mature, unmated females feeding on fresh Douglas-fir phloem produce a sex pheromone which is excreted with the frass and attracts both males and females (*436, 671, 876, 924, 1226, 1230, 1231*). The aggregating principle is concentrated in the hindgut of female beetles after they emerge from old brood trees, and the pheromone is released by defecation when they reach a new host (*671*). In contrast to other species of this genus, these females are capable of releasing the pheromone in exceedingly small amounts immediately after emergence. Upon initial invasion, odors released from the penetrated plant tissue arrest the aggregating population (*671, 1226, 1231*). The pheromone content of female hindguts declines rapidly as feeding and gallery construction progress. Live, feeding, virgin females are much more attractive than live, feeding, reemerged females, but mating does not irreversibly inhibit pheromone production in female beetles (*312*).

Crushed, emergent *D. frontalis* females, *D. brevicomis* males, and *D. pseudotsugae* females were all attractive to flying *D. frontalis* (*924, 1225*). Predators (*Thanasimus dubius, Heydenia unica*, and *Medetera* spp.) also responded to the test material (*227a, 1231, 1233, 1267*).

*Dendroctonus obesus* (Mann.), spruce beetle

One or more females, boring alone in a log, create an attraction to both sexes of the natural population during their attack flight. Males always predominate in responses to these logs, but after the initial flight, about equal numbers of each sex respond to logs without introduced females (*362*). The pheromone will also attract both sexes of *D. pseudotsugae* (*294*).

*Dendroctonus ponderosae* Hopkins, mountain pine beetle

Females boring into white pine logs release aggregating pheromones that attract both sexes (*44, 876, 1226, 1230*). Emergent males crushed at dry-ice temperature also attracted *D. brevicomis* in appreciable numbers (*876*).

*Dendroctonus pseudotsugae* Hopkins, Douglas-fir beetle

Pheromones released when virgin females bore into the wood of Douglas-fir attract and arrest both sexes, with males predominating (*185, 618, 619, 773, 924, 994, 995, 998, 1230*). The first beetles to invade freshly cut or damaged Douglas-fir are attracted by the terpenes present in the oleoresin; $\alpha$-pinene, limonene, and camphene are more attractive than $\beta$-pinene, geraniol, and $\alpha$-terpineol, but $\beta$-pinene and geraniol attract large numbers of ambrosia beetles (*Gnathotrichus sulcatus*). Such attraction is referred to as "host or primary attraction" and precedes the "secondary or beetle attraction," which occurs in response to the pheromones produced by the beetles (*994*). In laboratory tests, significant responses were obtained using low concentrations of oleoresin compounds and pine oleoresin fractions; the largest responses occurred on 2.5% Douglas-fir oleoresin (*619*). High concentrations of Douglas-fir oleoresin were repellent to both males and females. Both sexes are also arrested by female-infested western white pine, western larch, ponderosa pine, grand-fir, and western hemlock, but not by poplar, oak, plasterboard, or plywood (*618*). The insect prefers freshly felled host material to that which has been felled 15 days or more.

Males of *D. pseudotsugae* require at least 90 minutes of flight exercise before an arrestment response to female frass occurs (*133a*). According to Borden (*185*), attractive frass was produced by females in laboratory tests within 2 hours after boring into fresh logs, and peak attraction was reached in 8 hours. Production declined after 2 weeks but increased again when the females were allowed to attack new logs.

Males and females of *Gnathotrichus sulcatus* and *G. retusus* also respond to the borings produced by female *D. pseudotsuga* and to alcohol extracts of such borings (*1001*). Likewise, *Hylastes nigrinus* (*995*) and *D. obesus* (*294*) also respond to invaded logs.

Investigations have confirmed a sudden decrease in the number of beetles attracted immediately after mating (*619*). The presence of a male in the gallery reduces the attractiveness (*773*). Attraction ceases abruptly, although sparsely infested parts of the same trees may be colonized for some time. This cessation is the result of masking by the female triggered by stridulation of the male (*996, 998*). This stops the aggregation of beetles in flight, but not the arrestment of males. Such masking, discussed later, is viewed by Rudinsky (*998*) as a mechanism for survival through the regulation of attack and as a critical part of mating behavior.

*Gnathotrichus retusus* (LeConte), ambrosia beetle

*Gnathotrichus sulcatus* (LeConte), ambrosia beetle

The males enter the host (Douglas-fir) first, then attract both sexes by producing pheromone in the frass (*995*). *Gnathotrichus sulcatus* attacks felled or dead trees only after the material has aged, and the compound in western hemlock logs responsible for the primary attraction to males has been identified as ethanol (*264*).

*Ips acuminatus* Gyll.

Field traps containing live males attracted great numbers of both sexes, but traps containing live females caught practically no beetles (*81*).

*Ips avulsus* Eichh.

*Ips calligraphus* Germ.

In olfactometer tests, both sexes of these species were attracted in large numbers by volatile materials emanating from log sections of southern pines recently infested by males of the respective species (*1225*). In field tests, caged longleaf pine logs artificially infested with *I. calligraphus* males attracted adults for at least 3 weeks (*1265*).

*Ips confusus* (LeConte), California five-spined ips

A volatile substance responsible for mass attraction of both males and females is produced by mature males boring into the wood of ponderosa pine logs. The male initiates an entrance gallery in the wood, producing frass that contains the pheromones attractive to both sexes, but much more so to the female (*864, 1107, 1218, 1219, 1223, 1224, 1226, 1284, 1290, 1293–1295*). The attractive components are concentrated in the hindgut area and discharged by defecation (*1226*). Bark and wood samples of ponderosa pine become attractive 4–6 hours after males are introduced into the entrance tunnels, with peak attraction occurring at 18–144 hours. Frass is not attractive until 9–12 hours following male introduction.

Although Wood and Bushing (*1290*) at first reported that the frass of males boring in Douglas-fir is not attractive to females, Wood *et al.* (*1289*) later stated that the frass from males boring in white fir and Douglas-fir is attractive, indicating that the pheromone does not function in initial host discrimination.

Trees and logs that have been fed upon may retain their attractiveness for approximately 10 days following infestation, but lose their activity with loss of moisture. Neither feeding females nor their frass will cause assembling (*432*).

Borden (*184*) reported that teneral males produced attractant after a long maturation period. Single, mature males produced attractant for up to 18 days, but productivity declined after mating in direct correlation with the number of females per male. Male frass was significantly less attractive 15 minutes after exposure to open air and very little attraction remained after 1 hour. The aggregating pheromone of this species appears to be responsible for some degree of attraction to 2 predators (*Enoclerus lecontei* and *Temnochila virescens chlorodia*) and a parasite (*Tomicobia tibialis*) (*1286*).

*Ips grandicollis* Eichh.

In olfactometer tests, large numbers of both sexes were attracted by volatile materials emanating from log sections of southern pines recently infested by males (*1225, 1265*). Hertel *et al.* (*546*), however, reported that the odor produced by males attacking caged loblolly pine bolts was not very effective in attracting marked beetles; only 2.4% of 9417 beetles released around the attracting logs were recaptured. A dispersal flight may be needed before response takes place. Infested logs are also attractive to a predator, *Medetera bistriata* (*1267*).

*Ips latidens* (LeConte)

Frass produced by attacking males is attractive to both sexes, as is the pheromone produced by male *I. confusus* in ponderosa pine frass (*1294*).

*Ips pini* (Say), pine engraver

Adults are not strongly attracted to uninfested host logs, but both sexes are attracted to these logs after a few males have entered and become established in the inner bark (*31*).

*Ips ponderosae* (Sw.)

Males release an attractant, after they have fed for several hours on ponderosa pine, that lures both sexes. The volatile pheromone is obtained from the hindgut of feeding males, their frass, and air drawn from a box containing beetle-infested ponderosa pine logs (*1223*).

*Ips typographus* L.

Fresh spruce logs infested with males attracted a larger number of both sexes than did logs without males (*82, 1002*). The pheromone produced occurs in both the spring and second flights. Three common predators (*Thanasimus formicarius, Medetera signaticornis*, and *Epuraea pygmaea*) are also attracted by the pheromone (*1003*).

*Leperisinus fraxini* Pz, ash-bark beetle

Males and females aggregate on their host, *Fraxinus excelsior*, in response to volatile substances emanating from male-infested host material (*1050*).

*Lycus loripes* (Chevrolet)

A volatile attractant produced by males causes both sexes to aggregate and then distribute themselves into individual mating pairs or small clusters of pairs (*374*).

*Orthotomicus erosus* Woll.

Live males start boring into the wood of various species of pine after being attracted by the oleoresins. Both sexes are then attracted to the host by the male frass. A new infestation usually reaches optimum numbers within 8 days. Extracts prepared of the frass are attractive to both sexes but quickly volatilize (*299*).

*Orthotomicus sabinianae* Hop.

*Pityogenes carinulatus* (LeConte)

*Pityophtherus annectens* (LeConte)

*Pityophtherus confertus* Sw.

In field olfactometers, *Pinus ponderosa* wood infested with live males of *O. sabinianae* or *P. confertus*, *Pinus taeda* wood infested with male *P. annectens*, and *Pinus jeffreyi* wood infested with male *P. carinulatus* all became attractive to both sexes of the respective species of beetle. Neither the pine wood itself nor wood infested with females was attractive (*1218*).

*Popillia japonica* Newman, Japanese beetle

Electrophysiological tests indicated that males probably produce a pheromone which attracts both males and females (*11*).

*Pseudohylesinus grandis* (LeConte)

*Pseudohylesinus nebulosus* (LeConte)

Fir logs infested with females of these species attracted large numbers of both sexes of their respective species (*995*).

*Scolytus multistriatus* (Marsham), smaller European elm bark beetle

Flying males and females respond to a pheromone produced by virgin females boring in suitable host material. The pheromone is probably contained in virgin-female frass (*855, 855a*).

*Trogoderma granarium* (Everts), khapra beetle

In 1965, Finger *et al.* (*405*) reported that crawling larvae leave traces of an odorous substance that attracts other larvae of this species, but the investigators later retracted their findings after they were unable to reproduce their results (88).

In the same year, Bar Ilan et al. (89) showed that virgin female adults release a scent capable of attracting both sexes of this insect. This was substantiated by Levinson and Bar Ilan (725, 726) and by Yinon and Shulov (1323, 1324) in 1967. Ikan et al. (570) also reported that adult males likewise produce an attractant for both sexes, but Levinson and Bar Ilan (727) could not substantiate this.

The assembling scent produced by the female is apparently released to the outside via the cuticle, since approximately 5–10 times more attractant can be obtained by ether extraction of the entire body than from its surface. Production appears to be confined to the posterior body tissues, except for the hemolymph. Attractant formation starts after pupation of the females and continues until the end of the oviposition period. After mating, the percentage of beetles responding to the assembling scent declines markedly in males and to an intermediate extent in females (725, 728). Although both sexes are attracted, the attractiveness of the scent for virgin males is definitely higher than for virgin females (728, 729).

The assembling scents produced by both virgin males and females are repellent to adult *Tribolium castaneum*, *Tenebrio molitor*, *Dermestes maculatus*, and *Oryzaephilus surinamensis* (570, 1324, 1325). The scent produced by the female is attractive to *Callosobruchus maculatus*. *Sitophilus oryzae*, *Lasioderma serricorne*, and *Cadra cautella* are indifferent to the pheromones (1325).

*Trypodendron lineatum* (Olivier), ambrosia beetle

Rudinsky et al. (287, 995, 1000, 1001, 1004) has shown that both sexually mature, fertilized and unmated females boring into the wood of *Pseudotsuga menziesii* and Douglas-fir produce a volatile substance in the boring dust to which both males and females respond in flight. An ethanol extract of the female borings is likewise attractive to both sexes (995). A log merely infested with females or males is not attractive. The results were substantiated by Francia and Graham (421) and by Borden and Slater (187), but Chapman (292), although verifying that attack by female beetles was followed by a marked increase in attractiveness of the logs, stated that attraction disappeared after mating of the females. Males of *T. lineatum* require at least 30 minutes of flight exercise before an arrestment response to female frass occurs (133a).

# CHAPTER V

# ANATOMY AND PHYSIOLOGY OF THE PRODUCTION GLANDS

A good review of the literature pertaining to the scent glands of male and female insects was published by Richards in 1927 (*932*); he stated, "scent organs are found to be of very wide and, in many groups, fragmental occurrence in insects, playing an essential part in mating. When they occur in the female alone their function is to bring the sexes together, while those peculiar to the males are used, in nearly all cases, to rouse the female to the state in which she is ready to copulate."

## In Females

ORTHOPTERA

Although Stürckow and Bodenstein (*1158*) originally reported that the sex pheromone of female *Periplaneta americana* is produced in differentiated areas of the integument, located mainly on the head and to a much smaller extent on the remainder of the body, this was subsequently retracted by Bodenstein (*176*). Attempts to locate an obvious source by examining secretions of the head with the light microscope were unsuccessful. Extracts of the crop, caeca, midgut, and hindgut of attractive females 3 weeks to

79

3 months old were highly active, but the greatest amount of pheromone was present in the crop and midgut. No pheromone could be extracted from the proventriculus, malpighian tubules, salivary glands, or rectum. Extracts of female fecal pellets were highly active. The actual site of production remains unknown.

Barth's investigations (*111*) with the cockroach, *Pycnoscelus surinamensis*, indicate that the production of sex pheromone in virgin females is under endocrine control. Females of both parthenogenetic and bisexual strains of *P. surinamensis* produce a sex pheromone that stimulates courtship behavior of the bisexual strain, but the removal of the corpora allata less than 24 hours after the imaginal molt in the bisexual strain cuts off pheromone production. There is no effect on production in the parthenogenetic strain after allatectomy. The production of sex attractant in *Periplaneta americana* is controlled by the corpora allata.

Emmerich and Barth (*381*) showed that juvenile hormone (which activates corpora allata) from male *Cecropia* abdomens was effective in inducing pheromone production and accessory gland secretion in 8 of 18 allatectomized females of *Byrsotria fumigata*, although it did not stimulate oocyte maturation. Allatectomized females injected with farnesyl methyl ether (a juvenile hormone-mimicking compound) showed sustained pheromone production. This was substantiated by Barth and Bell (*115*) who also found evidence that oocyte maturation is likewise controlled by juvenile hormone.

It has been shown that adult females of *Byrsotria fumigata* fail to produce sex pheromone if their corpora allata have been removed shortly after the imaginal molt, and that the implantation of corpora allata into previously allatectomized females can induce pheromone production. The subject of mating behavior and its endocrine control in cockroaches has been adequately reviewed by Barth (*112*), who also showed that injecting *Cecropia* DL-juvenile hormone or farnesyl methyl ether into allatectomized female *Byrsotria* enhanced pheromone production (*131*).

LEPIDOPTERA

By far the greatest amount of work on the function of the production glands has been conducted with the insects in this order. Urbahn (*1212*) published a comprehensive treatise in 1913 on the abdominal scent organs of female butterflies. These organs, in *Phalera bucephala*, consist, in their simplest form, of merely a saddle-shaped field in the last intersegmental membrane between the eighth and ninth abdominal segments. When the abdomen is stretched, the intersegmental fold is simultaneously stretched to the outside. Deep ring-shaped scent glands are present in the noctuids *Cuculia verbasci* and *C. argentea*. Dorsally situated scent glands of the last intersegmental fold are present in *Hypogymna morio* and *Dasychira pundibunda*. Scent sacs are formed either singly or in pairs on each side of the body; they are situated either dorsally or ventrally and are withdrawn by means of special muscles. Such scent sacs are found in *Bombyx mori*, *Argynnis paphia*, *Saturnia pavonia*, and *Brenthis euphrosyne*. The glandulae oliferae originate by the transformation of the last intersegmental fold to a pair of glandular hoses which extend deep into the cavity of the eighth abdominal segment on both sides of the body. These hoses are found in *Vanessa urticae*, *Acronicta psi*, and *Argynnis paphia*. The dorsal odor glands of female *Orgyia ericae* are very highly developed. The female does not leave the pupal case; she merely extends the abdominal tip through an opening in the case to await the male's arrival.

Urbahn (*1212*) also studied carefully and at length (80 pages) the abdominal scent glands of female *Hypocrita jacobaeae*, *Callimorpha dominula*, *Caligula japonica*, *Aglia tau*, *Pterostoma palpina*, *Colocasia coryli*, *Porthesia similis*, *Dasychira fascelina*, and *Orgyia gonostigma*.

Freiling (*427*) described, in considerable detail, the male and female odor organs of numerous Lepidoptera. A detailed discussion of the release of sex odors (and other odors as well) by Lepidoptera is given by Hering (*545*). The excised female glands lure males who attempt to copulate with them, no heed being paid to the mutilated female. The female glands are usually modified cells between the eighth and ninth abdominal segments, although in *Argynnis* species they occur between the seventh and eighth rings. Drops of the liquid obtained from the sacculi laterales placed on filter paper lure males who attempt to copulate with the paper. The hairs on the underside of the female abdomen serve to fan and distribute the attractant particles.

Bronskill (*208*, *209*) has described a method for preparing permanent whole-mount ledidopterous genitalia for complete visibility of the female sex-pheromone gland. The genitalia are fixed in such a way that the eighth abdominal segment is completely visible by flooding with Carnoy's fluid (absolute ethanol-chloroform-glacial acetic acid, 6:3:1), held *in vacuo* for 8–12 hours, and hydrated with 95% and then 70–80% ethanol. The preparation is stained by Feulgen's method and then mounted. This procedure was used to prepare mounts of the genitalia of female *Choristoneura rosaceana*, *Argyrotaenia velutinana*, *A. quadrifasciana*, *Archips argyrospilus*, *A. mortuanus*, *Grapholitha molesta*, *Laspeyresia pomonella*, *Paralobesia viteana*, and *Hedia nubiferana*.

According to George (*446*), the female production gland of *Galleria molesta* is an invagination of the integument and consists of a single layer of enlarged columnar epithelial cells along the outer edges; it opens to the outside ventrally and posteriorly at the base of, and posterior to, the ovipositor lobes. The male has no comparable gland.

According to Barth (*111*), the production of sex pheromone in virgin female *Antheraea pernyi* and *Galleria mellonella* is not under endocrine control. Allalectomized female *A. pernyi* tied to wooden and wire frames and placed outdoors at dusk called normally and lured males released 25–30 yards away.

Riddiford and Williams (*937, 938*) have shown that a constituent of red oak (*Quercus rubra*) leaves, the host plant of *Antheraea polyphemus*, is necessary to trigger sex-pheromone release in the female insect. This so-called "oak factor" acts exclusively on the antennae of the female moth, stimulating her abdominal scent glands to release the male attractant. The stimulant has been identified as *trans*-2-hexenal (*934, 935*), a constituent common to the green leaves of many plants, but volatile substances also present in leaves other than those of the host mask or block the action of *trans*-2-hexenal on the females. Chanel No. 5 and formaldehyde also act as masking agents. The vapors of aqueous solutions of 2,4-hexadienol, heptanal, cycloheptanone, and octanal were inactive as a pheromone stimulant, whereas hexanal and *cis*-3-hexenol were occasionally active. The activity of the latter can be explained by its oxidation to *trans*-2-hexenal on exposure to air. The response of the female to oak factor is eliminated by prior excision of the corpora cardiaca but not of the corpora allata. In high concentration, the vapors of *trans*-2-hexenal elicit ovipositional behavior, even from virgin female moths. Riddiford and Williams (*939*) have shown that the corpora cardiaca must have intact connections with the brain for calling to occur in wild female *A. polyphemus* and *Platysamia cecropia* moths. Evidently, in response to the environmental signal of *trans*-2-hexenal, the brain stimulates the release of a hormone from the cells of the corpora cardiaca. This hormone then acts on the abdominal nervous system to provoke protrusion of the female genitalia, with release of pheromone. *Antheraea pernyi*, which is semidomesticated, shows no overt calling, and apparently releases pheromone continuously.

Rahn (*902*) has also shown that the presence of the host plant is necessary in attraction between the sexes of the leek moth, *Acrolepia assectella*, and the presence of corn plants in laboratory cages containing *Heliothis zea* significantly improved the percentage of mating in this insect (*1118*).

Kellogg (*657*) reported in 1907 that the female silkworm moth, *Bombyx mori*, protrudes paired scent organs from the hindmost abdominal segment. This gland is withdrawn into the body immediately after being touched by a male. The excised scent glands, but not the mutilated female, are highly attractive to the male. The structure of these glands was studied with the aid of the electron microscope by Steinbrecht and Schneider (*1144, 1147*) as well as Bounhiol (*195*). An electron microscopic study by Waku and Sumimoto (*1240*) of the ultrastructure and secretory mechanism of the female gland showed that the pheromone is synthesized in characteristic myelin figures that form from the mitochondria and appear in the pupal cell. The pheromone is then released into the cuticle through the plasma membrane of the microvilli lining the cell surface, finally being transported to the outside through pore canals characteristic to the cuticle of the gland.

The morphology and histology of the sex pheromone glands in female *Diparopsis watersi* has been described by Brader-Breukel (*199*).

Studies by Barnes et al. (*91*) of the sex pheromone gland of female *Laspeyresia pomonella* showed that the pheromone originates in a glandular structure opening dorsally in the intersegmental fold between abdominal segments 8 and 9. Ether extracts eliciting a male response were obtained from abdominal segments 8 and 9, but not from the remainder of the abdomen.

Roelofs and Feng (*962*) studied the pheromone scent glands of a number of Tortricinae and Olethreutinae. Sections of the female caudal tip revealed enlarged, columnar, nucleated gland cells lining the invaginated integument of the last intersegmental fold in *Laspeyresia pomonella*, *Grapholitha molesta*, *Hedia nubiferana*, and *Paralobesia viteana* (all Olethreutinae), and *Argyrotaenia velutinana*, *A. quadrifasciana*, *Archips argyrospilus*, *Choristoneura rosaceana*, and *Pandemis limitata* (all Tortricinae). In general, the production glands in the Tortricinae possess numerous large folds in the invaginated area, whereas those found in Olethreutinae appear as a thick, compact layer following the outline of the invagination. Only *Ancylis comptana fragariae* showed atrophied, underdeveloped glands.

Jefferson et al. (*624*) described the morphology of the female sex-pheromone glands of 8 species of Noctuidae. The gland is situated dorsally in the intersegmental membrane between abdominal segments 8 and 9 in *Autographa californica*, *Pseudoplusia includens*, and *Rachiplusia ou*. In *P. includens*, the gland is an eversible sac, but in *A. californica* and *R. ou* it may be an eversible sac or fold. In *Spodoptera exigua* and *Feltia subterranea*, the gland is an eversible sac situated ventrally in the intersegmental membrane between segments 8 and 9. The gland in *Heliothis phloxiphaga*, *H. virescens*, and *H. zea* is a complete ring of epithelium between segments 8 and 9, which is more highly developed ventrally in *H. virescens*.

The sex-pheromone gland in female *Choristoneura fumiferana* is located between the eighth and ninth abdominal segments. Once the gland is extruded by pulsation, it remains extruded for long periods, in contrast

with the intermittent protraction and retraction of the gland in a number
of other Lepidoptera. The period of extrusion is associated with photoperiod
(*1014*). Percy and Weatherston (*856a*) reported on the morphology and
histology of *C. fumiferana*, *C. pinus*, and *Malacosoma disstria*, and reviewed
the morphology and histology of the females of 35 other species of
Lepidoptera.

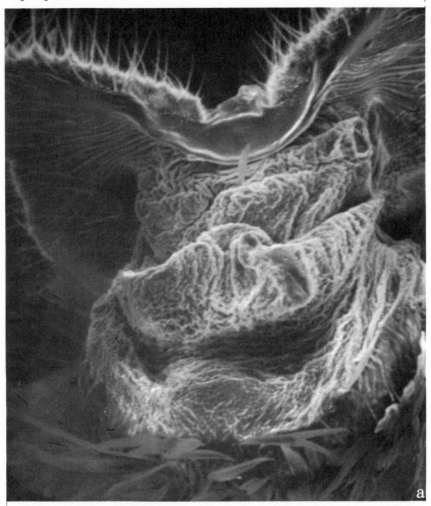

*Fig. V.1.* Everted intersegmental membrane of *Choristoneura fumiferana* female
abdomen. (a) Partial extrusion, 60X; (b) greater extrusion, 59X; (c) maximum extrusion,
59X. [Reproduced by permission of the National Research Council of Canada from
Weatherston and Percy (*1251*).]

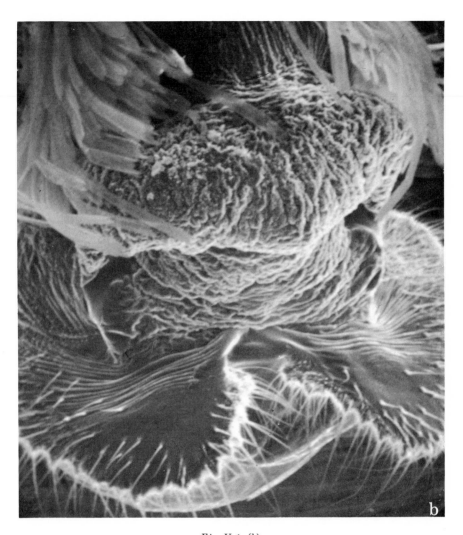

*Fig. V.1.* (*b*)

Weatherston and Percy (*1251*) have studied, by means of scanning electron microscopy, the sex pheromone-producing gland of *Choristoneura fumiferana* in relation to the mode of release. The gland is a modified intersegmental membrane situated dorsally between abdominal segments 8 and 9 (*1014*). Extrusion of the membrane by artificial means is shown in

Figure V. 1a, b, and c. The membrane has 3 distinct types of surfaces. The anterior portion has a surface similar to that of undifferentiated inter-segmental membrane, the glandular area shows numerous spikes over its surface, and the posterior area surface is similar to the glandular surface but lacks the spikes. Although the exact manner of pheromone release could not be deduced from the studies, a theory may be postulated in which the

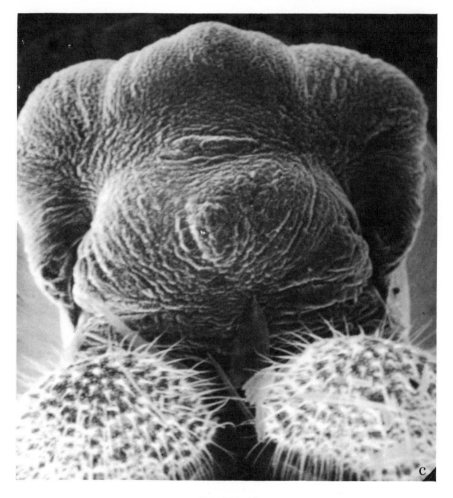

*Fig. V.1. (c)*

pheromone penetrates the cuticle in the manner suggested by Steinbrecht (*1144*) for *Bombyx mori* and is retained on the surface in the deep invaginations of the nonextruded gland until the latter becomes everted during calling. However, the function of the spikes is not yet clear.

The sex pheromone of the female *Estigmene acrea* is produced by glandular epithelium on the dorsal surface of the ninth abdominal segment. The padlike structures of the epithelium are spread apart by eversion of segments 8 and 9. The epithelium consists of a thin cuticle covered with hollow outgrowths and a procuticle; no pores are evident in the cuticle through which the pheromone can pass. Glucose-6-phosphate dehydrogenase and succinate dehydrogenase activity is high in newly emerged females and decreases with age (*740*).

Virgin, sexually mature females of *Ephestia* and *Plodia* commence calling by sitting with the wings folded and the apical half of the abdomen bent over the back between them. Meanwhile the apical abdominal segments are alternately extended and retracted, so that the intersegmental membranes are widely exposed. There is no doubt that during this process a scent attractive to the males is emitted. The segmental membranes, especially in the neighborhood of the orifice of the ductus bursae, have an appearance strongly suggesting the presence of secretory tissues (*933*).

A detailed study was made of the sexual scent organs of male and female *Anagasta kühniella, Cadra cautella, E. elutella,* and *Plodia interpunctella* by Dickins (*349*). In the males, scent glands are situated dorsally on the eighth abdominal segment. In the female *E. cautella,* the organs show gland cells and are much more complex than those in *A. kühniella* or *P. interpunctella;* the organs are present on the ninth abdominal segment. *Cadra cautella* shows typical glandulae odoriferae not shown by any of the other species. Barth (*101*) found that the head, thorax, wings, and the front half of the abdomen of female *P. interpunctella* did not evoke a courting dance in males, whereas the abdominal tips (segments 7–9) did evoke such a response, except in males whose antennae had been removed. Microscopic examination of the segments showed no odor scales or odor hairs, but 2 glands from part of the hypodermis of the intersegmental membrane were easily seen between segments 8 and 9. Tension of the cuticula and a resulting increase in blood pressure cause the secretion to be released.

Male *Plodia interpunctella* never become excited in the presence of noncalling females. The assumption of the calling position, with resultant stretching of the intersegmental membrane of the ventral body wall, exposes the mouths of the glands so that the scent is emitted. It is also possible that the glands are under nervous control (*829*). Calling by the virgin female is not continuous. Very rarely females failed to call, and dissection of these and of females that would not pair with males, although they were calling, showed no gross abnormalities. Müller (809) also postulated that the female ready for copulation stretches the ninth abdominal segment vertically upward, eventually stretching the last intersegmental membrane to expose the scent glands.

Calling by the female *Pectinophora gossypiella* occurs between 1:00 and 4:00 AM. The pheromone gland, incorporated into a folding structure between the ostium bursae and the papillae anales, is everted dorsally as the ovipositor becomes extended beyond the abdomen. It attains full size when the ovipositor is retracted ventrally (724a).

Amputation of the wings or antennae of a female gypsy moth (*Porthetria dispar*) does not affect her attractiveness to males, but males do not attempt to copulate with females whose abdomens or genital organs have been removed, and they are not successful in attempting to mate with those females deprived of abdominal scales and wings. Female abdomens excised at 9:00 PM and placed in a vessel containing 3 freshly emerged males caused great agitation in these males by 10:00 PM (893). Nolte (828) could not find the sex attractant-production glands of the female nun moth (*P. monacha*) in the area of the ovipositor; the excised abdomen lured males, but the wings and abdomen-free bodies were unattractive.

The sex-pheromone gland of female *Phthorimaea operculella* is a bulbous structure lying in the dorsal intersegmental region between the eighth and ninth abdominal segments. The histology of these glands has been described (9).

In 1958, Hammond and Jarczyk (504) reported that the scent organs of the female Egyptian cotton leafworm, *Spodoptera littoralis*, are tufts of modified scales or hairs with gland cells at their base, covering the ninth abdominal segment. During the resting period this segment is invaginated into the eighth segment. Large droplets of oily secretion cover more than 1 cell and penetrate the overlying cuticle. These investigators found no special secretory ducts in the glandular epithelial cells and concluded that the hairs have the function of increasing the evaporating surface of the secretion. Hammond and Jarczyk did not find any similarity in form or position between the sexual scent glands of *S. littoralis* and those previously described by Götz (468) for other noctuids. However, Jefferson and Rubin (621), in 1970, examined the sex-pheromone glands of this insect and reported that the females have an eversible sac of glandular epithelium situated ventrally in the intersegmental membrane between the eighth and ninth abdominal segments. The gland is similar in structure and location to the gland in the yellow-striped armyworm, *S. ornithogalli* (624), the granulate cutworm, *Feltia subterranea*, and the beet armyworm *S. exigua*.

Applying pressure to the abdomen of a female *Pectinophora gossypiella* causes the terminal abdominal segments to extrude. The sex-pheromone gland is an eversible saclike structure situated dorsally in the intersegmental membrane between the eighth and ninth segments. The glandular epithelium is composed of flattened cells with bulges near the nuclei (*625*).

Histological study of the terminal abdominal segments of the female *Orgyia leucostigma* reveals a modified intersegmental membrane located dorsally between the eighth and ninth abdominal segments. The glandular area is crescent-shaped and identical to that described earlier for *Dasychira pudibunda* by Urbahn (*1212*). The glandular cells are goblet-shaped, instead of columnar or cuboidal (*856*).

The attractive substance in the lesser peach tree borer, *Synanthedon pictipes*, is located somewhere on the terminal 3 segments of the female abdomen. Females are attractive to males only when these segments are above the plane of their bodies but parallel to the longitudinal axis (see Fig.II.2) (*302, 1080*). The duration of exposure varies from a few seconds to at least 90 minutes. Some females resume the calling position within seconds after retraction of the organs.

Female webbing clothes moths, *Tineola biselliella*, could be observed in a calling pose, protruding and retracting their ovipositor and vibrating the tip of this organ. The excised female abdomen was attractive to males but the head and thorax was not (*986*).

The gland producing the female sex pheromone of *Trichoplusia ni* is an eversible sac of glandular epithelium situated dorsally in the intersegmental membrane between abdominal segments 8 and 9. Histological studies revealed a vacuole-laden cytoplasm characteristic of secretory cells, whose apical region consists of vertical striations. The pheromone is apparently continuously secreted by females 1–7 days old (*623*). A study of the ultrastructure of the apical region showed that the vertical striations consist of parallel villi, each with a diameter of 500–700 Å and a length of 2–3 $\mu$m. Fibers are present along the length of each villus and extend into the glandular cytoplasm. No obvious ducts are discernible in the endocuticle immediately adjacent to the villi, and the manner of penetration of the pheromone through the endocuticle is unknown (*788*). Sower *et al.* (*1125a*) developed a gas chromatographic and a bioassay method for measuring the pheromone emanating from single females. Laboratory-reared females released pheromone for about 40 minutes per night at an average release rate of $7 \times 10^{-3}$ $\mu$g/minute.

In the protruded eighth and ninth abdominal segments of a calling female *Vitula edmandsae*, the intersegmental membrane is composed of columnar epithelial cells forming the attractant-producing tissue, as in all female Lepidoptera studied. The cells form a ring gland extending farther anteriorly on the ventral side of the abdomen. No glandulae odoriferae are present. The glandular cells are columnar, with vertical striations on the apical ends; no vacuoles are discernible. The endocuticle is stratified, and no canals are seen for passage of secreted material to the outside; the pheromone may diffuse through the endocuticle to the epicuticle (*1250*).

COLEOPTERA

Happ *et al.* (*511*) have shown that in the first 7–10 days following emergence of female *Tenebrio molitor* from the pupal cuticle, oocyte length and emission of the sex attractant increase. Females exposed to male scent showed greater growth of their terminal oocytes and a higher level of pheromone emission than females exposed to unscented air or to female scent. Females exposed to female scent emitted more sex pheromone than those exposed to unscented air. It is concluded that both male and female scents contain primer pheromones. Attempts by Tschinkel *et al.* (*1204*) to locate the organs or cells secreting the sex pheromone in the female have indicated the region of the metathoracic sternum and first 2 abdominal terga. The pheromone does not remain in the vicinity of the source, but gradually spreads out over most of the beetle's surface.

An anatomic and histochemical study of the spermathecal gland of female *Tenebrio molitor* indicated that, although this gland is not the source of the female sex pheromone, a glycoprotein secreted by the gland functions in mating itself or immediately thereafter (*510*). The secretion of the corpus allatum appears to be directly involved in the control of sex-pheromone secretion by female *T. molitor* (*783*). Sex-pheromone activity is significantly reduced by allatectomy, brain removal, or decapitation. Reimplantation of the corpora allata or brain does not restore the pheromone level, but topical treatment or injection with juvenile-hormone substitutes such as farnesyl methyl ether or *N*,*N*-diethyl-3,7,11-trimethyl-10,11-epoxy-*trans*-2,*trans*-6-dodecadienamide counteract the effect of the operations.

Fundamental differences exist in the mechanisms triggering production and release of aggregation pheromones in the genera *Dendroctonus* and *Ips*. In *D. frontalis*, *D. brevicomis*, and *D. ponderosae* extensive feeding by the female in new host material soon inhibits attractant release, whereas feeding triggers and sustains the generation of attractants in *Ips confusus* and related species (*1221*, *1226*). The inhibition of pheromone production by female *D. frontalis* is, however, reversible, since reemerged females again produce significant attraction (*312*, *313a*). The attractiveness of trees and logs infested with *Dendroctonus* is highest during the initial phase of attack and ceases with extensive feeding. The physiological condition of the host material determines the rate of feeding, which in turn affects production and release of the attractant (*1221*). Trees infested with *Dendroctonus* are characterized during the attractive stage by the formation of pitch tubes containing small amounts of bark, wood particles, and feces (*1226*).

Jantz and Rudinsky (*619*) and Zethner-Moller and Rudinsky (*1333*) described the digestive tract, malpighian tubules, and reproductive system of the male and female Douglas-fir beetle, *Dendroctonus pseudotsugae*. Feeding virgin females producing the sex pheromone possessed longer malpighian tubules than feeding mated females or those not feeding; these differences were not observed in males. Males were arrested only by the female rectum, except after borings from fresh phloem were added to the hindgut and malpighian tubes. It was concluded that the malpighian tubes of unmated, feeding females are associated with the production of the pheromone.

A histological study of the abdomen of female *Trogoderma granarium* showed that the epithelium of the ventral intersegmental fold between the fifth and sixth abdominal sternites is composed of thick columnar cells with large nuclei and a granulated cytoplasm; the fold is shrunken and atrophied in older beetles. This suggests that the fold serves as a multicellular gland for secreting the sex pheromone (*1133*).

Graham (*421, 473*) has shown that healthy, living host trees for *Trypodendron lineatum* contain neither olfactory attractants nor repellents, but a felled and dying tree contains a primary attractant for both male and female beetles. Oxygen deficiency is probably a causal factor in attractant formation, since detectable attractancy can be induced by keeping fresh sapwood at 20°C for 4 hours under anaerobic conditions (*473*). Once females have been attracted to the host by these volatile substances and feed, production of the assembling scent begins. Searching for the source of this pheromone in the female beetle, Schneider and Rudinsky (*1048*) tested various parts of the hindguts of attractive females. Strong arrestment was shown by the ileum, cryptonephridium, rectum, and posterior ileum, whereas none was shown by the pylorus or anterior ileum. The investigators concluded that the pheromone is produced by a layer of secretory cells at the posterior ileum and anterior rectum of females in the presence of digested bark and wood particles, and under the influence of a substance (hormone or enzyme) or process activating or activated by the maturation of the fertilized egg. This has been corroborated by Borden and Slater (*187*) whose data suggest that the pheromone of *T. lineatum* is probably a true secretion.

Hoyt *et al.* (*563*) studied the colleterial glands of female *Costelytra zealandica* which lie beneath the seventh sternite ventral to the vagina. Each gland is oval-shaped, consists of 2 rounded lobes, and arises as an outpocketing of the vagina. The interior wall is lined with thin projections around which many bacteria are found. These bacteria were isolated and cultured on agar in petri dishes. The resultant colonies were cultivated and the culture filtrate was tested for attraction to male beetles in field traps. Males were attracted to these traps, and subsequent analysis of the filtrate revealed the presence of phenol. The bacteria normally found in the glands apparently act in producing the female's sex pheromone, which has been identified as phenol. (See Chapter X.) Sectioning tests conducted by Henzell *et al.* (*543*) have shown that the production glands lie within the first 3 abdominal segments on the dorsal side of the female body.

HEMIPTERA (HOMOPTERA)

In the homopterous insect, *Schizaphis borealis*, males are unable to recognize females whose hind legs have been removed, but the males are attracted to the excised legs. The female's hind tibiae show pseudorhineria; the membranes are densely covered with small grooves or pits possessing a wavy relief pattern. It is concluded that the emitting organ for the sex pheromone is located on the hind tibiae of the oviparous female (*861*).

HYMENOPTERA

Queen honeybees, 1–2 weeks old, exude a pleasant aromatic odor from a complex of subepidermal cells in abdominal tergites 2–4 only at the time of swarming. The substance may serve to either attract or excite drones at swarming (*920*).

Casida *et al.* (*286*) were unable to find specific gland openings in the female *Diprion similis* that might be associated with production of the attractant. However, Mertins and Coppel (*783b*) have described a pair of abdominal glands present in females, but not in males, that are probably involved in pheromone production. The glands lie in the anterolateral margins of the abdomen, one on either side, and open via a short duct through a vertical, slitlike orifice in the intertergal membrane.

DIPTERA

Exposure of male *Dacus tryoni* to 8000 or 20,000 rad of cobalt on the ninth day of pupation had no effect on pheromone production. However, exposure to 5000 rad on the fifth day of pupation caused a significant decrease (50%) in pheromone production by the adult males (*412c*).

## In Males

ORTHOPTERA

Ishii (575–577) has shown, by anatomic and histological studies, that the aggregation pheromone of male *Blattella germanica* nymphs is probably secreted from the rectal pad cells into the lumen of the rectum, and then excreted with the feces. The secretion is apparently absorbed by the lipids in the body surface, and thus appears on the surface.

Whereas 90% of normal female *Leucophaea* cockroaches accept a male within 26 days after emergence, only 30% of allatectomized females accept a male in this time (382). Mating in this insect apparently depends on the presence and activity of the corpora allata (383). The failure of allatectomized females to mate may depend on an alteration of their ability to perceive a chemical odor produced by the male which stimulates the female to feed on his tergal gland; unless the female responds in this way, mating does not occur. This chemical odor is perceived by the female through her antennae. Tergal feeding stimulates the male to copulate. Although allatectomized female *Diploptera* are courted and mate as readily as normal females, the females of this insect play a passive role during courtship (988).

Roth (984) has shown the importance of stridulation by males of *Nauphoeta cinerea* when courting nonreceptive females. During gestation, females remain unreceptive to the male pheromone, apparently because mechanical stimuli resulting from the ootheca in the uterus prevents the receptivity center from being reactivated. If the nerve cord is transected in the pregnant female, inhibitory signals from the uterus are interrupted, the center becomes activated, and after a few days the female mates again despite the fact that she is carrying an ootheca (976).

LEPIDOPTERA

The scent glands of male lepidopterous insects are present on the body or the wings (1252). The simplest type of gland is hairy, pierced by a single canal (as in *Eurytides protesilaus*) or numerous canals (as in *Argynnis*), or the surface is covered with openings as in a sieve. In Pierids and Lycaenids the secretory organs are located on the wings (545).

In 1902, Illig (572) published a masterly treatise describing his work on the anatomy and physiology of the odor glands of male butterflies; he also reviewed the subject. Beautiful color plates illustrated his findings which are described in detail. The wings are the usual location for these organs in the male. However, in those butterflies that ordinarily fly by vigorous wing movements, such as the sphingids, the organs are situated in the abdomen and lead to scent brushes (hair-pencils). In many cases the odor is the same as that of the flowers frequented by these insects, but there is no doubt that they are meant for use in courtship. The male spreads the odor by vibrating his wings vigorously. Among those butterflies that fly during the day and mate in daylight, color may play a considerable part in sexual attraction, but among those that fly at night and mate in darkness, odor is the prime attractant. Exotic butterflies, such as *Danaus* and *Euploea*, that fly during the day but have very highly developed odor organs on both wings and abdomen, probably must compete with the highly odorous flowers in their hot, tropical world.

Eltringham (*378*) has described scent glands and brushlike organs in male *Xylophasia monoglypha* and *Phlogophora meticulosa*. These are situated in a long narrow pocket in the abdominal cuticle, starting in the first segment and running along either side for about 6 mm. It is slitlike, with the edges pressed closely together. Inside it is housed a brush, which can be opened like a fan by muscle fibers. For an example of a brushlike organ see Figure V.2. It is composed of numerous, long narrow scales.

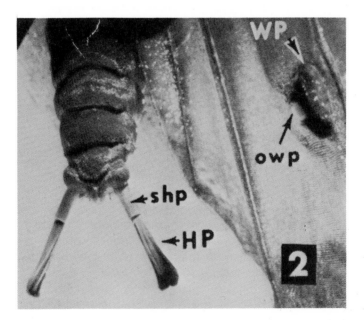

*Fig. V.2.* Dorsal view of posterior abdomen and right hindwing of male *Danaus gilippus xanthippus* with hair-pencils partially extruded. HP, right hair-pencil; shp, partially evaginated membranous sheath; WP, right wing pocket; owp, opening of wing pocket through which hair-pencil will be inserted. [From Brower and Jones (*211*).]

Each scale is penetrated by a minute tube from a gland cell at its base. The brush is evidently a distributing organ for the stimulating scent, presumably used when flying in proximity to a female.

Reichenau (917), investigating the scent of the male sphinx moth, *Sphinx ligustri*, supposed the scent material to be pressed by muscular action from the gland into the hairs or scales and diffuses from there into the atmosphere through their apices. He also maintained that the scales were provided with capillary tubes in which the scent substance was held in the form of bubbles. These bubbles undoubtedly make up the chitinous reticulum that is quite evident in the scales of *Ephestia cautella* and that probably holds the secretion within its meshes.

The anatomy and physiology of the scent gland of male *Hepialus hectus* was described in 1902 by Deegener (341). The pineapplelike scent of the secretion of these glands is an attractant for the female.

Barth (102) compared the anatomy and physiology of the male sex-pheromone organs in *Argynnis paphia*, *A. adippe*, and *A. aglaja*. These organs, situated on the wings, secrete an odorous chemical substance that excites the female in preparation for copulation.

The hair-pencils of *Danaus gilippus berenice* are glandular structures, with the surface of the individual hairs irregularly covered with tiny cuticular spherules ("dust") that detach readily and stick to the female's integument by means of a coating of liquid secretion that comes from specialized gland cells associated with the bases of the hairs. The dust is transferred to the female antennae during normal courtship. The hair-pencils are paired organs lying laterally inside the male abdomen at the end of the body and have arisen through invagination of the intersegmental membrane between sternites 8 and 9. The term hair-pencil refers to the cylindrical bundle of individual hairs; when partially extruded, the organ looks like a small brush, hence the term "abdominal brush" is also used. See Figure V.3 for an example of an extruded hair-pencil. The term "duftpinsel" has likewise been used because the brush is scented. They have also been erroneously called "anal scent glands," although they are neither connected to nor derived from the digestive system. Both of the hair-pencils are extruded simultaneously by the male butterfly through an increase in the pressure of its abdominal body fluids; retraction occurs by means of a retractor muscle attached to the base of the hair-pencil. They can also be forced out by carefully squeezing the posterior part of the male abdomen (210, 879). On natural extrusion, the hair-pencils dip into glandular wing pockets lined with small, flat scales arising from cells arranged in alternating rows with smaller cells. There is no doubt that the wing glands are active secretory organs. The interaction between the

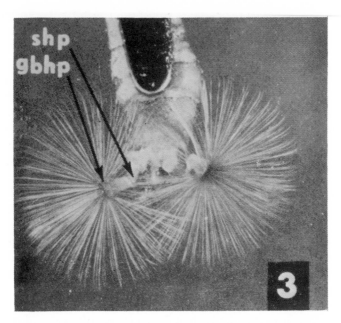

*Fig. V.3.* Hair-pencils of *Danaus gilippus xanthippus* fully extruded and splayed. The individual hairs arise from the glandular base (gbhp) of the hair-pencil, shown completely evaginated; shp, fully evaginated membranous sheath. [From Brower and Jones (*211*).]

hair-pencils and the wing glands in this butterfly has been considered in detail by Brower *et al.* (*210*).

In the related species, *Danaus gilippus xanthippus*, Brower and Jones (*211*) have shown that the males evert their hair-pencils and insert them into the wing pockets for an average of $7\frac{1}{2}$ seconds several times during each day on successive days prior to courtship. Although it was popularly believed that the hair-pencils obtain their scent from the wing pockets, experiments in which the latter were sealed shortly after male emergence showed conclusively that the hair-pencils produce scent independently. A secretion from the wing pockets may interact with the product of the hair-pencils.

The male pheromone brushes of *Leucania impura, L. conigera,* and *Phlogophora meticulosa,* all of which are noctuids, were studied by Aplin and Birch (*72, 162*). The brush is supported on a lever attached to the base of the first abdominal sternite and lies in a lateral pocket stretching from sternites 3 to 5. The lips of the pocket are tightly pressed together to restrain the hairs and limit evaporation of the scent. Although Eltringham

(378) had suggested that the scales lining the pocket might produce the scent, the pocket is separated by a pair of small glands giving rise to a thread of intertwined hairs leading into the brush. The brush is everted by specific muscles pulling the lever from the pocket and away from the abdomen. A muscle across the back of the basal plate of the brush contracts to spread the hairs. The brushes may remain expanded for only 1 or 2 seconds during calling immediately prior to copulatory attempts. Stobbe's glands in the second abdominal segment are essential for development of the scent, as are probably the pockets. The gland cells, greatly swollen with secretion in the pharate adult, discharge after emergence of the male from the pupa (162).

Males of *Achroia grisella* exude a strong aromatic odor that attracts and excites the female. The substance possessing this odor was described by Kunike (702) as being formed mainly by the thorax and, to some extent, by the abdomen, and dispersed into the atmosphere by tufts of hair present on each side of the end abdominal segments. However, Dahm *et al.* (324) have recently shown that the pheromone is produced in glands located at the base of each forewing. They have also shown (325) that the excited and searching females are guided by sound perceived by their abdominal tympanic organs and are able to find males from both upwind and downwind positions. Females whose tympanic organs were punctured could still locate males, but only after considerable searching. Removal of the antennae and labial palps reduced attractiveness of live males and suppressed the reaction of females to the pheromone. Destruction of the tympanic organs plus removal of the antennae and palps abolished the responses of females to males. The attraction is thus mediated by olfactory as well as auditory cues.

Wing glands in the males of *Aphomia gularis*, *Galleria mellonella*, *Plodia interpunctella*, and *Ephestia elutella* were discussed by Barth (100) in 1957. More recently, Röller *et al.* (973) studied the large glandular complex located at the base of each forewing in the initial angle between the costa and the subcosta of *G. mellonella*.

Hair-pencils associated with pheromone production in male *Trichoplusia ni* consist of 2 tufts of hollow scales, 4 mm long, arising from membranous tubes on the ventral surface of the eighth abdominal segment. Large flask-shaped cells bound into lobes by nonglandular interstitial cells are located at the base of the scent brush. A long, narrow reservoir lined with microvilli communicates with the scent scale. The gland cell is 150–200 $\mu$m long in newly emerged males but is completely atrophied within 2 days. Other glandular cells, located on the eighth abdominal sternite, possess minute fan-shaped scales lining the dorsal wall of the fold containing the

hair-pencils. These cells do not atrophy and remain continuously active
(*474, 476*).

## COLEOPTERA

Pitman *et al.* (*869*) have amassed data implicating the hindgut region,
including the malpighian tubules, in pheromone production by recently
excavated males of *Ips confusus.* However, their claims (*871*) to the
discovery of secretory areas in the hindgut were retracted (*1293*). According
to Renwick *et al.* (*923*), unfed females and males and parts other than the
hindgut are not attractive. The same group (*870*) report that males produce
the pheromone within 2 hours of infestation followed by a general increase
in the next 96 hours and than a decrease; by the fourteenth day the
pheromone was nearly undetectable. However, according to Wood (*1290*),
bark and wood samples of ponderosa pine become attractive 4–6 hours after
males are introduced into entrance tunnels, with a maximum production
of pheromone at 18–144 hours. Frass is not attractive until 9–12 hours after
introduction. Attractancy is initiated by males feeding in phloem-cambial
tissues of the host and is delayed until food is passed through the gut. Wood
and co-workers (*1289, 1290*) feel that attraction is associated with male
excrement rather than with the gut itself. Individual males have been
shown to produce 9.5 mg of frass per day for about 15 days. Since pheromone
production commences only after feeding, it is believed that either a
precursor is ingested and metabolized to the attractant or that metabolism
of the food material causes secretory activity in specialized cells (*1289*).
The attractant is found in frass produced by males boring in white fir and
Douglas-fir, indicating that the sex pheromone does not function in initial
host discrimination.

Pitman (*864*) reported that glucose-supplemented diets appeared to favor
pheromone production in *Ips confusus,* but he could not obtain definite
results with maltose, sucrose, glucose, raffinose, fructose, or potato starch.
Ponderosa pine with phloem low in nonelectrolytes was less favorable as a
dietary medium for pheromone production. Tests with billets of wood taken
from different trunk sections indicated that beetles feeding on billets higher
in carbohydrate produced more pheromone.

In a similar manner, male *Ips confusus* beetles treated topically with
100 μg of a juvenile hormone substitute (methyl 10,11-epoxyfarnesenate)
were stimulated to produce sex pheromone in the hindgut-malpighian
tubule region just as did males introduced into fresh pine logs. However,
boring males produced more pheromone than males treated with the
hormone (*186*).

A study by Hardee (*513*) showed that a constant supply of adequate food, especially cotton squares or small bolls, is essential to the continued production of pheromone by male *Anthonomus grandis*. Removal of food resulted in a 50% reduction in pheromone production after 1 hour and a reduction of more than 90% after 24 hours.

## HEMIPTERA (HETEROPTERA)

The metasternal scent glands of the male *Lethocerus indicus* consist of a single, blind-ending tubule. The male glands are much larger than those in the female, forming loops extending into the anterior part of the abdomen. Pattenden and Staddon (*849*) have estimated, from the dimensions of the gland, that the 2 male glands can accumulate a total of approximately 50 μliter of secretion, roughly 25 times as much as in the female.

Gupta (*494*, *495*) described the anatomy of the abdominal scent glands of Heteroptera and suggested that their function is primarily defensive in the nymph and both defensive and sexual in the adult.

## DIPTERA

Lhoste and Roche (*731*) found that the 2 glands that probably produce the pheromone in the male Mediterranean fruit fly are located in the last (seventh) abdominal segment. The attractive substance diffuses from these glands to the surface of an erectile anal ampul that is formed by pulsating pressure from the posterior portion of the rectum.

The pheromone produced by male *Dacus tryoni* has been shown by Fletcher (*411*, *412*) to be secreted by a gland complex developing from the posterior ventral wall of the rectum that, in mature males, consists of a secretory sac and a ventral reservoir to store the pheromone before release. The reservoir is initially lined with secretory epithelium, but, with the onset of sexual maturity, much of the secretory epithelium breaks down and the reservoir takes on a storage function. The pheromone is released through the anus by muscular contraction of the reservoir and rectum. Although the pheromone can be detected in flies 2 days after emergence, the reservoir usually does not become full of secretion until 12–14 days after emergence.

Similar rectal gland complexes in male *Dacus cucurbitae*, *D. dorsalis*, and *D. oleae* were studied by Schultz and Boush (*1051*), and in *D. oleae* by Economopoulos *et al.* (*369a*), who found the gland in *D. dorsalis* and *D. oleae* to consist of a reservoir in the right lateral posterior of the rectal sac and a bulbular secretory sac opening into the base of the reservoir. In *D. cucurbitae*, the gland is situated ventrolaterally off the rectal sac, forming a distinct chamber separated from the lumen of the sac by longitudinal and circular muscles; the epithelial lining consists of large cells greatly folded upon one another.

Males of *Rioxa pornia* secrete their volatile sex pheromone from glandular cells situated in the pleural regions of abdominal segments 3–5, where the thin epidermis is modified as a band of well-defined, columnar cells that are inflated during attractant production. These columnar cells are about 24 μm long and 6.4 μm wide. Passage of the pheromone to the outside is apparently directly through the cuticle overlying the cells, which, according to Pritchard (*888*), are consistent with those described by Jefferson *et al.* (*623*).

## MECOPTERA

Bornemissza (*189, 191*) has shown that males of *Harpobittacus australis* and *H. nigriceps* attract the females by means of a scent produced in reddish-brown vesicles located between tergites 6–7 and 7–8 of the abdomen, which expand and contract rhythmically. These vesicles do not appear to have any associated scent reservoirs and seem to consist of a specialized intertergal membrane, the cells of which presumably discharge their secretion directly onto the membrane surface. The ultrastructure of the secretory gland in *H. australis* has been described in detail by Crossley and Waterhouse (*319*). The vesicles consist of cuticle-secreting squamous cells, secretory duct cells, and paired secretory cells. The secretory duct through which the pheromone passes to the cuticular surface is secreted by an enveloping secretory duct cell bearing short microvilli adjacent to the duct. The secretion oozes out onto the surface of the everted gland through duct openings on the cuticular surface.

# CHAPTER VI

# THE MECHANISM OF ATTRACTANT PERCEPTION

It has long been known that chemical insect attractants, especially the sex attractants, are detected by means of sense organs located mainly in the antennae. In Lepidoptera, the antennae are probably the sole organs of chemoreception. Hauser (*526*), in 1880, reviewed the subject of insect olfaction, beginning with the work of Lefebvre (*718*) in 1838. Hauser described, in detail, the anatomy of the antennae of species in the orders Orthoptera, Neuroptera, Hemiptera, Diptera, Lepidoptera, Coleoptera, and Hymenoptera. He also reported that males of *Saturnia pavonia* and *Porthetria dispar* deprived of their antennae never mated. In such species the male is often able to locate a female from a considerable distance, and his olfactory organs, situated on the plumelike antennae, are highly sensitive (*588, 1179*). The fact that olfactory receptors are usually located on or in the antennae has been substantiated by many investigators since Hauser's work was published.

More recently, Wilson (*1269*) has reviewed the overall subject of chemical communication in the social insects, and Kwiatkowska (*707*) has reviewed the problem of modes of communication in the animal kingdom. The following references are to other recent reviews of insect olfaction and orientation: *129, 273–276, 284, 550, 684, 751, 757, 799, 862, 948, 1025, 1034, 1035, 1079, 1087, 1180, 1269–1271,* and *1313*.

In 1900, Mayer (*758*) reported that male promethea moths immediately flew 100 feet to a clear glass battery jar, covered with mosquito netting, containing 5 females. When the jar was inverted and sand was packed around its mouth to prevent the escape of air, males were no longer attracted. Males were also attracted to females wrapped loosely in cotton to make them invisible; these males grasped the cotton with their abdominal claspers in typical copulatory attempts. Males were attracted by and mated with females whose wings had been replaced by those of males. Mayer concluded that sight was not involved in male attraction. He arranged a small wooden box containing females in such a way that air blown into the box emerged through a small chimney, and found that males were attracted to the top of the chimney, regardless of the presence of carbon disulfide and diethyl sulfide fumes in the immediate vicinity. Males released 5 feet from severed female abdomens flew directly to them and ignored the remainder of the female body. Males deprived of their abdomen or whose spiracles were covered with glue were still attracted to females. However, males whose antennae were covered with shellac, glue, paraffin, Canada balsam, celloidin, or photographic paste did not seek females and displayed no excitement even if held within 1 inch of virgin females. Once the paste was washed away, the males again responded to the female. Abbott (*1*) found that male promethea moths whose antennae had been coated with shellac flew irregularly and were unable to find a female in their immediate vicinity, even at a distance of only a few inches.

The great French naturalist, Fabre (*389*), in 1904, imprisoned a female emperor moth, still damp with the moisture of metamorphosis, under a gauze cover in his study. The same evening males "seemed to take possession of the house; about 40 male moths were flying round the gauze cover." This was repeated for the next 8 nights. The males appeared to fly with certainty to the house but, having arrived, were uncertain as to the precise location of the attractive object, and final discovery was left to a vague and hesitating search. The position of the cage could be changed, the female was even imprisoned in a drawer, out of sight, without thwarting the males in their quest. Conditions were unfavorable for flight in that the weather was stormy and the dark house was surrounded by bushes and shrubs. Even if the female was surrounded by dishes of strong odorants such as naphthalene and oils, she was still located by the males. Fabre also found that a spot on which a virgin female had recently rested, but from which she had been removed, was attractive to the males, particularly if the object was absorbant, i.e., cardboard, dust, or sand, but less so if it was hard and smooth like marble or metal. Excising the antennae rendered the males much less skilled at finding the female.

Mayer and Soule (*759*) showed that male promethea and gypsy moths whose antennae were covered with flour paste would not mate with their respective females until the paste had been washed away with water. Normal males flew toward the females against the wind, frequently passing alongside and beyond the females. Under these conditions, the male would often remain poised on his wings, drifting back with the wind until he came to leeward of the female, then a few vigorous strokes would bring him toward her again.

Kellogg (*657*) reported in 1907 that male *Bombyx mori* with intact antennae and blackened eyes found females immediately, whereas those lacking antennae but able to see could not locate a female. Males possessing only a right antenna and exposed within 3–4 inches of a calling female circled repeatedly to the right until coming in contact with her; those lacking a right antenna circled to the left. This behavior was confirmed by Sengün (*1065*) who also showed that, in the absence of air movement, normal males sometimes detected females from a distance of 5 cm, but not 7 cm, whereas females could be located from a distance of 25–150 cm with moving air. Besides confirming the foregoing, Nakazema (*824*) reported that male *B. mori* with both antennae removed responded to females 76% less often than normal specimens. Coating the antennae with Vaseline reduced the response somewhat. Minor olfactory regions were given as the basal portions of the wings and the labial palpi. Chemoreceptors of the male silkworm moth are evenly distributed throughout the antennae, as determined by removing different numbers of antennal joints (*301*). The perceptive distance decreases with the removal of increasing numbers of segments.

In describing the courtship of the polyphemus moth, Rau (*910*) reported that during sexual excitement the antennae of males are erect and alert, while on other occasions they are drooping and limp. Male cecropia moths with one-half of each antenna removed will mate, but mating does not occur with males completely lacking antennae (*908*).

According to Prüffer (*893*), male gypsy moths deprived of their wings, wing scales, abdominal scales, 1 antenna, the tips of both antennae, or whose eyes are covered are able to find females and mate, although those deprived of their wings have difficulty getting next to the female. Males deprived of both antennae show no interest in females.

Male pink bollworm moths with both antennae removed do not respond to the female's attractive scent, but males with 1 antenna or with only part of the antenna clipped off respond readily (*140*).

Osmani and Naidu (*837*) have shown that females of the red cotton bug, *Dysdercus cingulatus*, confined in a cardboard box with pinhole perforations and kept 15 inches away attracted males. Males lacking antennae were not attracted to females.

Male *Grapholitha molesta* without antennae are neither attracted to, nor sexually stimulated by, the female's pheromone (*446*).

Although Riley (*942*), in 1894, conceded that the olfactory organs in lepidopterous insects are located in the antennae, he stated that there is good evidence (without citing such evidence) that in some hymenopterous insects the olfactory organ is localized in an ampulla at the base of the tongue.

Valentine (*1213*) found that antennaeless male *Tenebrio molitor* never responded to a female and had to be placed directly on the female's back before copulatory reflexes asserted themselves. Application of paraffin oil to male antennae also prevented the response. A male with 1 antenna removed reacts by swerving off in an arc to the side bearing the remaining appendage, but he still responds to the female's odor; however, he is incapable of following the lure when it is set in motion. The same circus motion was elicited by coating 1 antenna with oil. Removal of the maxillary palpi had no effect on male response to the lure. Removal of various lengths of the antennae showed that the male organs operative in the discovery of the female are located chiefly in the terminal 4 segments of the antennae. Valentine concluded that it is reasonable to suppose that the peg organs found on these terminal segments are operative in the response of the male to the female odor. These results are in direct contradiction to the conclusions reached by McIndoo (*771*) in 1915. He reported that, beyond a doubt, none of the antennal organs of beetles serve as olfactory sites, and that the olfactory pores on the wings and legs are well adapted anatomically for receiving odor stimuli, since the peripheral ends of their sense fibers come into direct contact with the external air. However, McIndoo's tests were not conducted with females or their scent, the odorous substances used were peppermint, thyme, and wintergreen oils, pennyroyal and spearmint leaves and stems, and decayed matter from *Harpalus pennsylvanica* beetles.

Among the Lymexelonidae (beetles), the male has either an extraordinarily developed second joint to the maxillary palp or branched antennae. If the palpal joint is removed, he is no longer attracted to the female. Male *Hylecoetus dermestoides* with their feathery maxillary palpi coated with a film of gum mastic became inactive and would not mate with females in their vicinity. After the mastic was removed with alcohol they quickly became active and began to mate. The females, on the other hand, have very simple maxillary palpi and antennae. Germer (*451*) concluded that the male detects the female with his maxillary palpi, which are well supplied with nerve stalks.

Numerous thin-walled sensoria present on the 8 distal segments of the antennae of male *Limonius californicus* apparently function as chemoreceptors for the female's pheromone. Response of males to the attractant was not affected by bilateral amputation of the 5 distal segments, decreased progressively with the amputation of segments 6 and 7, and was eliminated completely by removal of the eighth segment (*734*).

Male *Attagenus megatoma* with antennal segments 2–8 removed failed to respond to female pheromone in the laboratory (*223*).

Gara (*432*) showed that both male and female *Ips confusus* beetles freed of 1 antenna showed a 50% decrease in response to frass made attractive by male feeding, as compared with the response of intact beetles; removal of both antennae prevented the assembling response.

The female antennae in *Byrsotria fumigata, Leucophaea madeira,* and *Nauphoeta cinerea* serve as distance chemoreceptors for perceiving the male pheromones. The ability of the female to mate can be correlated with the distribution of thin-walled chemoreceptive types of sensilla on the antennae. In *B. fumigata* only the antennae perceive male stimuli, whereas in the other species sensilla on the last segments of the maxillary and labial palps are also capable of detecting the male, though not over a great distance. Although the male antennae bear the receptors for effecting a rapid response to females, the maxillary and labial palps and the cerci (except in *B. fumigata*) also play a role. In *Pycnoscelus surinamensis* the male antennae serve as distance receptors for sex odor; however, after antennectomy, sense organs on the last segment of the maxillary palps can detect a female on contact (*977, 982*).

Male wasps (*Bracon hebetor*) whose abdomens had been removed or covered with lacquer could no longer locate females (*813*) and failed to show the premating excitement characteristic of normal males in the presence of females (*487*). Such abnormal males wandered aimlessly about and bumped against one another, although they were in the immediate vicinity of a female. When such a male accidentally came in contact with a female he would mount and copulate by reflex reaction, without exhibiting wing flipping. No sexual excitement was shown by males whose antennae had been coated with celloidin, while males blinded by coating the eyes with asphaltum black behaved normally, showing excitement in the presence of females and mating successfully. In another investigation, males whose abdomens had been removed responded normally to females and attempted to mount (*488*).

Abbott (*2*) tried to determine if the antennae of male *Megarhyssa lunator* function in bringing the sexes together. Males whose antennae had been frozen by spraying with ethyl chloride were clipped on a wing to mark them, and released. Over a 2-week period, 11 of 16 released males returned to find the females, leading Abbott to conclude that the antennae are not used in bringing the sexes together. Males deprived of their antennae by excision also managed to locate the females (1).

*Drosophila victoria* males produce an attractant (or aphrodisiac) for the females. Virgin females deprived of their antennae failed to accept male overtures owing to their failure to receive olfactory stimuli; nevertheless, males attempted to copulate with antennaeless females despite the females' objections (1132). For the female to be normally responsive to the display of wing vibrations by males, the arista and funiculus of her antenna must be intact and able to move freely. The arista probably acts as a sail, twisting the funiculus and thus stimulating units of Johnston's organ at its base (750).

Minnich (791) and Abbott (1) published reviews of the location and physiology of insect olfactory organs, reporting that the antennae are not the only sites of these organs, although others had not definitely been located. A thorough search of a number of detailed reviews of insect olfactory chemoreception prepared by Marshall (752), Dethier and Chadwick (347), Dethier (345, 346), Götz (468), Hecker (529), Chauvin (300), and Wright (1307) shows, beyond a doubt, that most insects use their antennae to locate the opposite sex, although the maxillary and labial palpi, legs, and ovipositors may also be used by some species.

According to Ford (419), the attractants of both male and female moths are perceived by the antennae, which contain great numbers of "end organs" capable of being stimulated by volatile substances (scents) carried in the air. Several types of these structures exist, but in general they are minute cups in the chitinous surface with a projection in the center that is connected to a nerve cell. Antennae of female moths are relatively simple, and are chiefly engaged in registering the aphrodisiac scent of the male produced when he is close at hand; they consist of jointed rods only occasionally provided with short side branches. Antennae of the males of some species are much more elaborate, with brushlike extensions along both sides so that their surface is greatly increased (Fig. VI.1). On the other hand, the antennae of butterflies are never "feathered" or pectinated; they always end in a knob, while those of moths rarely do and never end in forms without the frenulum.

The sites of insect olfactory reception in the antenna have been found to be various types of sensilla. Schneider (1032), in his detailed review of insect antennae, defines a sensillum as "a specialized area of the integument, consisting of formative cells . . . , the sensory nerve cells and, in some cases, auxiliary cells." The numerous sensilla may be classified into 15 main

*Fig. VI.1.* Close-up of head of male gypsy moth, showing the feathery antennae used to detect the female sex pheromone. [By permission of the U.S. Department of Agriculture.]

groups, with sexual chemoreception, particularly in the lepidopterous insects, being assigned mainly to sensilla basiconica (sensory pegs or cones possessing one to several nerve fibers) and sensilla coeloconica (sensory pit-pegs or thin-walled cones situated on the floor of depressions in the antennal cuticle, and innervated by a bundle of nerve fibers) (*1024, 1030, 1031*).

Schenk (*1022*), in 1903, described the anatomy and histology of sexual differences in the antennae of 4 Lepidoptera and 10 Hymenoptera. The nervous system of the antennae in male and female *Saturnia pyri* was studied by Prüffer (*894*). The principle of the nervous system is the same in both, but the female antenna is characterized by a smaller number of nerve cells shown by a reduction of the antennal length.

In 1937, Barth (*101*) reported on a study of the anatomy of the antennae in *Plodia interpunctella* and *Aphomia gularis*. The antennae of both male and female *Plodia* consist of 48 segments plus a ring and a shaft segment, each of which is provided with sensilla coeloconica and sensilla styloconica. Excision of the palpi in *Plodia* and *Aphomia* does not prevent sexual excitement in the male. Removal of the entire right antenna and half of the left one still permits the male to respond, since it leaves him with 81 coeloconica and 9 styloconica sensilla, but removal of the entire right and three-fourths of the left antenna prevents reception (35 coeloconica and 2 styloconica left). Removing three-fourths of each antenna prevents reception (68 coeloconica and 4 styloconica), but a response is still elicited after one-half of each antenna is removed (leaving 162 coeloconica and 18 styloconica). Barth found that the typical male sexual dance is elicited within 1–2 seconds in a container that had previously held a female; apparently the sensilla styloconica come into play at first, followed by the sensilla coeloconica. Male *Plodia* antennae possess 320 sensilla coeloconica, whereas female antennae possess 274.

A detailed study by Boeckh *et al.* (*180*) of the antennal sensilla of the saturnids *Antheraea polyphemus, A. pernyi, Platysamia cecropia,* and *Samia cynthia* described sensilla coeloconica, basiconica, trichodea, chaetica, styloconica, campaniformia, and squamiformia. The polyphemus moth antenna had 70,000 sensilla and 150,000 sensory cells in the male, and 14,000 sensilla and 35,000 sensory cells in the female. Both sexes showed the same types of sensilla, except that the female showed no sensilla trichodea and larger numbers of sensilla basiconica.

The antennae of the noctuid moths *Heliothis zea, Spodoptera exigua, S. ornithogalli,* and *Trichoplusia ni* are setiform with many flagellar segments. Sensilla found in all species are chaetica, trichodea, coeloconica, styloconica, Böhm bristles, and ear-shaped sensilla for which the name "sensilla auricillica" has been proposed. Sexual differences appeared as slight differences in the number and distribution of the sensilla (*622*).

Myers (*818*) found 3 types of sensilla with perforated walls on the antennae of *Danaus gilippus berenice*. The most common are short, thin-walled pegs over most of the antennal surface. Long, curved, thin-walled pegs occur in patches on the inner medial surface. Multiple sensilla coeloconica are present, having up to 50 pegs in one sensillum. Long, thick-walled hairs on the antennae are mechanoreceptors and contact receptors. Sunken pegs of unknown function are also found. Myers (*819*) found that 2 of the 3 major types of sensilla (long, curved pegs and coeloconica sense organs) can be completely blocked in the female without reducing courtship success, showing that the widely distributed short, thin-walled sensilla alone are capable of receiving the male sex pheromone. Approximately 1000–5000 of these sensilla (about 5% of the total) must be exposed for normal courtship success.

Borden and Wood (*188*) observed at least 5 types of sensilla on the antenna of both sexes of *Ips confusus*. Short, thin-walled pegs (basiconica) and long, thin-walled hairs (trichodea) are permeable to crystal violet dye and are likely to be olfactory receptors. Removal of the antennal clubs eliminated a positive response of the female to male sex attractant. Responses of females after removal or covering (with collodion or india ink) of various portions of the club indicated that the sensilla trichodea are the receptors of the sex pheromone.

Vogel (*1235*) reported that the male honeybee antenna contains many more sensilla than that of the female or worker. Kaissling and Renner (*638*) were able to identify 2 types of olfactory receptor cells responding to pheromones in this insect. These cells are associated with the poreplates on the antennae of all 3 castes. One of the cell types in the drone is specialized for queen substance and the other responds to the scent of the Nassanoff gland.

In 4 reports published between 1956 and 1959, Schneider *et al.* (*1024, 1041–1043*) described in great detail the morphology of *Bombyx mori* antennae and their function in chemoreception. Using electrophysiological methods, wherein tiny silver chloride-coated electrodes (*222*) were inserted into a freshly excised male antenna that was further connected through an amplifier to an oscilloscope, Schneider (*1024, 1040*) succeeded in recording action potentials from the antennae. On stimulation with natural *Bombyx* sex attractant, a characteristic response (EAG or electroantennogram) appeared, whereas other olfactory stimuli or drugs resulted in different forms of electroantennograms; the amplitude of the EAG is directly related to the concentration of the attractant tested. When the olfactory stimulus is extract of female *Bombyx* sex attractant, electrical activity increases in the male antenna but not in that of the female, whereas activity increases in both sexes when the stimuli are cycloheptanone or sorbyl alcohol. Many compounds in high concentrations elicit an EAG (ether, ethanol, propanol, butanol, or xylene), but only the sex attractant elicits an EAG at extreme dilution. Electrophysiological measurement of attractants was reviewed by Schneider (*1027*) in 1961, and the technique was utilized by Morita and Yamashita (*802*) in 1961 to record receptor potentials from sensilla basiconica in the antennae of *Bombyx mori* larvae.

A detailed study of the *Bombyx mori* antennal morphology was conducted by Steinbrecht (*1146*) with the aid of light and scanning electron microscopy. He reported that the sensilla trichodea and sensilla basiconica are olfactory in function. In the male, the long trichodea contain the receptors for the female sex pheromone, bombykol, and cover the free space between the branches of the bipectinate antenna. It is probable that most of the odor molecules are adsorbed to the sense hairs. The number of highly sensitive receptor cells for bombykol on the male antenna was estimated as about 25,000. Using the number to calculate the sensitivity threshold of a single receptor cell, it is estimated that a single molecule of bombykol can elicit a nerve impulse in the cell. Excellent reviews of insect chemoreceptors and receptor sites are those by Wolbarsht (*1276*), Slifer (*1111*), Boeckh et al. (*181*), Kafka (*634*), Kaissling (*636*), and Schneider (*1034, 1036, 1038, 1047*).

Schneider's assumption that the EAG is essentially the sum of many olfactory receptor potentials comes from data recorded more or less simultaneously by an electrode located in the sensory epithelium; it has received experimental support (*1028*). The electrophysiological procedure has been extended to a number of Lepidoptera, such as *Callosamia promethea, Hyalophora calleta, H. euryalus, Antheraea pernyi, Rothschildia orizaba,* and *Samia cynthia* (*1026, 1028, 1029, 1045*), *Porthetria dispar* (*637, 1037, 1039*), and *Choristoneura fumiferana* (*14*). Priesner (*886*) has used it to determine the degree of interspecific effects of chemically unknown female sex attractants for 1900 species combinations between 104 saturniid species, and Schneider (*1028*) determined the species specificity of sex attraction in 7 saturniids. The electrophysiological technique has also been applied to the carrion beetles (*Necrophorus* spp.) by Boeckh (*177, 178*), pine beetles (*Dendroctonus brevicomis* and *D. frontalis*) by Payne (*851*), the American cockroach (*Periplaneta americana*) (*182*), and the honeybee (*Apis mellifera*) (*628*).

There were no significant differences in the EAG's of a male antenna using the female attractant-producing glands of other species as stimuli, but the effect of the related saturniids on the male *Bombyx* antenna was much smaller than that of the female *Bombyx* gland; neither the female *Bombyx* gland nor the synthetic *Bombyx* sex attractant (bombykol) elicited a response in the antenna of any of the saturniid males checked. Although the antennae of males and females responded to concentrated essential oils (wintergreen, clove) with an EAG, no antenna of any of the female saturniids or *Bombyx* responded to any of the glands (*1028*). Bombykol showed an electrical response threshold between $10^8$ and $10^{14}$ times lower than those of the other 3 cis-trans isomers of this structure. Contrary to earlier work with concentrated lures (*1024*), the EAG threshold has been shown to be much higher than the behavior threshold using the pure substance.

Kaissling (*636*) has applied the mass-action law to prove the connection between the odor concentration and the receptor potential's amplitude of the bombykol receptor of *B. mori* and the queen substance receptor of *A. mellifera.*

Schneider (*1029*) found that the exposure of excised male antennae to a sex attractant held on the tip of a glass rod could be greatly refined by the use of a living mounted male exposed to the test substance impregnated on a fluted filter paper in a short glass tube. Under these circumstances it was possible to record EAG's for many hours or days, instead of only 1–2 hours, and the cartridges were easily exchanged during the course of an experiment. Initially the reaction of a male *Bombyx* to bombykol was practically constant from a threshold below $10^{-10}$ $\mu$g to about $10^{-4}$ $\mu$g; with higher concentrations the reaction intensity first rose slowly and then very rapidly. From this response, Schneider deduced that the insect could not distinguish between different molecular densities with great sensitivity in the lower concentration range, but the sensitivity is much greater in the higher range. In support of this hypothesis was his finding that dissected glands of female *Bombyx* were electrophysiologically as effective as filter paper containing $10^{-2}$–10 $\mu$g bombykol; this concentration range was exactly where the slope of the curve was steepest and where odor-intensity discrimination was expected to be optimal. The female *Bombyx* antenna showed no EAG response to the gland or to synthetic bombykol, indicating that the female moth does not possess the specific receptor for detecting its own attractant. Although the antenna of female *Periplaneta americana* gave an EAG response to its own sex attractant that was approximately 50% as large as that given by the male antenna (*182*), Boeckh *et al.* (*183*) have reported that the female antenna does not contain receptors for that odor.

In subsequent electrophysiological experiments by Schneider *et al.* (*1039*), the EAG response of male *Bombyx mori* antennae to bombykol and its isomers increased from $10^{-3}$ to 100 $\mu$g, where it reached a plateau. EAG's elicited by an odor source containing less than $10^{-2}$ $\mu$g were not significantly different from the control EAG's. Significant EAG's were elicited by $10^{-1}$ $\mu$g, 1 $\mu$g, and 1 $\mu$g, respectively, of the cis,trans, cis,cis, and trans,trans isomers of bombykol. In behavioral tests with bombykol using live male moths, 50% of the moths responded with wing flutter to $10^{-4}$ $\mu$g of bombykol; significant reactions were observed with $10^{-5}$ $\mu$g and nearly 100% reacted to $10^{-3}$ $\mu$g. At the EAG and behavior thresholds, the air stream contained $1 \times 10^{7}$ and $2 \times 10^{2}$ molecules/cm$^3$, respectively. More accurate estimates made possible by the use of tritiated bombykol have resulted in a revision of the EAG threshold to $10^{4}$ molecules (*1037*). Intrinsic limitations in the use of EAG's to bioassay male pheromones among the Lepidoptera have been discussed by Birch (*164*), and Adler (*10*) has described a number of physical conditions important to the reproducibility of EAG's.

A more exact (and more difficult) method for measuring insect response to a sex pheromone involves the use of single-cell recordings, which show that the sensilla trichodea of the male lepidopterous antennae are the specialized pheromone receptors. Consequently, single olfactory receptor cells react only to the pheromone and not to other olfactory stimuli. The effectiveness of different compounds can be compared with their physical-chemical properties. The use of the single cell technique has been discussed recently by Boeckh (*179*), Priesner (*887*), and Kafka (*633, 635*). Olfactory receptor cells respond to qualitatively different odor stimuli with either excitation (increased impulse frequency) or inhibition (depressed impulse frequency). Receptor potentials and nerve impulses are recorded simultaneously with the same extracellular electrode. The time courses of excitatory and inhibitory receptor potential are different, and the electrophysiological reaction threshold of the inhibitory stimulus is higher than that of the excitatory threshold. Recordings obtained from single cell receptors appear as spikes rather than curves.

A fascinating study described very recently by Kasang (*652*) involved the chemophysiology of tritiated bombykol reception on the male *Bombyx mori* antenna. Bombykol has been shown to undergo metabolic transformation on the antenna. Although the amount of material extracted from the antenna at different time intervals after bombykol application was too small for direct chemical analysis, thin-layer chromatographic comparison of the eluted material with 100 different lipids showed that the metabolic products are fatty acids, esters, and alcohols. Apparently, bombykol metabolism is the result of enzymatic activity by 2 different enzymes and is dependent upon pH and temperature. It was concluded that bombykol is first adsorbed on the antenna, then diffuses from an outer lipid phase into an inner aqueous phase (sensillum liquor and receptor dendrite cytoplasm). Also, the metabolism of bombykol into fatty acids and esters is not related to the excitation of the bombykol receptor cell.

Riddiford (*936*) has obtained strong evidence that receptor proteins present in the antennae of male *Antheraea pernyi*, *A. polyphemus*, and *Platysamia cecropia* adsorb (or react with) pheromone molecules, resulting in male response. (See references *112, 181, 316, 979, 1036,* and *1276* for reviews of electrophysiology and the role of receptor cells.)

According to Kettlewell (*663*), male moths assemble or fly to females only upwind, gauging the direction by contrasting the number of molecules striking each antenna per unit of time. The volatility is greater at higher temperatures, with considerable loss of scent by convection currents occurring in hot sunlight. Immediately to the windward side of the male is a "negative zone"; males flying directly into this zone show behavior varying with species. In *Endromis versicolora*, a male striking this zone wheels around and returns several times close to the ground; in *Parasemia plantaginis*, the males alight on the ground and proceed for the rest of the journey on foot or by fluttering through the undergrowth.

Laboratory and field tests conducted by Schwinck (*1054, 1056, 1057*) and reported in 1954, 1955, and 1958 showed that a pure odor attraction of male *Bombyx mori* by the female could only be demonstrated for close orientation, resulting from trial and error as successive differential perceptions. At a greater distance the attractant has only an excitatory effect; unoriented distant searching is initiated by the odor stimulus. Estimation of the concentration gradients for open-room diffusion, supporting the test results, showed that this gradient was not the orientating factor for distance attraction. Air streams containing the attractant showed a much greater orientation for males than did the attractant in the absence of air streams; the males move in proportional linear orbits against the wind. The odor stimulus is only the cause of the streaming orientation, and the directional factor is merely the air stream and not a concentration gradient. Odor stimulation acts as a constant stimulator of streaming orientation; a strong diminution of attractant concentration leads to elimination of stream orientation. It was concluded that the sequence is (1) random searching flight by the male, (2) orientation against the wind when the female odor is detected, and (3) upwind flight. If the male strays from the wind stream, he resumes random searching flight until the odor is again detected. The female sexual odor is thus not an attractant for distance orientation, but only an excitant to release another orientation mechanism; it becomes a true attractant only over a short distance from the male. Laboratory tests with *B. mori*, *Porthetria dispar*, *P. monacha*, *Orgyia antiqua*, and *Lasiocampa quercus* males with partial antennal amputation showed that the sensilla styloconica played no part in male orientation to the female odor.

Schwinck's theory of a guidance mechanism directing the course of the male toward a receptive female is supported by laboratory observations with *Trichoplusia ni* (*1077*). Males remain in the typical resting position until they are exposed to air carrying the sex pheromone; then the antennae are "raised and brought slightly forward of a plane perpendicular to the body axis." The wings are extended and vibrated and the male flies toward the source of the odorous air stream. The positive orientation of the male to the air current may be facilitated at close range by air movement from a receptive female's vibrating wings. According to Sower *et al.* (*1125*), at a maximum female sex-pheromone release rate of $1 \times 10^{-8}$ gm/minute, a male olfactory sensitivity to pheromone concentrations as low as $1 \times 10^{-17}$ gm/ml of air, and a wind velocity of 1 m/second, a communication distance of less than 100 m is indicated for adult $T.$ $ni.$ Vision can be used in short range orientation of a pheromone-stimulated male to a female, and the male may orient to a visual model 2 cm away from the source of the volatile pheromone (1089).

Orientation of male and female $Trogoderma$ $granarium$ in a gradient of assembling scent was studied in a circular arena by Levinson and Bar Ilan (728, 729). In the vicinity of the odor-emitting zone, the behavior of males is characterized by vibration of the antennae and a zigzag pattern of approach with intermittent stops.

Casida $et$ $al.$ (286) reported that the approach of the male introduced pine sawfly ($Diprion$ $similis$) to the attractive female is characterized by a zigzag pattern decreasing in amplitude as the "point" source is approached. Within several feet of the attractant source, the males usually proceed slowly in locating the attractant or occasionally go directly to the active material.

Wilson and his associates (193, 1268, 1272) deduced the shape and size of the ellipsoidal space within which male moths can be attracted under natural conditions. With a moderate wind blowing, "the active space has a long axis of thousands of meters and a transverse axis parallel to the ground of more than 200 at the widest point" (1268). These investigators have described a general method for estimating the threshold concentrations of odorant molecules (1273).

In a discussion of the olfaction of various wild silk moths, Collins and Weast (305) stated that the female sex scent dissipates in the air to such an extent that it cannot be detected by a male farther away than about one-half mile. The male may pass by several females before a scent trail is detected, when the male flies upwind. After he nears a female, he may use a zigzag pattern to "home in."

Male $Anagasta$ $kühniella$ released in a wind tunnel downwind of a calling female flew upwind, but on losing the scent they made crosswind casts until the female was located. In still air males approached females from below (1193).

The manner in which a moth finds a mate has inspired biologists and biochemists to propose a number of theories about the mechanism of olfaction in both moths and man. In 1894, Riley (942) reported experiments in which he liberated a male cynthia moth in a park $1\frac{1}{2}$ miles away from a female moth in his window; the following morning the 2 were together. He tried to account for the attraction of insects for one another from a distance by a sort of telepathy; he stated, "this power would depend neither upon scent nor upon hearing in the ordinary understanding of these senses, but rather on certain subtle vibrations as difficult for us to apprehend as is the exact nature of electricity." Although Fabre's experiments (*389–391*) certainly indicated strongly that female moths attract the males by odor, he could not bring himself to believe that odor could draw moths from hundreds of yards or even miles away. He therefore postulated the existence of another sense, unknown to us, which by a vibrational stimulus warned the males from afar. He felt that something about the moth vibrates, causing waves capable of propagation to distances incompatible with an actual diffusion of matter.

In 1913, Teudt (*1177*) presented a summary of several theories of olfaction, comprising (1) odor particles dissolved by the mucus, (2) intramolecular vibration of odorous substances when the molecules come in contact with the nerves, and (3) rhythmic axial revolution of the molecule, dependent on the number, position, and quality of the combined atoms in the molecule. As a result of his own observations, Teudt explained the attraction of female moths to males in the presence of very odorous materials, such as naphthalene, by saying that the vibrations in the male's odor receptor organs do not react to the vibrations of the naphthalene odor; he concluded that odor detection takes place by electron vibrations.

Poulton (*885*), in 1928, discounted a "wireless" theory of moth assembling by calling attention to the fact that female moths may impart their attractive scent to inanimate objects with which they have been in contact. This theory had been explained as an "assembling" of males responding to a "wireless" call sent out by the antennae of the female despite the fact that it was well known that the attractive part of the calling female was the abdomen.

Observations over more than 10 years led Dyson (*364–366*) to conclude that odor must be related to a characteristic molecular vibration pattern, and he assigned certain odors to certain Raman frequencies ("osmic" frequencies) in the general range of 1500–3000 cm$^{-1}$. This was followed in 1950 with a hypothesis by Duane and Tyler (*358*) in which the attractant of the female moth emits infrared radiations picked up by sensitive receivers located in the male's antennae. These investigators measured the radiation from female polyphemus and cecropia moths using a recording infrared spectrophotometer and found a definite pattern in the region from 3 to 11 $\mu$m. By means of a small thermocouple buried in the fine fur of the female's thorax, it was determined that an active female raised her temperature as much as 11° above that of the room by vibrating her wings or moving her legs; thus, she radiated energy at a greater rate than her surroundings. Duane and Tyler also measured the lengths of antennal hairs of male cecropia moths and determined them to be between 40 and 80 $\mu$m; all variations in the length of the hairs appeared to be close to 6 $\mu$m or multiples thereof. This led the investigators to speculate in the following way: "It is noteworthy that four microns is one-half the wave length of eight microns which is well within the emission band of the female. Does this mean that the male Cecropia moth has a tuned antenna array which is his receptor for locating the female?"

Callahan (267) supports an electromagnetic radiation force of attraction between the sexes, based on flight behavior and configuration of the antennae in flight, as measured in the infrared region. He has described detailed studies (268, 269, 271), claiming that night-flying moths generate considerable far-infrared radiation in the 9–11 $\mu$m region and that they are able to locate minute thermal sources of this radiation in total darkness by means of their compound eyes (270); they are thus able to detect each other as thermal points of far-infrared radiation against the cool ambient background of nighttime temperatures. Callahan (269) also claims that these moths (most of his studies were carried out with *Heliothis zea*) have antennae bearing organs with measurements and configurations of far-infrared resonators. The male *H. zea* responds to the female over a long distance by detecting her vibrational signals and only senses her sex pheromone when he is very close to her. Moth flights are reduced on nights with a full moon when emission in the 8–14 $\mu$m range from the moon interferes with moth behavior. Callahan's theory of optic-microwave assembling has been reviewed in several detailed papers (38, 273–276).

Laithwaite (710), in 1960, proposed a radiation theory of moth assembling. Pointing out that air turbulence is practically continuous, he claimed that it is difficult for the followers of an olfactory theory of assembling to explain that "males will assemble to a virgin female both up and down wind," flying in a direct line. Laithwaite claimed to have substantiated this by experimental observation with released males of the common vapourer moth (*Orgyia antiqua*) and to have found that males released from a distance of 10 feet immediately converged on an empty box, a fertilized female, a dead female, eggs, and an empty female pupal case, whereas those released 100 yards away were not attracted by any of these objects despite the fact that a virgin female will attract males from this distance. The male's antenna was compared to an electromagnetic aerial, with the spacing (0.2–0.02 mm) of the pectinations indicating an operating wavelength in the far-infrared band. Females placed in an extremely fine-mesh wire-gauze box, which allowed scent particles to pass through but would effectively screen electromagnetic waves, attracted males from a short distance only.

In a detailed reply, Kettlewell (*666*) pointed out that Laithwaite had cited no positive evidence to support his radiation theory and had been misled, by low wind velocities and countereddies, into believing that males are capable of downwind assembling to a female. The total substance of Laithwaite's evidence, according to Kettlewell, was his observation of the similarity between radar antennae and assembling male moth antennae.

Recent scanning electron microscopy and conventional electron micrographs of the *Heliothis zea* antenna indicate that these moths may sense temperature and humidity electromagnetically, but they detect the sex pheromone by olfaction (*486*). Ultrasonic emissions by *H. zea* combined with acoustic properties of the ears may provide an echo-locating capability which could be used for the detection of large objects at a distance of more than 2 m and the resolution of details down to 7 mm at distances up to 24 cm (*654*).

Nolte (*828*), in 1940, had attempted to discount a radiation theory of insect olfaction by citing Prüffer's tests in which live female gypsy moths placed in a lead cylinder impervious to radioactive rays still lured males, whereas females kept under a glass did not.

By means of a panel of observers used to select by smell 16 compounds with an odor resembling that of nitrobenzene, Wright (*1298, 1304*) set up a theory of odor based on its correlation with a pattern of molecular vibrations below 1000 $cm^{-1}$ wave number as studied by means of Raman frequencies. He claimed that this refuted Dyson's theory of molecular vibration by osmic frequencies in the range 1500–3000 $cm^{-1}$. In a reply, Dyson (*367*) admitted an error in assuming the frequencies concerned to be those in the higher Raman ranges, but claimed that Wright's results proved that his (Dyson's) original hypothesis correlating odor specificity with molecular vibration was sound in principle. Wright (*1299, 1302, 1312, 1313, 1315, 1316*) extended his theory of low-frequency molecular vibration to insect olfaction, claiming that the nerve cells of the male's olfactory end organ contain a pigment of unknown constitution (*1299*). The theory presupposes some kind of direct interaction between the vibration of the odorous molecules and the male's receptor organs. It predicts that in a group of dissimilar chemicals that can act as sex attractants for the males of a certain insect, certain frequencies will be present that will be absent in biologically inactive substances, other things being equal (*1300, 1305*).

Attempting to explain how the male insect follows a scent, Wright (*1301*) claimed that an insect in free flight before it enters an odor cloud probably searches by flying a series of rather long zigzag paths. When it enters the cloud, its tendency to turn is inhibited as long as the interval between pulses tends to decrease. If it starts to move out of the cloud or away from the source, the interval between pulses will increase, releasing the inhibition on the tendency to turn. This would cause abandonment of a fixed flight path to make a series of short, violent zigzags until it once more locates a path in which the pulse interval tends to decrease. "Where sex attractants are involved, and these are the scents which operate over great distances, the emitting insect, usually the female, could actively assist the guiding process by emitting the scent in a series of short puffs."

In 1964, Wright (*1308*) speculated that insects probably need a combination of primary odors to evoke a response, rather than a single odor. He theorized that the sex-attractant scents, in most cases, are probably due to a mixture of substances which must all be present to compete with indifferent molecules on the sensory surface of the insect's receptors.

Experimental data obtained very recently by Doolittle *et al.* (*355*), Friedman and Miller (*428*), and Russell and Hills (*1005*) do not support Wright's far-infrared vibrational theory. Doolittle and his associates systematically altered the structure of cue-lure [4-(*p*-hydroxyphenyl)-2-butanone acetate] by replacing hydrogen atoms with deuterium in various parts of the molecule. These substitutions did not affect the attractiveness of the compound to the male melon fly, despite the fact that shifts in far-infrared spectral position did occur; no relationship between olfactory response and absorption in the far-infrared region was found. Friedman and Miller and Russell and Hills firmly established that chiral isomers (enantiomers) of carvone and related compounds possess distinct odor differences. The experimental data indicate different physiological responses to isomers differing only in their steric enantiomorphism and not in their vibrational (far-infrared) energies. The foregoing work with the enantiomeric carvones has been corroborated by Leitereg *et al.* (*721*).

A stereochemical theory of olfaction proposed by Amoore (*23–25*) may be of considerable assistance in the future determination of the olfactory mechanism in insects (*549*). The following are put forth as primary odors: camphoraceous, pungent, etherlike, floral, pepperminty, musky, putrid. Certain definite molecular properties characterize all compounds with the same primary odor and distinguish them from all compounds with different primary odors. Thus, the camphoraceous odor is exhibited by spherical molecules about 7 Å in diameter, the musky odor by disk-shaped molecules about 10 Å in diameter, the floral odor by kite-shaped molecules, pepperminty odors by wedge-shaped molecules with a polar group near the point of the wedge, etherlike odors by very small or thin molecules, pungent odors by electrophilic molecules, and putrid odors by nucleophilic molecules. Other characteristic odors are considered to be complex odors due to the molecules fitting 2 or more different primary odor receptor sites. If a chemical is volatile and its molecules have the appropriate configuration to fit closely into the receptor site, then a nervous impulse is initiated, possibly through a mechanism involving disorientation and depolarization of the receptor cell membrane. The theory, which shows good agreement with actual experience (27, 55, 157, 684, 686, 1019), is discussed in detail by Amoore et al. (26, 28, 29). Theories of chemical olfaction have been reviewed by Schneider (1035), Roderick (948), and more recently by Klopping (685).

It is now well known that on prolonged exposure to female sex phero- mones, males of a number of insect species sooner or later become adapted or habituated and cease to respond. The normal response may be attenuated or otherwise modified if the males are preadapted to some part of it, as has been shown with Epiphyas postvittana (95), Anagasta kühniella (1194), Trichoplusia ni (443, 1090, 1194), Pectinophora gossypiella (597), Porthetria dispar (166), and Ips confusus (184).

The response of males to the female extracts may also be inhibited or masked by the presence of foreign contaminants, as in Porthetria dispar (149, 895, 1248), Samia cynthia (614), Heliothis virescens and H. zea (139) (this has been denied by Shorey and Gaston, 1088), Diprion similis (286), and Periplaneta americana (613). Masking of food attractants and pheromone triggers has been shown for Drosophila melanogaster (1310), and Antheraea pernyi (936), respectively. These masking agents have been given the name "metarchons" by Wright (1310). Although the masking of attractants may be a significant problem in attempting to demonstrate the presence of a sex attractant in a particular insect species, this phenomenon offers the possibility of insect population control by releasing masking agents for natural attractants with resultant disruption in communication between the sexes (595, 1314).

An interaction of chemically and sonically induced behavior produces a natural pheromone mask of the aggregation pheromone in Dendroctonus pseudotsugae. The attraction of both sexes to the female-produced phero- mone under natural conditions ceases abruptly upon the arrival of stridulating males next to receptive females. Presumably the sound of the stridulation is perceived by a phonoreceptor in the female, so that the release of the masking substance by the female is controlled by the central nervous system (996, 998, 999). The result is to evenly distribute the available males, while preventing overcrowding.

The recent isolation, identification, and synthesis of a number of naturally occurring insect sex pheromones has made it possible to study the effects of "secondary substances" with closely related structures on response. For example, potent inhibitors of response have been discovered for the natural attractants of Grapholitha molesta (960), Argyrotaenia *velutinana* (*958, 963*), *Choristoneura rosaceana* (*966*), *Pectinophora gossypi-ella* (*597*), and *Plodia interpunctella* (*202*), and to hexalure, a potent synthetic attractant of male *P. gossypiella* (*155*). These inhibitors presumably react with the male antennal receptor sites to modulate or compete with the attractant's sensory input to the brain. Roelofs and Comeau (*958*) have also discovered a number of compounds which synergize the phero-mone of *A. velutinana*. Some of the inhibitors can delay and reduce male responses to the attractant if they are presented for a period of time before the attractant is released (*597, 958*). The structural requirements for activity of these inhibitors and synergists are closely related to those of the attractant, suggesting the possibility of molecular interactions with the binding sites on identical receptors, with varying degrees of affinity. (See Chapter XI for the chemical names of the synergists and inhibitors of the respective attractants.)

Attempts to demonstrate inhibition of the response with structures related to the natural attractants have been unsuccessful with *Ostrinia nubilalis* (*688*) and *Apis mellifera* (*157*). Male *Cadra cautella* are inhibited from responding to their own females when in the presence of calling female *Plodia interpunctella* (*431*).

It is evident that it may be possible to use pheromones and their inhibitors in insect control by at least 2 methods: (a) disruption of premat-ing communication between the sexes by permeating the atmosphere with a pheromone (*443*) and (b) prevention of mating by masking the natural female scent through the release of specific inhibitors in the atmosphere. Several good reviews of attractant masking have appeared (*595, 1079*).

# CHAPTER VII

# RESPONSES TO SEX PHEROMONES

Responses of insects to sex pheromones are either behavioral (in the laboratory or field) or electrophysiological (in the laboratory). A variety of bioassay methods have been developed for measuring the behavioral responses. Among the laboratory behavioral methods are the following:

1. *Open container*, in which live insects confined to a jar or beaker are exposed to puffs of air containing the pheromone
2. *Small cage*, in which the test insects kept in small cages are exposed to puffs of air or small filter papers impregnated with the test substance
3. *Olfactometers*, which may be (a) straight tube, (b) Y-tube, (c) arena, (d) revolving wheel, (e) pinioning rack, (f) chromatographic effluent, or (g) cage trap

Among the field behavioral methods are the following:

1. *Tethered balloon*
2. *Trapping*, which may be accomplished in (a) small field cages, (b) large field cages, (c) greenhouses, or (d) open fields

Examples of these methods are given in the descriptive paragraphs that follow for numerous species of insects.

## Behavioral Responses

ACARINA

Although not insects, ticks and mites have been studied by entomologists and found to give many of the responses shown by insects. Berger *et al.* (*138*) found that sexually mature males of 3 species of hard ticks (*Amblyomma americanum, A. maculatum,* and *Dermacentor variabilis*) attached to rabbit hosts, attempted to mate with other males when exposed to air puffed from droppers whose inner walls had been coated with solutions of the extracts of their respective virgin females (7 or 8 days old). Only males that had been fed exhibited the response.

Clumps of absorbent cotton impregnated with crude ether, water, or methylene chloride extracts of deutonymphs of two-spotted spider mites (*Tetranychus urticae*) and placed on leaf discs attracted males (*307, 308 308a*). Finely ground polyvinylpyrrolidone was also used as a substrate on leaf discs (*307*).

ORTHOPTERA

*Blattella germanica,* German cockroach

Filter papers over which nymphs had crawled, as well as those treated with methanol extracts of nymph feces, elicited aggregation when placed with nymphs in glass containers (*576, 577*). Papers conditioned with roaches whose abdominal tips had been amputated did not elicit the response. A modification of this method was used by Volkov *et al.* (*1237*) to bioassay extracts of virgin females on males. Filter papers treated with female extracts and placed in glass tumblers closed with a metal screen in the form of a funnel were placed in glass chambers containing 100 males. Attracted males were caught in the traps; females were not attracted.

*Byrsotria fumigata*

Females caged with males mounted and fed on the male's tergum in response to the pheromone released by the male (*982*).

*Eurycotis floridana*

Partitioned chambers segregating males from females were used to demonstrate initiation of mating by females (*113*). Removal of the partition allowed females to approach the wingless males, mount, and attempt to feed on the tergal gland. Receptive males showed a side to side rocking movement termed "lateral vibration." Tactile stimulation of the male abdominal dorsum releases copulatory thrusts.

*Locusta migratoria migratorioides,* African migratory locust

Extracts of air from a locust rearing room were bioassayed on solitary fourth instar hoppers by moistening a pad of cotton wool with the extract and placing it at the bottom of a 1-liter jar underneath a wire-mesh floor. The pheromone volatilized from the wad into the air of the jar. An active fraction caused the hoppers to aggregate into swarms (*827*).

*Nauphoeta cinerea*

A pheromone produced by males that stimulates receptive virgin females to mount and feed on the male tergal glands was tested by impregnating an aliquot of male extract on a small piece of filter paper and placing it in a little depression of a dummy male made of glass. The tests were run in red light during the afternoon with 10- to 18-day-old females in beakers. An active material attracted females to palpate the paper (Fig. VII.1). The sex pheromone produced by males which elicits arrestment and tergal feeding by females is called "seducin." It keeps the female in the proper position long enough for connection to be made and stimulates receptive virgin females as well as females shortly after parturition. Extracts from adult male *N. cinerea*, *Blatta orientalis*, *Blaberus discoidalis*, *Byrsotria fumigata*, *Diploptera punctata*, *Eublaberus posticus*, *Gromphadorhina portentosa*, *Leucophaea madeira*, and *Periplaneta americana*, as well as *Blaberus*

*Fig. VII.1.* Bioassay of male pheromone of *Nauphoeta cinerea*. Females, attracted to a dummy, palpate the filter paper impregnated with pheromone. [From Roth and Dateo (*983*), Pergamon Press, New York.]

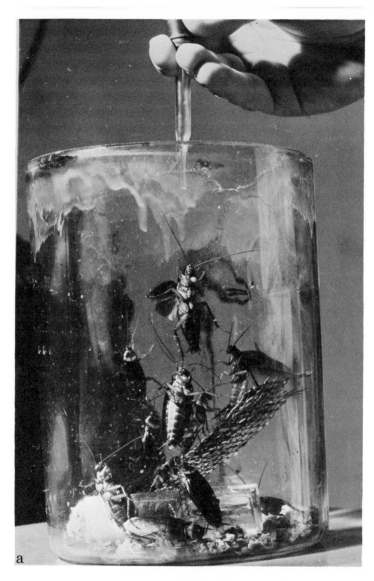

*Fig. VII.2.* Response of male American cockroach to the female lure. (a) Air blown from pheromone-treated dropper into container excites the males; (b) males converge on glass plate treated with lure; (c) with wings raised in characteristic sexual excitement, males attempt to mate with the plate. [By permission of the U.S. Department of Agriculture.]

*craniifer* larvae all induced female *N. cinerea* to sexually mount and feed. Pregnant females did not respond (*982, 983*).

*Periplaneta americana*, American cockroach

Males maintained in glass jars responded with wing flutter and abdominal extension to parts of virgin female bodies waved close to them (*1158*). The same behavior, as well as copulatory attempts, was elicited by blowing air from pheromone-treated droppers into the containers (*600, 985*)

*Fig. VII.2 (b)*

*Fig. VII.2 (c)*

(Fig. VII.2a, b, c). McCluskey *et al.* (*764*) used a modified **Y**-tube olfactometer, allowing starved males a simultaneous choice of food odor or female sex-pheromone odor; preferential response to the sex pheromone was shown.

## Hemiptera

*Cimex lectularius,* bedbug

Filter paper discs scented with the odor of males or females and placed in choice arenas containing both sexes were used by Levinson and Bar Ilan

*(730)* to measure aggregation response. Methanol extracts of such papers were also effective.

*Aonidiella aurantii*, California red scale

Activity in virgin female extracts on filter paper was demonstrated initially in the laboratory in petri dishes by males attempting to mate with the paper at the spot of application and by their repeated return to the spot *(1173)*. The attractant, whose production is not continuous and can be controlled, could also be impregnated into drops of paraffin resembling dummy females. Later, a moving turntable olfactometer was developed by Tashiro *et al.* *(1172, 1174)* in which free-flying males in a room responded to Munger cells containing virgin females on lemons or to their homogenates or condensates held on sand.

*Lygus lineolaris*, tarnished plant bug

Tanglefoot traps made of rigid vinyl plastic sheets, holding cages of virgin females, were used in the field to capture males *(1020)*.

*Matsucoccus resinosae*, red pine scale

Virgin females in circular screen cages and an ultraviolet lamp placed in a flight chamber with free-flying males attracted males, who landed on the screen and attempted to copulate with it. Filter papers on which virgin females had rested also attracted males in a petri dish *(350)*.

*Nezara viridula*, southern green stink bug

Laboratory studies with a Y-tube olfactometer showed that virgin females were attracted to males within 1 hour of exposure. Females released in large, outdoor screen cages were attracted to virgin males confined in mesh cages, and wild females were attracted to sticky board field traps holding caged males. The tachinid parasite, *Trichopoda pennipes*, was also attracted to males *(792)*.

*Planococcus citri*, citrus mealybug

Males in petri dishes were attracted to virgin females (12–15 days old) or to their extracts placed on circles of filter paper. The males clustered on the spot of application and exhibited recognition by holding their antennae out at right angles to the body *(477)*. Extracts prepared from females with nonpolar solvents were more effective than those prepared with polar solvents, but loss of activity occurred during freezer storage.

*Rhodnius prolixus*

Males and females kept in the dark in a cheesecloth-covered petri dish placed in a wire chamber were observed and photographed with the aid of a timed camera shutter and electronic lights. Free-flying males in the chamber were observed to congregate on the dish near copulating pairs. The pheromone is not produced if males and females are separated *(83)*.

DIPTERA

*Cochliomyia hominivorax*, screw-worm fly

Benzene or chloroform extracts of virgin males were presented to virgin females in flasks on filter paper strips, and the number of "strikes" made during a 15-minute period was recorded (*414*). Mated females and virgin males were not affected. Strangely, although virgin females of a Florida strain responded to males of both that strain and a Mexican strain, females of the Mexican strain did not respond to either pheromone (*413*).

*Culiseta inornata*, mosquito

In excitant experiments, an observation chamber containing glass flasks holding males was used in the dark. Air drawn from a container of females did not give a consistent male response, but filter paper strips on which virgin females had rested gave a characteristic male searching response. Olfactometer tests utilized a central chamber consisting of a disposable gallon carton to which glass jars with inverted cones were attached thereby forming traps; air was drawn through the screen-bottom traps into the chamber, which was slowly rotated to minimize position effects. Male catch by virgin female extract was consistently high (*683*).

*Dacus tryoni*, Queensland fruit fly

Discs of filter paper impregnated with male anal-reservoir secretion were placed in a cage containing virgin females and elicited an increase in locomotor activity, rapid vibration of the wings, and probing of the discs with the ovipositors. Female responses were confined to the late afternoon and dusk period, with the maximum response occurring at dusk (*411, 412, 412a, 412b*).

*Drosophila melanogaster*, vinegar fly

An olfactory bioassay device consisted of a glass cylinder to observe male behavior and a two-way stopcock to regulate passage of a stream of air through another cylinder containing an odor source or directly into the cylinder of males. The odor source (live females or males) was placed in the cylinder in small polystyrene cylinders with copper-screen ends. Air passed over virgin females elicited male courtship behavior (wing vibration and touching movements) (*1081*).

*Harpobittacus* spp., scorpion fly

Male and female pairs were placed in transparent cylindrical containers and their behavior was observed. In another method air was passed through a cage of males and directed into a cage of females. With the air-stream method, females congregated on the side of the cage nearest to the males. In the field, free-flying *H. australis* females responded to caged males of the same species 10 m away, but they ignored cages of male *H. tillyardi* (*190*).

*Lucilia cuprina*, sheep blowfly

Males held in a wire-screen cage enclosed in a glass assay chamber were exposed to a blast of air that had been passed through a glass arm containing caged females remote from the males. The males became sexually stimulated and tried to mount other males (99).

*Musca domestica*, housefly

Simulated fly models (pseudoflies) fashioned from black shoestring and treated with an extract of virgin females were exposed to males in a petri dish; this caused the males to jump onto the model (strike) (314, 968, 969). Under natural conditions, a male may strike a female while in flight, but copulation never occurs in the air. If a virgin female accepts a male, she thrusts her ovipositor into the genital opening of the male (815). Murvosh et al. (816) and Mayer and Thaggard (762) used a large cage-type olfactometer in which virgin females or their extracts were confined in a glass trap and a stream of air was passed through the trap toward free-flying males; stimulated males entered the glass trap.

## COLEOPTERA

Wright (1311) has described a laboratory olfactometer in which an attractive zone is established and maintained in a precisely determined location on the floor (or wall) of an observation chamber and also in the air adjacent to it. Insects must therefore manifest their normal behavior in finding their way to the zone. Attenuation of the attraction by "background odor" can be studied quantitatively. Sex attractants can be secured for testing by placing insects of the appropriate sex in an enclosure and passing air through it. The olfactometer can be used for many species of insects; it gave excellent results with *Trypodendron lineatum* beetles. Field observations with this insect also prompted Chapman (293) to stress the need for a study of "odor meteorology," concerned with the rates and types of odor distribution near the ground.

*Anthonomus grandis*, boll weevil

A laboratory bioassay procedure utilized an inverted glass funnel arrangement in which weevils were released and then subjected to the odor of live males (or females) or their extracts contained in aerated flasks. Responding weevils entered the flask of their choice (519, 1208). Females older than 2 days responded positively to males, and 5-day-old insects responded in the greatest numbers. Sex attraction in the field was studied by using traps composed of adhesive-coated 15 × 15 cm plywood backboards placed on a 15 × 10 cm plywood stage. Caged weevils were placed in a holding chamber, and attracted females or mixed sexes were trapped on the adhesive (200).

*Attagenus megatoma*, black carpet beetle

Burkholder and Dicke (*224*) used a test chamber made from a glass desiccator with a glass-plate lid. Filter paper discs on which virgin females had rested were exposed to males, who responded by extension of the antennae, zigzag approach, and copulatory attempts on one another. Subsequently, Burkholder (*223*) developed a multiple-choice olfactometer with air flow, a single-choice olfactometer, and a Y-tube olfactometer as well as Styrofoam field traps.

*Costelytra zealandica*, grass grub beetle

A laboratory clear-plastic Y-tube olfactometer with a wire-screen trap at the end of each arm was used to observe the response of 10- to 15-day-old males. Observations were made at 6:00–9:00 PM. Crushed female beetles on filter paper were placed in one of the traps and air was passed through toward the males, who proceeded to crawl into the trap and tried to copulate with the crushed females (*542*). In another laboratory method, test fractions were placed on a paraffin dummy which was then dropped among 10 males confined under an inverted funnel. Males responded within a few minutes to 0.1–10 $\mu$g of phenol by attempting to copulate with the treated dummies. Field tests at dusk involved the use of simple open tins into which phenol-water mixtures had been placed; attracted males flew into the tins (*541, 543*).

*Dendroctonus* spp., bark beetles

A laboratory bioassay apparatus allows tethered beetles to rotate 560° on the horizontal plane in the center of 4 air streams converging at right angles on the same plane as the beetle. Test materials are contained in capillary pipets, and turning tendencies and wing-beat frequencies are recorded electronically (*565*). Another method uses a walkway constructed over a female boring in a log or over screened vials recessed in a board and used for male arrestment tests (*618, 998*). The walkway has also been modified so that air is pumped into a stainless steel tube containing a test fraction and then into a Teflon tube under the walkway; males are arrested with excited motion and attempt to penetrate the mesh over the outlet of the Teflon tube (*670*). Arena and tube-type olfactometers have also been used in the laboratory in darkness and in light; males moved upwind to reach female frass (*185*). In another laboratory method, walking beetles were exposed to the effluent from a gas chromatograph or to the air from medicine droppers to which gas chromatograph effluent collection tubes had been attached (*924*).

A number of field methods have been used to measure attraction to beetles boring into logs in field cages (*618, 773*) or to baits contained in muslin sleeves, paperboard cylinders, or metal cylinders (*868*). One of the most popular and effective field methods utilizes tree trunk-simulating olfactometers made of canvas sleeves inflated by a fan and wrapped around trees into which beetles are boring (*433, 876, 924, 925, 1227, 1231*). Beetles responded to attractive trees.

*Diabrotica balteata*, banded cucumber beetle

Free-flying males are attracted to baited, adhesive-coated traps hung in a greenhouse in winter or spring. Males in the natural population are attracted to these traps placed in soybean plantings in the summer (*322, 1052*).

*Ips* spp., bark beetles

A number of laboratory bioassay methods have been used for measuring attraction to *Ips confusus*. One of these uses a small flight chamber containing a petri dish with beetles of both sexes; these are exposed to male frass (*184*). In an arena-type laboratory olfactometer, walking female beetles responded to male frass, males feeding on logs, and male fecal pellets. Although male frass mixed with host or sap resin was attractive in this olfactometer, the mixture was not attractive in the field (*877*). The 3 compounds making up the *I. confusus* pheromone were exposed in the laboratory to flying populations by placing filter papers treated with the material on adhesive-coated boards or under screens of plastic mesh in petri dishes (*1294*). A laboratory multiple-choice olfactometer used by Wood *et al.* (*1289*) was based on the movement of females to the source of the attractant and their klinotactic behavior (excited circling and milling about). A detectable response was evoked by dilutions of benzene extracts of male frass equivalent to $3 \times 10^{-8}$ gm of frass, approximating that produced by one male in one-third of a second. Bark and wood samples of ponderosa pine become attractive 4–6 days after males are introduced into entrance tunnels, with peak attraction at 18–144 hours (*1290*). The ratio of females to males responding positively is approximately 8:5. Pitman *et al.* (*869*) forced a stream of air, regulated by a flow valve, through a vial containing the test material and into a plastic tube containing walking beetles. Raw pine oleoresin, pinene, myrcene, and 3-carene exerted a repelling effect on female *I. confusus*, who were indifferent to an attractant-free air stream and expressed phloem sap from *Pinus ponderosa*. Male fecal pellets elicited a strong response.

Vité and co-workers (*1222, 1226*) tested logs infested with male *Ips* in the field by placing them in large-diameter aluminum tubes and blowing the attractants against a Tanglefoot-lined screen, a glass pane above a Tanglefoot-lined paper, or a trough containing water and a mild detergent. Wood *et al.* (*1288*) used traps made of hardware cloth cylinders fastened to the end of a rod driven into the ground. The cylinders were dipped in adhesive and air was passed through a charcoal filter into an aluminum tube containing the attractant. Bakke (*81*) used traps made of frame cages covered with plastic sheeting to attract both sexes of *I. acuminatus*. Small glass barrier traps were mounted above the edge of the top opening. Males and females responded similarly to caged males.

*Limonius californicus*, sugar beet wireworm

A laboratory bioassay (*733, 734*) utilized a simple tube olfactometer with the test materials, impregnated on filter paper strips, pressed to the floor of a compartment and exposed to a stream of air. In the field (*609, 1075*) microscope slides or cotton wads moistened with ethanol extracts of females attracted males.

*Limonius canus*, Pacific Coast wireworm

Male beetles became sexually excited in the laboratory when they were offered glass rods that had been dipped into female extracts containing as little as 0.005 female equivalent per milliliter of solvent. In the field, extracts placed on watch glasses attracted large numbers of males (*833*).

*Tenebrio molitor*, yellow mealworm

Tschinkel (*1203, 1204*) has assayed the sex pheromone by exposing males under petri dish covers to glass rods that had been dipped in active extracts. Males attempted to copulate with the rod by bending the abdomens downward and anteriorly around the end of the rod. Happ (*509, 512*) used Lucite chambers, containing virgin males, which were mounted on a turntable; air was passed through a central box containing virgin females into the chambers. Males responded by rushing upwind and attempting to mount one another. A similar choice-chamber olfactometer was used by August (*79*) to test pheromones emitted by both sexes.

*Trogoderma* spp.

Discs of filter paper impregnated with test material and placed in petri dishes were used by Bar Ilan *et al.* (*89, 725*) to measure responses of the khapra beetle to the sex attractant. Beetles of both sexes 1–3 days old were released in the center of the dish and the numbers of insects attracted to treated discs as compared with a control disc were counted at 15-minute intervals. An ether extract of dried bodies of either sex, especially the female, attracted both sexes. The same procedure was used by Yinon and Shulov (*1323–1325*) to assay the assembling scent. They found that extracts of either sex were attractive to both sexes of *T. granarium* and repellent to *Triboleum castaneum*. A modification of this test procedure, using treated filter paper discs placed in shell vials containing males of *T. granarium*, was used to measure the response to the sex attractant produced by females (*949*); interspecific responses of males of 7 species of *Trogoderma* (*T. glabrum, T. granarium, T. grassmani, T. inclusum, T. simplex, T. sternale, T. variabile*) were measured to extracts of virgin females of these species.

A positive response consisted of antennal extension toward the disc and stretching of the male during circular running beneath an active disc (*1217*). Stanić et al. (*1133*) and Adeesan et al. (*8*) also used impregnated filter paper discs to bioassay female attractant on males of *T. granarium*. Levinson and Bar Ilan (*726, 728, 729*) measured olfactory responses of both sexes of this insect to the emanation from live females in a filter paper arena placed over the females on a hot plate maintained at 30°C. All of the foregoing tests were conducted in controlled darkness.

Burkholder and Dicke (*224*) used the laboratory olfactometer made from glass desiccators, described for *Attagenus megatoma*, as a bioassay method for the female attractants of *T. glabrum* and *T. inclusum*.

*Trypodendron lineatum*, ambrosia beetle

Walkway olfactometers and flight chambers were used by Schneider and Rudinsky (*1048*) and Francia and Graham (*421*), respectively, to bioassay for responses of both sexes to female-infested wood. Moeck (*793*) used an olfactometer consisting of a glass runway and air-inlet chamber base fitted to a wooden frame. Conditioned air was passed through a Y-tube. Attracted beetles walked through the olfactometer arm toward the stimulus.

## HYMENOPTERA

*Apis mellifera*, honeybee

A queen in a small wire-screen cage tethered to a helium- or hydrogen-filled balloon raised to a height of 10–15 m by a nylon cord (see Fig. II.1) attracts drones (*438, 450, 843, 1007*).

## LEPIDOPTERA

*Achroia grisella*, lesser wax moth

Virgin females in a flight cage (50 × 50 × 75 cm) exposed to cotton wicks treated with an ether extract of male wings responded with wing fanning and a circling dance, but they found the odor source only after much searching. Live males, however, were easily found. Adhesive-coated cardboard squares on the cage floor exposed to air blown across males trapped numerous females (*325*).

*Adoxophyes orana*, summerfruit tortrix

A laboratory bioassay method consisted of counting the number of males that exhibited a mating dance after exposure to a methylene chloride extract of whole female bodies (*1167, 1169*). A concentration as low as $1 \times 10^{-6}$ female equivalent may be detected. Response was at a maximum after the males, who were reared under continuous light, were held in darkness for 7 hours. Field evaluations were conducted in a tea garden using tub-type traps fitted with a blacklight lamp and baited with virgin females, or cylinder-type sticky traps baited with females (*1168*).

*Alabama argillacea*, cotton leafworm

Caged males 5–7 days old who were exposed to vapors of an active extract expelled from a glass pipette lifted their antennae, vibrated their wings, flew in a hovering position, and attempted to copulate with one another (*136*).

*Anagasta kühniella*, Mediterranean flour moth

Males released in a wind tunnel downwind of a calling female flew upwind. If they lost the scent they made crosswind casts and located the female. Red light was used to make the observations. Males also vibrated the wings with circus movements and made copulatory attempts (*1193*).

*Argyrotaenia velutinana*, red-banded leaf roller

Glass choice olfactometers were used in the laboratory to observe the response of males to female extracts placed on filter paper discs. The number of males responding in the first 30 seconds was recorded (*961*).

*Autographa californica*, alfalfa looper

Males in bioassay cages elicit wing vibration when exposed to filter paper discs impregnated with active extracts of virgin females. (*442, 1084, 1093, 1096*). Extracts were prepared from the abdominal tips that were collected at midnight of the fourth day after female emergence. Males exhibited a cyclic rhythm of responsiveness. Maximum response occurred during the hours of darkness when males were synchronized to a 12-hour-light, 12-hour-dark cycle (*1084*).

*Bombyx mori*, silkworm moth

Filter papers impregnated with an active material were placed in a glass tube 7 mm in diameter, and a stream of air was blown through the tube over the head of a live male, who responded with wing vibration (*228, 232, 637, 849, 1054*). The amount of bombykol (the pure, natural pheromone) released from a filter paper was determined by using tritium-labeled bombykol and a liquid scintillation spectrometer. It was found that a paper disc impregnated with $3 \times 10^{-5}$ $\mu$g of bombykol and subjected to an air stream of 50 ml per second for 2 seconds released $1.4 \times 10^6$ molecules of bombykol per second (*1044*).

An assay apparatus composed of glass compartments, cages, and a closed air-flow system was used by Takahashi and Kitamura (*1163*) to quantitatively determine the sex pheromone activity of this insect as well as the eri-silkworm moth and the almond moth. The responsiveness of males (wing vibration and circling dance) was affected to a great extent by the size of the cages used and the air-flow rate.

Bayer (*122*) and Bayer and Anders (*123*) used caged live males as test animals to bioassay eluates from the exit port of a gas chromatograph through which a female extract was passed. The criterion for activity was wing flutter.

*Bucculatrix thurberiella*, cotton leaf perforator

Males performed a typical mating dance when exposed to virgin females in Y-tube and "swastika" olfactometers (*918*). Attraction of 100% was not obtained in either olfactometer.

*Cadra cautella*, almond moth

Tests were conducted in an isolated, dark room using modified Frick traps baited with female extract, and released males. Males in cages also responded by wing vibration when exposed to female extract. The same methods were also used with *Plodia interpunctella* (*206*).

*Choristoneura fumiferana*, eastern spruce budworm

*C. pinus*, jack-pine budworm

Caged males were subjected to a stream of air passing over a filter paper disc treated with an ether rinse of flasks that had contained 2- to 4-day-old virgin females and showed sexual excitement by wing vibration and a circling dance. Males responded only to rinses of flasks that had contained females of their own species. In the field, Tanglefoot-coated board traps baited with virgin females attracted males (*1015, 1016*). A laboratory olfactometric method for *C. fumiferana* involved counting the number of males buzzing (rapidly beating their wings while circling on the substrate) on exposure to the female pheromone (*1015a*).

*Crambus* spp., webworm moths

Males were lured to field traps made of Stikem-lined, 1-gallon ice-cream cartons, from which the lids and bottoms had been removed, which contained small screen cages of live virgin females. In the laboratory, caged males exposed to filter paper discs treated with extracts of virgin-female abdominal segments responded with frantic flight, spreading of the wings, and extrusion of the genital claspers (*85*).

*Cryptophlebia leucotreta*, false codling moth

Caged males responded to active substances by vigorous flight and a circling dance (*533*).

*Danaus gilippus berenice*, queen butterfly

The male pursues the female, overtakes her in the air, and induces her to alight by rapidly brushing her anterior with scent-disseminating hairpencils. The pheromone acts as a chemical arrestant of the female's nonspecific escape flight from the pursuing male (*210*).

*Diatraea saccharalis*, sugarcane borer moth

Although test materials were originally bioassayed by using them to bait sticky field traps (*857*), such materials were subsequently tested by a laboratory method (*506*). Caged males exposed to the vapors of active solutions from a pipette responded with excitement, wing vibration, and flight; clasper extension occurred but was difficult to observe. Bioassays were conducted in diffuse light between 8:00 and 12:00 PM; best results were obtained within 2 hours after the beginning of a dark phase.

*Dioryctria abietella*

Male responses to live calling females were observed in an elaborate laboratory olfactometer by passing air through a chamber of virgin females into a chamber containing males. Typical responses included rapid crawling while dragging the abdominal tips over the surface of the test chamber, antennal waving, and wing fluttering in short bursts one-fourth of a second long (*396, 397*).

*Diparopsis castanea*, red bollworm

Initially, the attractiveness of females was tested by placing them in field traps and counting the number of males caught overnight (*1210*); this procedure was then made the basis for a laboratory method used to test chemical fractions (*76*). Extracts and other fractions are absorbed on cotton-wool rolls (dental rolls) that are placed in traps in the test room; 30 male moths are released in the room at 5:00 PM and the males captured are counted at 8:00 AM the next morning. Using virgin females as bait, catches of up to 90% of the liberated males have been recorded, but catches were much lower with female extracts.

*Epiphyas postvittana*, light-brown apple moth

Male response to the female sex pheromone consists of 4 behavioral steps: (1) antennal elevation, (2) activation and flight, (3) upwind orientation, (4) copulatory movements. One assay method involved the use of a glass compartmented olfactometer into which caged males were placed; air was passed over an applicator disc treated with an active material and caused the males to respond (*97*). A subsequent type of olfactometer consisted of straight glass tubes. The upwind ends of the tubes had mixing chambers for the introduction of the test sample (*98*); 2 unmated males 2–10 days old were used in each tube. All assays were conducted at the appropriate period of maximal male responsiveness in a diel cycle under constant illumination of 3.5 lux. Males were maximally responsive by the second night after emergence.

*Euxoa ochrogaster*, red-backed cutworm

Air passed over a filter paper disc treated with female extract in a glass compartmented olfactometer under constant illumination elicited antennal extension, wing vibration, flight toward the source, clasper extension, and copulatory attempts in males (*1155*). The response was considered positive when at least 50% of the males either extended their claspers or attempted to copulate.

*Galleria mellonella*, greater wax moth

Virgin females were placed singly in the compartments of muffin pans under wire-screen covers and exposed under red light at 5:00–10:00 PM to test materials impregnated on aluminum foil strips over which air was passed. Females responded to active materials by raising the antennae, vibrating the wings, and performing a circling, running dance (*972, 973*).

*Grapholitha molesta*, Oriental fruit moth

Males in waxed-paper cages exposed to female pheromone approach the source fanning their wings in great excitement, pausing with the abdominal tips curved upward, and opening and closing their claspers. From 20 to 30% of laboratory-reared males responded to as little as 0.01 female equivalent, while maximum response was obtained with 0.1 female equivalent (*446, 962*). Although traps baited with synthetic pheromone in the field captured males in Australia, field-collected males failed to respond to the synthetic extracts or to crude extracts of females in laboratory bioassays (*991a*).

*Heliothis virescens*, tobacco budworm moth

*H. zea*, corn earworm

Although caged males did not respond to air puffed from a pipette contaminated with female extract, Berger (*139*) reported that male *H. virescens* exposed to a female component from a gas chromatograph responded by vibrating their wings, extending their claspers, attempting to mate with one another, and flying toward the source of the gas. However, Shorey *et al.* (*867, 1084, 1088, 1093*) subsequently reported that males of these species responded with wing vibration to filter paper discs treated with an ether extract of virgin females. McDonough *et al.* (*768*) have reported that male *H. zea* will respond with wing flutter to a glass rod that has been dipped into an attractant solution or to air from a medicine dropper treated with the attractant. Agee (*13*) has reported on the premating behavior of *H. zea*, which includes wing flutter and clasper extension by the male. Starks *et al.* (*1134*) have described an olfactometer used to measure the reaction of male *H. zea* to chemicals; it can be applied to many noctuid moths. It consists of a wooden box with an electrical system used to detect flying insects directly by light reflected from the insect. The number of flights of the insects toward a filter paper treated with the test material is monitored.

*Laspeyresia pomonella*, codling moth

Caged males exposed to the air from droppers treated with extracts of virgin females or to small hardwood applicator sticks or filter paper discs that had been dipped in such extracts performed a circling dance, extended their claspers, and attempted to copulate with other males. The discs were hung in 1-quart ice-cream carton traps lined with sticky polymerized butane (*262, 767*). Field tests were conducted in 8-foot field cages erected

over apple trees; each cage contained 50 released 1-day-old males. A carton
trap containing 50 female equivalents and another containing 5 live females
were hung in the cage. The extract caught 11% of the males and live females
caught 31% of the males in 6 days (*262*).

*Lobesia botrana*, grapevine moth

Laboratory tests conducted in a straight glass tube olfactometer showed
that males given a choice between 1 or 3 virgin females went directly to the
three females (*461*).

*Malacosoma disstria*, forest tent caterpillar

Pieces of filter paper moistened with pheromone solutions were placed in
a glass T-tube olfactometer and air was passed over the paper toward males.
Such males migrated toward the filter papers (*1154*).

*Orgyia leucostigma*, white-marked tussock moth

A laboratory bioassay apparatus consisted of Pyrex glass tubes 4 feet
long and 4 inches in diameter. The tubes were covered at 1 end with a
piece of cheesecloth and the other end was left open to receive a cardboard
container with males. Virgin females, which are naturally wingless, were
placed on the cheesecloth; a current of air was passed over them, into the
tubes, and through the carton of males. The males were stimulated to fly
and attempt copulation with the cheesecloth cover. Over 60% of the males
exhibited strong responses (*856*).

*Ostrinia nubilalis*, European corn borer

Caged males were exposed to the air from droppers treated with virgin
female extracts, and they responded by performing a circling dance with
wings extended upward and vibrating, genitalia extended, and claspers
opening and closing. The excited state sometimes lasted as long as 5 minutes
after exposure of the males (*690*).

*Pectinophora gossypiella*, pink bollworm moth

Caged males respond to a mating pair by rapidly vibrating their wings
and curving their abdomens upward while stationary or crawling (*840,
841*). Caged males exposed to air blown from droppers treated with female
extracts responded with excited flight and copulatory attempts (*597*). The
same type of behavior was elicited in males held in glass flask-type or
T-shaped olfactometers (*489, 625*). Cup-type traps baited with filter paper
wetted with female extract were hung in wire-screen cages containing males;
the males were attracted when a stream of air was passed through the
trap (*840*).

*Plodia interpunctella*, Indian meal moth

Males introduced into petri dishes that have contained a virgin female
show rapid wing vibration (*1053*). Males placed in a container with live
females vibrate their wings and antennae and attempt to mate (*349*). Caged

males likewise responded with wing flutter when exposed to stirring rods with bulbed tips that had been dipped into female extracts (*205, 206*). This response to the female pheromone is undoubtedly regulated in some manner by the reproductive system, since after mating the males exhibit a complete lack of response to the pheromone for several hours (*203*). Males surviving after treatment with Bidrin (a phosphate insecticide) responded as well as normal males.

*Porthetria dispar*, gypsy moth

A laboratory bioassay method involves exposure of filter paper (squares or rolled cartridges), glass rods, or vials previously in contact with the substance, or gas chromatographic effluent, to males held captive by their wings in plastic, spring-type clothespin mounts hung from pins or a wooden rack. Tests were conducted in still or moving air directed toward the males. An active substance elicited flicking movements of the antennae and curvature of the abdomen in the direction of the test object (Fig. VII.3) (*166, 167, 612, 1249*). Laboratory results appeared to be in general agreement with field data obtained with baited traps. Individually caged males were exposed at 4:00–8:00 AM to filter papers treated with active material and vibrated their wings for at least 30 seconds (*1157*). Caged, virgin, calling females lured males in the field, but sight apparently assists the males in locating the females (*351*).

*Prionoxystus robiniae*, carpenterworm moth

Field traps are used to bioassay test canidates for male attraction. These traps, measuring 12 × 12 × 8 inches, are constructed of wire screen and are fitted with 4 cone-shaped entrances (*1120*).

*Pseudoplusia includens*

*Rachiplusia ou*

Caged males of *P. includens* responded with sexual excitement to pheromone-treated filter papers. Caged *R. ou* males responded similarly to air blown from pheromone-treated droppers (*1096*).

*Rhyacionia buoliana*, European pine shoot moth

The female sex pheromone was bioassayed by observing, in the laboratory, the number of males displaying a directed upwind orientation in an air stream containing the pheromone. A pheromone quantity equivalent to 0.005–0.01 female was necessary to induce directed male orientation, and repeated exposure to the pheromone caused a rapid decrease in male response (*331a*).

*Samia cynthia*, cynthia moth

Adult males were placed in individual wire-screen cages (25 × 25 × 25 cm) and kept in a darkened room at 25–27°C for at least 20 hours before being used for bioassay tests at 3:00 PM in dim light. Four such cages were

*Fig. VII.3.* Male gypsy moth exhibits sexual excitement while held by wings on rack in laboratory bioassay of female sex attractant. [From Bierl *et al.* (*158*), Copyright 1970 by the American Association for the Advancement of Science, Washington, D.C.]

placed in a row, and a cage containing a virgin female was put at the head of the row. A gentle stream of air was directed through the female's cage toward the other cages. Within 10 seconds, 3 of the 4 males began to flutter their wings violently and almost immediately flew onto the wall of the cage through which the air stream was entering. The tip of the abdomen was curved characteristically downward. The sexual excitement ceased within 3 minutes and the males required a recovery period of at least 30 minutes before another positive response could be elicited (614).

*Sitotroga cerealella*, angoumois grain moth

In a Y-choice olfactory maze, males exposed to active materials darted about, showed rapid wing whirring, and attempted to copulate with nearby objects, including other males. In the field, sticky traps made of plastic jars with Stikem-coated ends were used to attract males (201, 667).

*Spodoptera eridania*, southern armyworm

Caged males were exposed to air puffs from a medicine dropper treated with an extract of virgin females and responded by raising the antennae, extending and vibrating the wings, flying or running to the source of the pheromone, performing a circling dance, raising the tip of the abdomen, extending the claspers, and attempting to copulate with other males. A 25–50% response was considered good (65, 913).

*Spodoptera exigua*, beet armyworm

Males exhibited a cyclic rhythm of responsiveness when placed in a glass olfactometer and exposed to ether extracts of virgin female abdominal tips that were collected at midnight. Maximum response occurred during the hours of darkness when the males were synchronized to a 12-hour light-dark cycle. Females used for extraction were 4 days old and males used for measuring response were 5–8 days old (442, 1084, 1096).

*Spodoptera frugiperda*, fall armyworm

Caged males exposed to air from pheromone-treated droppers react within 5 seconds by orienting their antennae toward the source, vibrating their wings, and flying to the source of the stimulant (Fig. VII.4). They repeatedly attempt copulation with the dropper and with each other in the vicinity of the stimulant source. The activity lasts for 2–6 minutes; in 50 trials, 85% of the males responded (43, 1062, 1063). Although the insects mate under laboratory conditions approximately 2 hours after sundown during a 4-hour period, adult males were kept under controlled conditions with programmed light so that sundown would occur at noon and they could be tested at 2:00–4:00 PM under the diffuse light of a fluorescent desk lamp (1062).

*Spodoptera ornithogalli*, yellow-striped armyworm

The bioassay method used was the same as that for *S. exigua* (above).

*Trichoplusia ni*, cabbage looper

Caged males were exposed to a filter paper strip treated with a female extract; they moved their antennae 90° from a position along the anterior margin of the forewing to an elevated position in a wide vee along the head. This was followed at once by rapid wing vibration, slight elevation of the

*Fig. VII.4.* Male fall armyworm moth responds to a medicine dropper treated with the female sex pheromone. [By permission of the U.S. Department of Agriculture].

abdomen, and eventual flight to the source of the stimulus (Fig. VII.5). Males repeatedly tried to mate with the strips and with nearby males, but not with untreated strips. Cylindrical ice-cream carton traps (8 cm in diameter and 23 cm long), with screen-funnel ends, containing extract-treated strips, were placed in a screen cage (60 × 80 × 60 cm) containing 50 virgin females who were 4 days old; a 6-cm fan provided air circulation in the cage. The trap caught 5 males within 50 minutes, whereas a check (empty) trap caught none. About 99% of 7- to 8-day-old males held under continuous light responded within 2–3 minutes to 0.05 female equivalent by rapid wing vibration and antennal excitation; the duration of response was 2–5 minutes. Males 17 to 18 days old could still respond to a female, but the largest numbers of males responded at 3–4 days of age (567).

The sex pheromone may be bioassayed at 1:00–5:00 AM in diffuse light by noting the number of males that start to vibrate their wings or fly after exposure to an extract of the female abdominal tips. The original bioassay procedure could be used to detect a concentration as low as that equivalent to $1 \times 10^{-4}$ female, but pheromone contamination of the laboratory air caused variable results. Subsequent development of a bioassay apparatus utilizing a completely closed flow system and an external air source resulted in a quantitative method sensitive to the $10^{-6}$ female equivalent level (442). The sensitivity of this type of bioassay depends on the effort taken to exclude males from extraneous contact with the pheromone (1083). The minimum male threshold of response to pheromone, determined by laboratory bioassay, was about 8 molecules/mm³ of air (1125a).

Caged males were exposed to air passed over virgin female extracts in the laboratory and exhibited strong wing vibration (135, 137, 441, 1084, 1085, 1093, 1096). Olfactometric methods (535, 616, 1184) measured the number of males attracted into compartments containing live females or their extracts. In a laboratory flight tunnel, males stimulated to orient

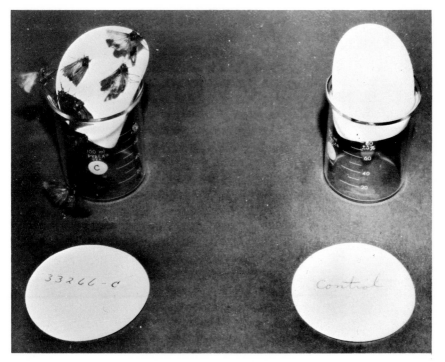

*Fig. VII.5.* Male cabbage looper moths are attracted to a pheromone-treated filter paper but not to an untreated control. [By permission of the U.S. Department of Agriculture.]

toward the source of a current of air containing female sex pheromone moved in a direction opposite from that of males stimulated to orient toward an incandescent light. When low-intensity light and pheromone odor were present in the tunnel at the same time, male orientation toward the pheromone sources was completely abolished and most of the males congregated adjacent to the light source (*1086*). Combination blacklight and pheromone traps were used in the field (*535, 539, 561, 1278*).

*Vitula edmandsae*, bumblebee wax moth

Filter papers on which virgin females had been crushed were placed in containers and air was passed over them; this caused the males to flutter their wings vigorously, expose their hair-pencils, and make clasper movements (*47*).

### Electrophysiological Responses

As discussed in Chapter VI, the basis for this method was first developed by Schneider *et al.* (*1024, 1041–1043*). It involves the insertion of tiny electrodes (usually silver chloride coated) into the freshly excised male antenna or the antenna of the live whole mount, further connected through an amplifier to an oscilloscope. Action potentials are produced (*1024, 1040*). On stimulation with the pheromone a characteristic response (EAG or electroantennogram) appears and can be recorded. The EAG is essentially the sum of many olfactory receptor potentials produced more or less simultaneously and detected by the electrode located in the sensory epithelium.

Electrophysiological measurement of pheromones has been very well reviewed by Schneider (*1027, 1033, 1036*), Wood *et al.* (*1292*), Yamada (*1319*), and others (*112, 181, 979, 1276*). Summed receptor potentials of many cells elicited by odor puffs are easily recorded by inserting 1 electrode near the tip of the antenna and another electrode at the base. The EAG gives information on the qualitative spectrum and the intensity-reaction relationship of a large number of sensory units. Extracellular recordings from single receptors may be made with a glass capillary electrode inserted into a hair sensillum of the antenna and an indifferent electrode placed in the hemolymph space of the antenna. The form of the EAG depends on the odor components of the air current, and the amplitude depends on the odor concentration or the rate of air movement. Adler (*10*) has described the physical conditions important to the reproducibility of electroantennograms as "(1) length and inside diameter of the glass tube (cartridge) holding the treated paper; (2) distance from the outlet end of the cartridge to the antenna; (3) approximate position of the treated paper inside the cartridge; (4) size and type of the filter paper being used; (5) solvent used (if any); (6) sample dose per cartridge; (7) flow rate of the air stream; (8) type of electrodes and electrolyte used; (9) time of day the test was made."

Since the sex attractants extracted from female moths elicit strong EAG responses in the antennae of the male of the same species, but not in the female, the electroantennogram is a useful index for bioassaying these pheromones. The fact that the sex attractant extracted from the female American cockroach, *Periplaneta americana*, elicited an EAG in antennae of males, females, and nymphs of this species indicated the lack of purity of the attractant (*182, 183*).

Schneider (*1033*) has shown the presence of sex pheromones in females of *Antheraea* spp., *Bombyx mori*, *Brahmaea* spp., *Hyalophora euryalus*, *Lasiocampa quercus*, *Rothschildia* spp., *Samia cynthia*, and *Saturnia pyri* by preparing their abdominal extracts and using these to elicit EAG responses in the corresponding males. Schneider (*1025*) has shown that still-air stimulation of *B. mori* male antennae by the female extract, cycloheptanone, and sorbyl alcohol increases the spontaneous activity. Only the male antennae respond to the sex attractant, but the antennae of both males and females respond to the other substances. Summated spontaneous spikes were obtained using large-surface iron or silver-silver chloride needle electrodes, whereas single spontaneous spikes were obtained with fluid-filled capillary silver-silver chloride microelectrodes. Priesner (*886*), in a monumental study, demonstrated the presence of sex pheromones in females of 104 species of Lepidoptera; he also determined the degree of interspecific effects for 1900 species combinations.

Among the Lepidoptera, the electrophysiological method has also been used to bioassay the sex pheromones produced by females of *Antheraea polyphemus* (*180*), *Bombyx mori* (*637, 1039, 1044*), *Choristoneura fumiferana* (*14, 1251a*), *Diparopsis castanea* (*801, 825*), *Platysamia cecropia* (*180*), *Porthetria dispar* (*1157*), *Samia cynthia* (*180*), *Trichoplusia ni* (*474, 853*), and *Zeiraphera diniana* (*953a*). Grant (*475*) has also used the method to determine that the flowers of *Abelia grandiflora* contain strong olfactory feeding stimuli for both male and female *T. ni*.

With the exception of the Lepidoptera, the order Orthoptera has received the most attention with regard to electrophysiological response measurement; all of the investigations utilized the American cockroach, *Periplaneta americana*. Although Boeckh et al. (*182*), in 1963, had reported that the sex attractant of the female *P. americana* stimulated antennal receptors of both males and females and preadult instars, they later reported (*183*) that the antennae of female adults and nymphs of any instar do not possess receptors for this attractant. During attempts to determine whether the olfactory center of the roach possesses neurons specific to the sex pheromone, Yamada *et al.* (*1318–1321*) found that an extract of the sex attractant elicited an electrical response in single cell receptors in the brain of both males and females. They have designated the neurons responsible as "sex pheromone specialists" (*1321*).

Among the order Coleoptera, the EAG's of females and males of both the southern pine beetle, *Dendroctonus frontalis*, and the western pine beetle, *D. brevicomis*, were recorded by Payne (*851, 852*). Antennae of both sexes of *D. frontalis* responded with equal intensity to dilutions of the female-produced aggregation pheromone, frontalin, presented in increasing concentrations. Male antennae were more responsive than female antennae to frontalin in *D. frontalis*, but in *D. brevicomis*, female antennae were more responsive to frontalin than male antennae and male antennae were more responsive to brevicomin (*D. brevicomis* aggregation pheromone) than female antennae.

Male and female Japanese beetles showed a greater electrophysiological response to male extract than to female extract. Females responded strongly to both male and female extracts (*11*).

Ruttner and Kaissling (*1007*) recorded spikes and slow potentials from the poreplates of antennae of *Apis mellifera* and *A. cerana* drones. The same type of olfactory sense cells respond to queen substance (the sex pheromone of *A. mellifera* queens) and to the secretion of the mandibular gland of *A. cerana*. Olfactory cells of the same type were found on antennae of queens and workers of both species. However, Kaissling and Renner (*638*) have identified 2 types of olfactory receptor cells responding to pheromones of *A. mellifera*. One of these is specialized for queen substance, while the other responds to the scent of the Nassanoff gland. Kaissling (*636*) has used the mass-action law to prove the connection between the odor concentration and the receptor potential's amplitude of the bombykol receptor and the queen substance receptor of *B. mori* and *A. mellifera*, respectively. Calculation of the number of hypothetical odor receptors per receptor cell showed it to be in the same order of magnitude as the number of fine-pore tubules in the wall of the antennal sensory hairs.

Schneider and Seibt (*1046*) recorded the EAG's from antennae of male and female queen butterflies, *Danaus gilippus berenice*. Both males and females responded equally strongly to the hair-pencil of males, its crude extract, and a ketone isolated therefrom. Responses to a second component, a diol, were weak. Hair-pencils of male *Lycorea ceres* (a related species), which also contain the ketone, are equally effective in eliciting EAG's from both sexes of the queen butterfly and the monarch butterfly, *D. plexippus*, another related species. However, *D. plexippus* hair-pencils, which lack the ketone, do not elicit EAG's in monarch or queen antennae. Grant (*474*) was able to record EAG's from antennae of male and female *Trichoplusia ni* exposed to excised male hair-pencil scales and methylene chloride extracts of such scales. He considered this as additional evidence for the production of an aphrodisiac by males for females. However, Birch (*164*) believes that the EAG technique is not a satisfactory approach to receptor systems when applied to male pheromones. He reasons that male pheromones of Noctuids are released in large quantity very close to the female and her target receptors; with so much pheromone present at the receptors, there is little relevance to EAG measurements of threshold responses. Birch recorded EAG's in response to male hair-pencil scent from *Leucania pallens, L. conigera, Apamea monoglypha, Phlogophora meticulosa,* and *Mamestra persicariae.*

Grant (*476a*) obtained EAG's from male and female *Trichoplusia ni, Heliothis virescens, Pseudaletia unipuncta* (Haworth), and *Leucania* sp. responding to extracts of the scent scales of their own males. *Anagrapha falcifera, Manduca sexta, H. zea,* and *Spodoptera ornithogalli* did not respond to the scent scales of their own males. All these species, as well as *Autographa precationis* Guenée, responded to the hair-pencils of *T. ni.* The apparent lack of species specificity may be explained by assuming that the nonspecific odor-generalist olfactory cells on the antennae were stimulated.

# CHAPTER VIII

# INFLUENCE OF AGE OF THE INSECT ON PRODUCTION OF AND RESPONSE TO SEX PHEROMONES

Although many species of insects may produce and emit sex pheromones during their entire life-span beginning at the time of emergence, many others do not attain sexual maturity until they reach a certain age and the pheromone production may cease some time before the natural death of the insect. Similarly, the responding sex may or may not be sexually mature at emergence. It is therefore necessary that such factors be taken into account in attempting to demonstrate the presence of chemical sex attraction or excitation in any insect, as the following survey clearly shows.

ORTHOPTERA

Little attractant and arrestant for female *Nauphoeta cinerea* is present in males at emergence, its production reaches a maximum at 3–6 days of age and remains constant up to 18 days of age (*983*).

HEMIPTERA (HETEROPTERA)

Female *Lygus hesperus* become attractive to males when 6–9 days old and remain attractive for up to 28 days (*1153*).

## HEMIPTERA (HOMOPTERA)

Tashiro and Moffitt (*1175*) reported that virgin female *Aonidiella aurantii* are attractive to males from the time the gray margin begins to develop for as long as 84 days thereafter, being most attractive during the first and second weeks after the gray margin begins. Males may mate from emergence until death. According to Rice and Moreno (*930*), females reared on lemons begin attracting on the twenty-second day of development, but those reared on potatoes begin to attract when 27 days old.

## LEPIDOPTERA

No pheromone activity was detected in female pupae or pharate adults of *Adoxophyes fasciata*, but activity rapidly increased from midnight (after emergence) until mating occurred. Virgin females retained their maximum titer of pheromone activity for about 10 days, but after mating there was a decrease of about 120-fold (*820a*).

Female *Anagasta kühniella*, 3–5 days old, were more attractive than those 1 or 7 days old. The response of males was maximal at dawn, increased with age for 5 days, and was greater in light of 1 lux intensity than of 200 lux (*1195*).

Male *Argyrotaenia velutinana*, 4–5 days old, respond to the female attractant to a greater extent than do those 14–15 days old (*961*). Females 1–2 days old are most attractive; attractiveness declines with age so that 9- and 10-day-old females are only half as attractive (*1281*).

Male and female *Bombyx mori* do not fly and are eager to mate immediately after emergence (*629, 744*). Steinbrecht (*1145*) has shown that the female is already attractive in the last days of its pupal stage. According to Prüffer (*893*), males and females of the related gypsy moth (*Porthetria dispar*) become sexually mature 2–3 days after emergence; males emerge several days before the females. Although newly emerged females are very weakly attractive, the amount of attractant emitted rapidly increases during the first day after emergence and remains constant until death (*1021*). Virgin females bioassayed in the laboratory were maximally attractive at the end of the first hours of adulthood (*167*); tests conducted in a garden showed that, of 300 male gypsy moths released, 156 were attracted to a newly emerged virgin female, 6 were attracted to a 7-day-old virgin female who had already oviposited, 2 were attracted to a mated female who had oviposited, and 52 were attracted to a female who had mated immediately before but had not yet oviposited (*895*). Female nun moths (*P. monacha*) were most attractive when 1–2 days old (*578*). Ambros

(21) and Hanno (507) claimed that newly emerged females are only weakly attractive, the attraction becoming stronger in females 1–2 days old; attraction almost disappears after mating and is completely gone by oviposition. However, Eidmann (373) reported that females are capable of mating immediately upon emergence, as are females of most of the spinners (including *Dasychira pudibunda* and *Dendrolimus pini*). The female continues to attract males for about 12 days, according to Hanno (507), with average maximum potency lasting for 8 days (394).

Female *Cacoecia murinana* moths placed in small cages in the field were attractive from the day of emergence to a maximum age of 11 days, luring large numbers of males but no females (424).

Female *Choristoneura fumiferana* mate most readily when they are less than 12 hours old and become progressively less attractive or less receptive with age. Mated females are not as attractive as virgin females (839). Maximum response was obtained from males 2–3 days old held under continuous illumination (1015a).

The sensitivity of male almond moths (*Cadra cautella*) to the female sex pheromone increases gradually with age up to the seventh day after emergence; maximum sensitivity occurs at 3 days of age. The pheromone is already detectable in female pupae which have begun eclosion; it increases in amount up to a maximum of 3 days after emergence and maintains this level until death if it does not mate (705).

Females of *Crambus mutabilis*, *C. teterrellus*, and *C. trisectus* begin to emit the attractant soon after emergence and remain most attractive for the first 2 or 3 days of life; attractiveness decreases with age. Most males are attracted during the first 2–5 days of life (85).

Female *Diatraea saccharalis* begin to emit their attractant soon after emergence and are most attractive during the first 3 days of life, after which attractiveness decreases with age (858).

The longest total duration of calling by virgin female *Epiphyas postvittana* occurred on days 2 and 6 of adult life and the shortest on day 4; the peak duration probably occurs on day 2. The pheromone content increased until day 5 and then dropped sharply (96a).

The sex pheromone of *Euxoa ochrogaster* was not extracted in detectable quantity from 4-day-old virgin females but was recovered in increasing concentrations from virgins 7–20 days old. Unmated males did not respond to the pheromone until they were 5–9 days old (1155).

Although Vöhringer (1236) reported that the male *Galleria mellonella* begins to release attractant 12 hours after emergence, copulation was observed between males $1\frac{1}{2}$–2 hours old and females about 12 hours old. Röller *et al.* (974) subsequently showed that the males develop a characteristic odor when they are 2–24 hours old; this odor reaches a maximum during the evening hours.

Male *Grapholitha funebrana* were attracted to females to the greatest degree when 1–3 days old (*1018*). Virgin female *G. molesta* increased in attractiveness with age at least until their fifth day of life (*448*).

Working with *Heliothis virescens, H. zea, Pseudoplusia includens, Rachiplusia ou, Spodoptera exigua, S. ornithogalli,* and *Trichoplusia ni,* Shorey *et al.* (*1095*) found that females of all species had a relatively high sex pheromone titer by one-half day after emergence; both species of *Heliothis* and both species of *Spodoptera* contained detectable sex pheromone at the time of emergence from the pupa. However, Gentry *et al.* (*445*) reported that female *H. virescens* must be at least 4 days old before they can attract males. Ignoffo *et al.* (*567*) used 4-day-old virgin females to test the response evoked by an extract of female *T. ni* abdominal tips. Although a small proportion of females was capable of mating on the night after emergence, most females mated for the first time on the second or third nights (*1091*); Shorey *et al.* (*1095*) reported that the proportion of females mating reached 50% of its maximum level by $1\frac{1}{2}$ days after emergence.

Although Fabre (*389*) reported many years earlier that female *Lasiocampa quercus* must be 2 or 3 days of age before they can attract males, Dufay (*359*) demonstrated that newly emerged females are quite attractive to males.

One-day-old females of *Lobesia botrana* and *Paralobesia viteana* are more attractive to males than are 2-day-old females, although virgin females remain attractive for their normal life-span (9–13 days) (*462*). Laboratory experiments showed that a 2-day-old female *P. viteana* immediately excited a male brought to her vicinity (*461*).

The majority of virgin female tobacco hornworm moths (*Manduca sexta*) do not become attractive to males until the second night after emergence, and even then some individuals are not attractive. Vigor and vitality of the females are important factors in the production of attractant as well as responsiveness of the males. Individual testing appears to be the only sure way of ascertaining attractiveness (*17*). According to Ouye and Butt (*840*), female *Pectinophora gossypiella* 1–6 days old were attractive to males, but low catches of 1- to 3-day-old males were obtained. Berger *et al.* (*140*) found that females yielded active extracts regardless of age.

Two-day-old female *Platynota stultana* were most attractive to males, followed by 3-, 1-, and 4-day-old females. Females 6 days old and older were almost unattractive (*14a*).

Richards and Thomson (*933*) reported that females of *Plodia inter-*

*punctella* are ready for mating a very short time after emergence, almost as soon as the wings are dry. Lehmensick and Liebers (*720*) reported that females are not attractive until they are at least 3 hours old, and males do not respond until they are at least 50 minutes old. One-day-old males placed in a dish in which a 1-day-old female had rested immediately began a courtship dance and attempted to copulate. Although females 5 days old were not attractive, a weak male response was evoked by 15-day-old females. According to Brady and Smithwick (*205*), newly emerged females contained a concentration of attractant comparable to that found in 1- to 5-day-old adults. Virgin females released attractant almost continuously during their adult life. Chemical fractions obtained from the crude extract of female abdominal glands were used for bioassay on 40- to 50-hour-old males (*1331*).

Females of *Prionoxystus robiniae* lived an average of 9 days and were most attractive early in life (1–3 days old) (*1120*).

Virgin female Nantucket pine tip moths (*Rhyacionia frustrana*) are most attractive early in their life-span, becoming unattractive as they age. The maximum age for an attractive female is 9 days (*1296*).

Female *Spodoptera littoralis* and *Agrotis ypsilon* do not release their sex attractants until they are at least 1 hour old (*407*). The cells of the secretory abdominal glands producing the scent are larger in 1- and 2-day-old females than in those 3–5 days old; indeed, these cells begin to decrease in size after the second day (*504*). The pheromone of *S. frugiperda* is not present in very young females, but it increases rapidly for 18 hours after eclosion, at which time the males become receptive to the stimulus (*1063*).

Newly emerged *Synanthedon pictipes* females can become attractive to males as soon as they are ready for flight, usually 25–35 minutes after emergence, but only about 59% of all females tested prove to be attractive (*302*). They remain attractive up to 5 days.

*Vitula edmandsae* virgin females produce increasing amounts of attractant during the first 2 days after emergence, then the emission decreases (*47*).

Most *Zeadiatrea grandiosella* females began to attract males on the night of emergence; attractancy dropped sharply after the fourth day and was completely absent in 8-day-old females (*334*).

## COLEOPTERA

In both sexes of *Anthonomus grandis* peak activity occurred when the weevils were 4–6 days old. Females responded to males and males attracted females at 2 days of age. Few females less than 2 days old responded to attractive males (*520*).

Female *Attagenus megatoma* were not attractive to males until at least 24 hours after emergence. Male response was highest 6–7 days after female emergence. Males aged 1 day or less did not respond, while 97% of males aged 6–7 days responded (*223*).

Although male *Costelytra zealandica* are probably at least 7 days old before mating occurs in the field, males aged 2 days responded to the sex attractant in the laboratory (*543*).

The scent attractive to both sexes is released only after emerging male *Ips confusus* have fed for several hours on new host material (ponderosa pine) and the gut has been filled (*1224*). Teneral males produced attractant after a long maturation period. Single, mature males produced attractant up to 18 days, but after mating productivity declined in direct correlation with the number of females per male (*184*).

Happ and Wheeler (*512*) have reported that pheromone emission by virgin female *Tenebrio molitor* reached a peak about 4 days after eclosion from the pupal skin. Tschinkel *et al.* (*1204*) state that both pheromone production and male response are undetectable in newly emerged adults but rise to their maximum extent within 1 week after eclosion. However, Valentine (*1213*) reported that adults are at least 10–12 days old before mating occurs.

## Hymenoptera

Male *Bracon hebetor* 24 hours old show a typical sexual response upon exposure to live or crushed females (*488*).

Female introduced pine sawflies (*Diprion similis*) emerging from ground level cocoons are usually mated by the time they have crawled part way up the stems of grasses; once mated, they no longer elicit a response from the males (*310*).

Adult *Macrocentrus ancylivora* usually copulate almost as soon as they emerge, although there are exceptions in which copulation does not occur until several hours afterward (*406*).

## Neuroptera

Male *Osmylus chrysops* begin calling females 2 days after emergence (*1275*).

## Diptera

Attractive pheromone is collected from 24-hour-old male *Cochliomyia hominivorax*. Females 6 days old are more responsive to males than younger or older females (*414*).

Pheromone is first detected in male *Dacus tryoni* 2 days after emergence, but in the majority the gland reservoir does not become full of secretion until 12–14 days after emergence (*412*).

Male *Lucilia cuprina* do not attain their maximal level of responsiveness to the odor of mature females until the fourth day after emergence, nor is maximal male response achieved until the females are 2 or 3 days old (*99*).

Male and female houseflies (*Musca domestica*) will not mate for at least 16 and 24 hours, respectively, after emergence. Most males mated when they were 26 hours old and most females were 48 hours old before insemination (*815*).

# CHAPTER IX

# INFLUENCE OF TIME OF DAY ON SEX PHEROMONE PRODUCTION AND MATING

Since it is safe to assume that sex attractants and excitants in insects are released only immediately before or during the period of the day in which mating normally occurs, it is extremely important that the effect of time of day on sex-pheromone production and release be considered in investigations designed to demonstrate the existence or collection of these pheromones. In at least 1 insect species—the sugar-beet wireworm, *Limonius californicus*—definite indications have been obtained that the pheromone is produced by the female and stored in her body in bound form until she wishes to attract the male (*609*). Other examples of this may be uncovered in the future. It is also known that many species produce their pheromones as they are needed. This chapter cites numerous cases in which the effect of time of day has been determined.

ORTHOPTERA

Adult males of the American cockroach, *Periplaneta americana*, exhibit daily rhythms in responding to the sex attractant of virgin adult females. Peak responses were noted at 2:00 AM and the lowest responses at 8:00 PM with normal photoperiod tests. In reversed photoperiod tests peak responses were at 2:00 PM and the lowest at 8:00 AM (*263*).

### LEPIDOPTERA

This order comprises many species (mainly moths) in which mating normally occurs during the hours of darkness and many others (mainly butterflies) in which assembling and mating occur during daylight hours. Attempting to learn as much as possible about the mating habits of these insects, Götz (*465*), in 1941, devised a simple yet ingenious device. This clock-operated turntable consisted of an iron tripod holding a clock mechanism whose hour axis moved a glue-covered aluminum dial 32 cm in diameter. Divided into 12 segments, the dial made a complete circle at the top with a lead disc containing a covered opening to allow entrance of the male. In each of a number of metal containers, which were fastened to the underside of the tripod rim, a virgin female was placed, whose odor could penetrate through a gauze covering over the opening. Males attracted to the females of their species remained fastened to that part of the turntable holding the female. Through the use of this apparatus, Götz found that this attraction occurs during evening hours (9:00 PM to midnight) with *Lobesia botrana* and in the early morning hours (2:00–6:00 AM) in the case of *Paralobesia viteana (Clysia ambiguella)* (*465–468*).

Fabre (*391*) found that hordes of males visited a captive female peacock moth (*Saturnia pavonia*) each evening between 8:00 and 10:00 PM, while a captive female lesser peacock moth (*S. pavonia minor*) was visited by a number of males between noon and 2:00 PM. A female oak eggar moth (*Lasiocampa quercus*) attracted males at about 3:00 PM.

Adult *Adoxophyes orana*, 1–4 days old, mated about 1 hour before sunrise (*1167*) or between 2:00 and 4:00 AM (*1168*).

Bioassay tests conducted by Roelofs and Feng (*961*) at various times of the day showed that 4- to 5-day-old males of *Argyrotaenia velutinana* held under continuous light showed their greatest response at 9:00 AM. However, Wong *et al.* (*1281*) reported that peak sexual activity among males in the field occurred during the hour immediately after sunset; extracts prepared from females killed at that hour attracted more males than extracts prepared at other times of the day.

The peak mating activities or presumably pheromone release by *Bucculatrix thurberiella* occurred between 5:00–8:00 PM, as recorded by a clock-operated turntable (*918*). A smaller peak activity occurred between 1:00 and 2:00 AM, and the number of males attracted thereafter decreased until no moths were caught at midnight.

Female *Choristoneura fumiferana* started calling earlier (by sunset) than female *C. pinus* (by $1\frac{1}{2}$ hours after sunset). Male *C. fumiferana* are also attracted by females to traps earlier, although the times of the peak catches were approximately the same in both species (*1015*).

The nocturnal calling behavior of female *Dioryctria abietella* and the time of sex-pheromone release was observed under light-dark cycles of 12 hours each (*396, 397*). Calling began 5–6 hours after darkness, was maximum after 9 hours of darkness, and ceased just prior to the onset of the light phase. Peak periods of mating and female calling coincided.

The response of male *Euxoa ochrogaster* to the female was greatest between $1\frac{1}{2}$ and 4 hours after the dark period began (*1155*).

*Holomelina lamae* and *H. nigricans* mate during the day whereas *H. aurantiaca, H. immaculata, H. rubicundaria, H. fragilis*, and *H. ferruginosa* call and mate only at night in the laboratory (*953*).

Field tests conducted by Wong *et al.* (*1282*) showed that only a few male codling moths (*Laspeyresia pomonella*) were caught in female-baited traps before or during sunset. Peak activity occurred during the hour immediately after sunset and steadily declined through midnight. Batiste (*121*) designed a timing sex-pheromone trap for collecting moths that utilized a moving sticky belt as a collecting surface for a modified carton trap holding live females. Male *L. pomonella* flight was initiated well before sunset and terminated when the temperature reached about 61°F.

Female *Prionoxystus robiniae* almost always began attracting males at midday (11:00 AM–1:00 PM) and continued attracting until dark (*1120*).

The mating flight of male *Thyridopteryx ephemeraeformis* is usually at its height between 3:00 and 6:00 PM; few males are attracted before 1:00 PM (*626*).

Female *Parasemia plantaginis* attract their males during the day and assemble at dusk. The normal calling time of *Callimorpha dominula* is from midnight to 5:00 or 6:00 PM, but a male was caught by a caged female at 11:15 PM. The normal mating time of insects is greatly prolonged in the case of caged females (*663*).

Rau and Rau (*911*) determined the following assembling periods: 3:30–6:00 PM (mainly 4:00–4:30 PM) for *Callosamia promethea*, dawn for *Platysamia cecropia*, 9:00–12:00 PM and again from 3:00 AM to dawn for *Samia cynthia*. Assembling of *Antheraea polyphemus* is apparently not confined to 1 brief period of the night. Moonlight influences the period of flight in some species. Mayer (*758*) reported that the male *C. promethea* is attracted to the female between the hours of 2:00 PM and sunset. Soule (*1124*) reported 4:00–4:30 AM as the hour of mating for *P. cecropia*, and Rau (*909*) stated that "mating in this insect starts between midnight and morning and ends at those hours on the following day."

Under field conditions, 98% of male *Platynota stultana* males were attracted to female-baited traps between 6:00 and 10:00 PM. The response began about ½ hour before sunset and continued until 1½ hours after sunset, with a peak at 45 minutes after sunset (*14a*).

The female *Porthesia similis* is ready for mating about 1½ hours after emergence, which usually occurs at 7:00 AM (*1212*). If unfertilized, she may continue protruding and retracting the ovipositor for days at the same rate until mating occurs.

Male *Heliothis virescens* are attracted to females only in the early morning hours, between 4:00 AM and daylight (*445*).

Although the male gypsy moth (*Porthetria dispar*) flies during the day, mating usually occurs during the evening and night hours (*459, 893*). Female abdomens excised at 9:00 PM and placed in a cage containing 3 newly emerged males caused great agitation among the males within 1 hour (*893*). Varying reports of peak male nun moth (*P. monacha*) activity are 6:00 PM–1:00 AM, beginning at dusk (*19, 21*), 7:00–9:00 PM (*828*), and late afternoon and evening (*459*).

Standfuss (*1132*) reported that a newly emerged female *Saturnia pavonia* moth attracted 127 males from a distance between 10:30 AM and 5:00 PM. A new female does not usually protrude her ovipositor until the following day in order to increase her drawing power, although attraction and mating may occur without calling. If unmated, she remains inactive all night until forenoon, then begins calling until evening (*1212*).

The period of highest attractiveness for *Spodoptera littoralis* is during the first 2 or 3 hours after darkness (*1266*), and from dusk to midnight (or 6:00 PM–2:00 AM) for *Cacoecia murinana* (*424*). Maximum mating activity occurs between 5:00 and 7:00 AM in *Solenobia triquetrella* (*1059*), 10:30 PM–3:00 AM in *Manduca sexta* (*17*), and midnight to 4:00 AM or dawn in caged *Trichoplusia ni* (*567, 1083*). Sower *et al.* (*1126, 1127*) report that female *T. ni* synchronized to a 12-12 light-dark cycle released sex pheromone only between 5 hours after dark and the end of the dark period, with peak release from 8 to 11 hours after dark. Mean time of mating in the laboratory and field was after midnight (*1078*). Sexual activity in *Diatraea saccharalis* is mostly confined to the hours between 1:00 and 4:00 AM (*858*).

Males are attracted to female webbing clothes moths (*Tineola bisselliella*) in the morning or evening (*1182*) and to female *Rhyacionia frustrana* and *R. buoliana* in the evening (*711a, 1296*).

Although male *Pectinophora gossypiella* are attracted to females in the field during darkness, those males reared in the laboratory under continuous illumination will respond readily at any time to an extract of the female abdomens (*840*).

Barth (*101*) reported that the production of sex attractant by females of *Plodia interpunctella*, *Anagasta kühniella*, *Achroia grisella*, and *Galleria mellonella* does not appear to be restricted to specific hours. Traynier (*1195*) states that virgin female *A. kühniella* emit the sex pheromone at the beginning of the light period of a 24-hour cycle of light and dark periods. The response of males was found to be maximal at dawn. It is obvious from the foregoing that lepidopterous insects exhibit specific but extremely varied times of courtship and mating.

## Coleoptera

The volatile substance responsible for mass attraction of both sexes to male *Ips confusus* is produced only during the initial period of attack on ponderosa pine logs, but the attraction shows a definite diurnal pattern at 8:00–10:00 AM and 2:00–6:00 PM (*432*, *1224*). Similarly, the response flight of both sexes of *Dendroctonus pseudotsuga* to the volatile substance produced by sexually mature, unmated females feeding on Douglas-fir phloem exhibit a distinct diurnal and seasonal pattern (*992*). Flight begins in the morning as soon as a threshold temperature of 68°F has been reached, attaining its peak around noon. Later in the season, high temperatures during midday always result in marked flight depression. Concentrated flight ceases at sunset, even though the temperature may remain optimal for flight.

Male *Xenorhipis brendeli* shows its greatest response to females between 12:30 and 3:00 PM daily, with more than 80% of all males being captured between 12:40 and 1:30 PM (*1253*).

## Hymenoptera

Attraction of male *Bracon hebetor* by females occurs during daylight (morning) hours (*488*).

A virgin female *Diprion similis* set in the field at 11:00 AM attracted its first males within 30 seconds. Activity continued until 4:00 PM, when a total of over 7000 males had been attracted. The greatest activity occurred between midnight and midafternoon (*310*).

## Diptera

In *Dacus tryoni*, mating is normally confined to the dusk period. Responses of females to the male pheromone were observed only in the late afternoon and dusk period (*412*).

The greatest activity of female hessian flies (*Phytophaga destructor*) to attract males occurs in the early morning (*285*).

# CHAPTER X

# COLLECTION, ISOLATION, AND IDENTIFICATION OF SEX PHEROMONES

## Collection

Insect sex pheromones may be collected by 2 main methods—volatilization and solvent extraction.

1. Volatilization
   a. Passage of a stream of air through a container of live insects
      i. Condensate collected at low temperature
      ii. Condensate adsorbed on a fat (enfleurage)
   b. Steam distillation of dead insects into an organic solvent
2. Solvent Extraction
   a. Extraction of insects
      i. Blending (mechanical) with a solvent
         (a) Mortar and pestle for small numbers
         (b) Electric blender for large amounts
      ii. Washing or rinsing with a solvent
      iii. In Soxhlet extractors
   b. Extraction of insect by-products (feces, organs)

The preferred solvents for extraction are methylene chloride, hexane, and ethyl ether, since they are sufficiently volatile for removal without exposing the extract to high temperatures. Methylene chloride has the added advantage of nonflammability.

## Isolation

Isolation of a sex pheromone from an extract or the solution of a condensate usually involves a combination of chromatographic techniques (column, thin-layer, or gas). The following steps are frequently used.

1. Evaporation of the solvent (under reduced pressure; e.g., 15 mm)
2. Low temperature precipitation of inactive waxes and other fats in acetone or methanol at $-20°$ to $-70°C$ for 24 hours
3. Column chromatography
   a. On Florisil
   b. On silicic acid
   c. On silver nitrate-impregnated silica gel (to separate saturated from unsaturated components)
   d. Hexane followed by increasing percentages of ether in hexane are commonly used as eluting solvents for the above adsorbents
4. Thin-layer chromatography
5. Gas chromatography
   a. Preparative (collection of fractions)
   b. Determination of purity
   c. Nonpolar and polar packings

## Identification

Identification of a pure pheromone involves a combination of chemical and physical methods.

1. Chemical methods
   a. Test reactions to determine functional groups in the molecule
   b. Hydrogenation of an unsaturated molecule, including measurement of the amount of hydrogen absorbed if possible
   c. Ozonization or permanganate oxidation of an unsaturated molecule
2. Physical methods
   a. Ultraviolet and infrared spectroscopy (to determine functional groups and unsaturation)
   b. Nuclear magnetic resonance spectroscopy (to determine number of protons and their grouping)

c. Mass spectrometry (to determine molecular formula and molecular fragmentation)

d. Optical rotation (to determine if molecule is optically active)

Excellent explanations or reviews of pheromone collection and isolation are given in references *436, 540, 598, 741, 1103, 1232*, and *1241*. Separation of cis and trans isomers are well explained in references *765* and *1243*. References *540, 598, 646, 700, 739, 800*, and *1099* are recommended as aids in identifying pheromones.

Methods for collecting, isolating, and (or) identifying specifix sex pheromones are given in this chapter.

## ACARINA

*Amblyomma americanum*
*Amblyomma maculatum*
*Dermacentor variabilis*

Partially engorged, unmated females were separated from their hosts and homogenized several times with methylene chloride. The extracts were concentrated and subjected to chromatography on a column of silicic acid and eluted with hexane containing increasing percentages of ether. The active fraction, eluted with 8–10% ether in hexane, was subjected to preparative gas chromatography on a column (4 mm × 3.7 m) of 12% diethylene glycol succinate (DEGS) on 60/70 mesh Anakrom ABS at 150°C (nitrogen flow rate 60 ml/minute); the retention time of the active fraction was 15 minutes. The retention time was identical in extracts of females from all 3 species. Although the pheromone was not identified, it appears to be a phenolic compound, soluble in sodium hydroxide and discharged from this solution with carbon dioxide. It shows ultraviolet absorption (in cyclohexane) at 260–285 m$\mu$, and its retention time is slightly longer than that of thymol (*138*).

## ORTHOPTERA

*Blattella germanica*

Volkov *et al.* (*1237*) claim that a sex attractant produced by virgin females may be collected by extracting, with ether, filter papers used to line containers holding these insects. From papers housed with 700 females for 21 days, they obtained 282 mg of a waxy extract with an unpleasant odor whose active component was volatile with water vapor. Addition of acetone to the extract caused a solid to separate which was filtered off, and the filtrate was freed of solvent. The 89 mg of active fraction remaining was found to be free of amino acids (negative ninhydrin test); it was treated with petroleum ether (bp 40°–60°C) and the solution was decanted off and washed with 5% sodium bicarbonate, 3% hydrochloric acid, and water, and dried. Removal of the solvent under reduced pressure left 68 mg of a yellow, mobile oil with a faint odor. Gas chromatography on a column (4 mm × 6 m) of polyethylene glycol adipate on Chromosorb W at 142°C (helium flow rate 80 ml/minute) showed the presence of 3 components by thermal conductivity detection.

Ishii (575) and Ishii and Kuwahara (576, 577) were unable to demonstrate the existence of a female attractant for males in this species, but they did show that male nymphs release a pheromone causing aggregation of nymphs of both sexes. The active substance is collected by washing the body surface, especially the abdomen, with ether or extracting the feces with this solvent or with methanol.

*Locusta migratoria migratorioides*

An aggregation pheromone produced by swarming hoppers can be collected by extracting, with petroleum ether or risella oil, air from the rearing rooms, hopper abdomens, or hopper feces (827). By heating at 110°C a risella oil extract of the room air for 1 hour under vacuum and collecting the distillate in petroleum ether, a solution was obtained whose infrared spectrum showed carbonyl absorption (1722 and 1212 cm$^{-1}$) and the absence of alkenes, alkynes, and aromatic groups.

*Nauphoeta cinerea*

A substance called "seducin," which attracts females and keeps them in the proper position for coupling, is extracted with methylene chloride from the abdomens of males 1–2 weeks old (983). Partition of the neutral portion of the extract into methanol-soluble and methanol-insoluble fractions and column chromatography of the former (0.097 mg per male) on Florisil (successive elution with pentane, methylene chloride, ether, and methanol) showed that the active substance was removed with a mixture of ether and methanol. Fractional precipitation of an extract of 4000 males with methanol at −10°C to free it of inactive lipids gave an active methanol-soluble fraction of 0.02 mg per male. Seducin is probably a stable, polar, neutral lipid of low volatility.

*Periplaneta americana*

Wharton et al. (1258) obtained solutions of the attractant by extracting, with petroleum ether at room temperature, papers on which virgin females had been kept. These extracts, concentrated by vacuum distillation at 22°C (distillates collected in a freezing bath were not attractive to males), were stable under refrigeration for several months. In order to obtain larger amounts of attractant, Wharton et al. (1257) exposed many papers to virgin females and extracted the papers with water. The extract was made

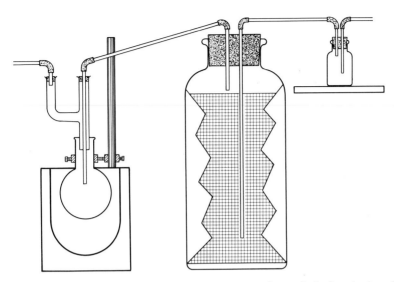

*Fig. X.1.* Apparatus for collection of sex attractant from virgin female American cockroaches. [Redrawn and modified from Yamamoto (*1322*), Entomological Society of America, College Park, Md.]

alkaline, acidified to pH 5.0–5.5, and heated until 60% had been distilled. The distillate was freed of fatty acids and repeatedly redistilled until the volume of the final distillate was convenient for complete extraction with isopentane. Purification and isolation of 28 μg of a "pure" attractant were accomplished by a combination of successive column chromatography (Florisil, silicic acid-Celite, Florisil, and alumina), eluting with increasing concentrations of ether in isopentane, and gas chromatography on a 4-foot column of 5% Apiezon M on 80/100 mesh Chromosorb W at 130°C (argon flow rate 110 ml/minute). Gas chromatographic eluates were collected in coiled traps at −80° or −160°C. The active fraction (about 28 μg) had a retention time of approximately 145 minutes. Infrared spectra indicated that the material was aliphatic and contained ester carbonyl (*1257*).

An improved method for collecting larger quantities of the attractant was reported by Yamamoto (*1322*). Virgin females (1000) were confined in a 10-gallon milk can, and a stream of air (2 cubic feet/second) was passed through the container and into a receiver immersed in an alcohol-dry ice bath, as shown in Figure X.1. A number of cans could be connected in series through a common air inlet.

The condensate obtained from approximately 10,000 females, collected by Yamamoto's method, was chromatographed on silicic acid; elution with 10% ether in hexane, after successive elution with hexane and 3% ether in hexane, removed the attractant from the column completely. Attractant was obtained by steam distillation as 12.2 mg of a pale yellow liquid with a characteristic sweet odor; it elicited a response from males at levels below $10^{-14}$ $\mu$g (605). Gas chromatography indicated that this material was pure (only a single peak obtained) and that it was obviously different from the substance with which Wharton's group had been working, since its retention time was much lower (6 minutes as compared with 145 minutes).

The attractant analyzed for $C_{11}H_{18}O_2$ was not optically active and showed no absorption in the ultraviolet region. Its infrared spectrum showed it to be an ester and indicated the presence of an isopropylidene group $[(CH_3)_2C=]$. Through the combined use of acidic alkaline hydrolysis, hydrogenation, hydrogenolysis, oxidation, and nuclear magnetic resonance spectroscopy, the structure 2,2-dimethyl-3-isopropylidenecyclopropyl propionate [1] was assigned to the attractant molecule (605). However, this structure was synthesized and proved to be inactive on males (335, 602, 1225).

$$(CH_3)_2C=C \underset{\underset{H}{\diagup}\overset{\diagdown}{\underset{OCOCH_2CH_3}{}}}{\overset{\diagup}{\underset{C}{}}} C(CH_3)_2$$

[1]

The true structure of the sex attractant remains open.

Wharton et al. (1256), in a critique of the preceding isolational and structure work (605), stated that collection of the attractant by extraction of filter papers gave 3–5 times more attractant per roach per day than did the air-collection method. They also stated that the attractant isolated by the latter method could not have been pure, but a reply (600) to this critique put forth additional evidence to refute Wharton's claims.

Recently, Bowers and Bodenstein (197) reported that an active extract prepared from virgin female midguts (176) withstood refluxing in 5% methanolic potassium hydroxide but not in 10% alkali. The activity was also destroyed by treatment with sodium borohydride, a reagent which specifically attacks aldehydes and ketones, but activity was quantitatively regenerated by oxidation of the product with chromic acid. Treatment with bromine destroyed activity but debromination with zinc restored about 50% of the activity. This has led to speculation (197) that the attractant may be an unsaturated ketone with a hindered ester group.

HEMIPTERA

*Aonidiella aurantii*
Tashiro and Chambers (1173, 1174) were able to collect the female's attractant by aeration of live virgin females and by extracting either the condensate or homogenized females with ether. Extracts of females prepared

with acetone, methylene chloride, chloroform, ethanol, hexane, and petroleum ether were also sexually attractive to males. These investigators have stated that the pheromone is continuously present in females, who can release or withhold it at will. Thin-layer chromatography of an ether extract on silica gel, using a developing solution of hexane-ether-ethanol (70:25:5) and rhodamine B under ultraviolet light to make the spots visible, resulted in 6 spots; the major active spot had $R_f$ 0.9.

Although steam distillation of the condensate or ether homogenate gave attractive material, this method of purification was discarded in favor of column chromatography (1246). When the homogenate or condensate was chromatographed on a column (2 × 47 cm) of Sephadex LH20, acetone-chloroform (1:1) removed high molecular weight material and all acids; removal of acids with sodium bicarbonate destroyed the activity. Column chromatography on silica gel impregnated with 25% silver nitrate gave attractive material through elution with 25% ether in hexane. Preparative gas chromatography of this attractant on a column of 15% Carbowax 20M on 60/80 Chromosorb W gave active peaks in the area between cis-7-dodecen-1-ol acetate and cis-9-tetradecen-1-ol acetate. This evidence and that supplied by carbon-skeleton chromatography indicate that if the pheromone is an acetate it is the acetate of an unsaturated branched $C_{13}$ alcohol or an unsaturated straight-chain $C_{12}$ alcohol.

*Lethocerus indicus*

The clear liquid with a cinnamon odor produced by the adult male and that probably acts as an aphrodisiac for the female is obtained by opening the abdomen dorsally, removing the white tubules, pressing them in a glass, and filtering the press liquid (242). The clear, water-white liquid boils at 168°–170°C and possesses refractive index ($n_D^{25}$) 1.4160, which is unchanged after distillation. The infrared spectrum showed carbonyl absorption at 5.80 μm and gave a positive hydroxamate test for an ester grouping. Saponification of the liquid with ethylmagnesium bromide gave acetic acid and an alcohol which was esterified with 4'-nitroazobenzene-carboxylic acid chloride; the ester formed showed mp 128°–129°C. Oxidation of the double bond of the natural attractant, together with the foregoing data, identified the attractant as trans-2-hexen-1-ol acetate [2]. A synthetic sample of this material showed bp 165°–166°C and $n_D^{25}$ 1.4173; its infrared spectrum lacked 2 small bands (8.50 and 9.25 μm) present in the spectrum of the natural product that were undoubtedly due to a small amount of impurity.

$$CH_3(CH_2)_2 \overset{H}{\underset{H}{C}} = CCH_2OCOCH_3$$

[2]

In a continuation of the foregoing investigation, Devakul and Maarse (*348*) isolated from the gland liquid (known as "Meng Da" in Thailand), by gas chromatography and thin-layer chromatography, compound [2] and the related compound *trans*-2-hexen-1-ol butyrate. More recently, Pattenden and Staddon (*849*) also isolated these compounds from the secretion of the male sternal scent glands by gas chromatography on Apiezon L at 75°C.

As a result of the investigations of Butenandt and Tam (*242*) and Devakul and Maarse (*348*), 2 closely related compounds, *trans*-2-octen-1-ol acetate [3] and *trans*-2-decen-1-ol acetate [4], were isolated by distillation and gas chromatography from the steam distillates of *Musgraveia* (*Rhoecocoris*) *sulciventris* and *Biprorulus bibax*, respectively; they are believed to be sex attractants (*844*). Identification was accomplished by infrared spectroscopy and ozonolysis.

$$\underset{H}{\overset{H}{CH_3(CH_2)_4C}}=CCH_2OCOCH_3 \qquad\qquad \underset{H}{\overset{H}{CH_3(CH_2)_6C}}=CCH_2OCOCH_3$$

[3]                                    [4]

*Planococcus citri*

Pentane extracts of sexually mature females were more attractive to males than extracts prepared with more polar solvents. Losses of activity during freezer storage indicated instability, possibly by oxidation (*477*).

LEPIDOPTERA

Aplin and Birch (*72, 73, 160*) have isolated and identified odorous compounds from the male abdominal scent brushes of 7 species of noctuid moths. It is believed, but not known definitely, that these compounds induce the female to copulate. The brush was carefully removed and extracted with methylene chloride or carbon tetrachloride, and the extract was analyzed by gas chromatography; components were identified by mass spectrometry and comparison of the spectral patterns with those of the synthetic compounds. Components from *Leucania conigera*, *L. impura*, and *L. pallens* were 80% benzaldehyde and 20% isobutyric acid. The following were identified from other species (name, compound, and percent content are given): *Apamea monoglypha*, pinocarvone, 95; *Mamestra persicariae*, benzaldehyde, 10; benzyl alcohol, 2; phenethyl alcohol, 85; *Phlogophora meticulosa*, 6-methyl-5-hepten-2-one, 63; 6-methyl-5-hepten-2-ol, 28; 2-methylbutanoic acid, 9; *Polia nebulosa*, benzaldehyde, 8; benzyl alcohol, 10; phenethyl alcohol, 70 (*163*).

*Achroia grisella*

Dahm *et al.* (*324, 325*) attempted to collect the sex attractant by passing a stream of air through a container of active males and cooling the condensate in cold traps, but the condensate was biologically inactive. Active material was obtained by extracting male wing glands with ether. Combined gas chromatography and mass spectrometry gave the pheromone components, which were identified as undecanal [5] and *cis*-11-octadecenal [6] by hydrogenation, ozonolysis, and behavior on silica gel impregnated with silver nitrate.

$$CH_3(CH_2)_9CHO \qquad\qquad CH_3(CH_2)_5\overset{H}{C}{=}\overset{H}{C}(CH_2)_9CHO$$

[5]                            [6]

A combination of 100 $\mu$g of [5] and 1 $\mu$g of [6] was necessary to attract more than 50% of virgin females.

*Adoxophyes fasciata*

*Adoxophyes orana*

A methylene chloride extract of whole female bodies was saponified with 5% methanolic potassium hydroxide and the free alcohols, purified through a Florisil column, were acetylated. Active materials obtained were chromatographed on a column of silicic acid impregnated with 20% silver nitrate, eluting successively with 2% and 5% ether in pentane, followed by ether. The active fraction (5% ether-pentane) was purified by gas chromatography on a column packed with 15% PEGA on acid-washed Chromosorb W, giving a mixture of *cis*-11-tetradecen-1-ol acetate (see compound [8], page 173) and *cis*-9-tetradecen-1-ol acetate (see compound [24], page 190). Identification was accomplished by reductive microozonolysis to 11-acetoxyundecanal and 9-acetoxynonanal, respectively. Approximately 4.4 ng of the mixed pheromone was obtained from each female *A. fasciata* (1:1.8 [8]/[24]) and 5.1 ng/female from *A. orana* (1:3.4 [8]/[24]) (*1169a, 1169b*).

Neither [8] nor [24] alone was attractive in the laboratory or field, but a mixture of both was attractive. The optimum [8]/[24] ratio in the field was 1:4 for *A. fasciata* and 1:9 for *A. orana* (*1169a, 1169b*).

*Alabama argillacea*

The sex pheromone was extracted from virgin female abdominal tips by grinding batches of 300–400 tips several times in a blender with 5 mg of sodium sulfate and 100 ml of methylene chloride, filtering the homogenates, and evaporating the combined filtrates to give 100 mg of mobile yellow oil (*136*). Elution of the active material from a silicic acid column with petroleum ether or hexane indicated that it might be a hydrocarbon, and this appeared to be confirmed by the failure of acid or alkaline hydrolysis, hydrogenation, bromination, and ozone to cause inactivation, and the absence of infrared bands for hydroxyl, amino, carbonyl, or ether groups.

Retention times for the active material relative to known compounds were obtained by gas chromatography on 3 different column packings; they showed that the sex pheromone probably contains about 20 carbon atoms.

*Anagasta kühniella*

Traynier (*1195*) obtained active extracts by grinding female abdominal tips with ether or benzene, filtering, and storing the extracts at −8°C in the dark. Kuwahara and Hara (*704*) succeeded in isolating and identifying the sex pheromone. A methylene chloride extract of 566,000 females and males (approximately equal numbers) was freed of solvent, the residual oil (770 gm) was extracted 4 times with 1-liter portions of methanol at −20°C to separate inactive lipids, and the combined methanol solution was evaporated. The oil was saponified with 4% methanolic potassium hydroxide, and then extracted with ether; the inactive extract was acetylated with acetic anhydride to give 21.4 gm of active oil. Repeated chromatography on silicic acid columns (eluting with 25% benzene in hexane and with 0.25–5% ether in hexane), followed by chromatography on silicic acid impregnated with 10% silver nitrate (eluting with 1–10% ether in hexane), gave 2.2 mg of active material that was subjected to preparative gas chromatography on 15% PEG-20M at 170°C. A total of 0.8 mg of pure pheromone was obtained and identified as *cis*-9,*trans*-12-tetradecadien-1-ol acetate [7] by micro-ozonolysis.

$$CH_3C\overset{H}{=}\underset{H}{C}CH_2C\overset{H}{=}\overset{H}{C}(CH_2)_8OCOCH_3$$

[7]

*Argyrotaenia velutinana*

Active extracts were obtained by homogenizing the last 2 abdominal segments of virgin females with acetone, methylene chloride, ether, benzene, chloroform, methanol, or 95% ethanol, but those prepared with methylene chloride were more potent than the others (*950–952, 961, 962*). Active material was obtained by elution with 15% ether in petroleum ether from a column of Florisil (*961*) and by elution with 5% ether in petroleum ether from a column of silicic acid (*951, 952, 961*). Structural data obtained by chemical and physical tests (*951, 961*) indicated that the attractant was an unsaturated acetate. The pure attractant (200 µg from 40,000 females) was obtained by preparative gas chromatography of the active fraction resulting from column chromatography; it was identified as *cis*-11-tetradecen-1-ol acetate [8] by retention time on a polar and nonpolar column, hydrogenation, ozonolysis, and nuclear magnetic resonance spectroscopy (*951, 952*).

$$CH_3CH_2\overset{H}{C}=\overset{H}{C}(CH_2)_{10}OCOCH_3$$

[8]

The attractant has been designated "riblure."

*Autographa californica*

Pheromone extracts were prepared by removing the terminal abdominal segments of 4-day-old virgin females, macerating the segments with ether, and filtering the solution (*1093*). Thin-layer chromatography of the ether extract on silica gel plates, using benzene to develop the spots, showed an $R_f$ of 0.3–0.5 for the pheromone (*441*).

*Bombyx mori*

Butenandt (*227, 228*) obtained 1.5 gm of benzene extractive from 7000 virgin-female abdominal tips. The active neutral fraction was esterified with succinic acid, and the ester was saponified and sublimed in a high vacuum at 60°–70°C. The 100 mg of impure, waxy, crystalline substance was analyzed for carbon, hydrogen, and oxygen and was thought to be a diol of approximate composition $C_{16}H_{30}O_2$.

Working with an alcohol extract of 60,000 fertilized females, Makino *et al.* (*744*) obtained a neutral fraction which they passed through a column of alumina to give 4.5 gm of orange wax. This was freed of sterols ("bombicestrol") to give a brown syrup that distilled at 100°–110°C (0.06 mm) as a thick, faintly yellow oil that elicited a circling dance in males at 0.0005 $\mu$g and caused wing vibration at 0.00025 $\mu$g. Chromatography of this oil on filter paper with butanol–acetic acid–water, 85% phenol, or benzene-methanol (6:4) as solvents showed spots at $R_f$ 1.0, 0.98, and 0.82, respectively. The attractant, designated "bombixin," was obtained by elution of the spot from the paper; it showed 86.41% carbon and 12.05% hydrogen, did not absorb in the ultraviolet spectrum, and contained a primary hydroxyl group according to its infrared spectrum. The minute amount of material obtained was active at 2–4 $\times$ $10^{-5}$ $\mu$g.

Although Hecker (*528*) reported in 1956 that the pure 4'-(*p*-nitrophenylazo)benzoate of the attractant had been obtained from an extract of 313,000 female abdominal glands, saponification of this ester gave a substance showing activity at "below $10^{-5}$ $\mu$g/ml," which was considerably lower than that finally determined ($10^{-12}$ $\mu$g/ml) for pure "bombykol" (*237*) by Butenandt and Hecker in 1961. The attractant was proposed as a doubly conjugated alcohol of 12–15 carbon atoms (*528*).

Using a petroleum ether extract of the sacculi laterales from 3428 females, Amin (*22*) obtained 13.5 mg of a *p*,*p*'-nitrophenylazobenzoate, mp 78°C, that gave, on hydrolysis, a solution highly attractive to male moths. On the basis of a mixed melting point determination with $N,N$-dimethyl-$p,p'$-nitrophenylbenzamide, Amin decided that the silkworm moth sex attractant was dimethylamine. This was vigorously denied by Butenandt and Hecker (236), who showed that the minimum active concentration of dimethylamine was 10 mg/ml, whereas the natural attractant was many times more attractive to males.

In 1959, the pure attractant was obtained as its 4'-nitroazobenzene-carboxylic acid ester and identified as 10,12-hexadecadienol (231, 232, 234). The extract prepared from 500,000 virgin female abdominal tips with ethanol-ether (3:1) was saponified, and the active neutral fraction was freed of sterols and esterified with succinic anhydride. Saponification of the succinates, treatment with 4'-nitroazobenzenecarboxylic acid chloride, and chromatography gave 12 mg of the attractant derivative from which the attractant was regenerated by saponification (238, 529, 530). Identification was accomplished by hydrogenation and permanganate oxidation. Although the configuration of the conjugation in bombykol was at first thought to be cis,trans (234), subsequent synthesis of the 4 possible geometric isomers showed bombykol to be *trans*-10,*cis*-12-hexadecadien-1-ol [9] (233, 239, 531, 1198).

$$CH_3(CH_2)_2\overset{H}{C}{=}\overset{H}{C}-\overset{H}{C}{=}\underset{H}{C}(CH_2)_9OH$$

[9]

Bayer (122), as a result of an independent investigation of the gas chromatography of a pentane extract of female glands, reported nearly 30 peaks to be present on the gas chromatogram. Of these, 3 substances emerging from the instrument showed attractiveness to males (30, 122, 123); 1 component was highly active. Schneider and Hecker (1040) had previously shown that several unsaturated alcohols were attractive to males in varying degree, but the most potent of these was only $10^{-6}$ times as strong as the natural attractant.

Oleic, linoleic, and 14-methyl-9,12-pentadecadienoic acids were isolated by Butenandt *et al.* (241) in 1963 from the female glands; they were not attractive. The entire, interesting story of this attractant has been reviewed by Butenandt (230).

*Cadra cautella*

Active extracts of female abdominal tips have been obtained with ether (775) and from the tips (206) or whole females (1292) with methylene chloride. A methylene chloride extract of 94,000 whole bodies, subjected to low-temperature crystallization from methanol and hexane, gave 7.3 gm of yellow oil that was further purified by chromatography on a column of silicic acid. Functional group tests and gas chromatographic behavior of the 240 mg of oil obtained suggested that the attractant could be the acetate of a $C_{14}$ unsaturated alcohol (*63, 1292*).

Kuwahara et al. (*706*) extracted 1,200,000 female bodies with methylene chloride, subjected the extract to crystallization from methanol at −20°C for 24 hours, and evaporated the filtrate to dryness to obtain 313 gm of an active yellow-brown oil which was chromatographed on a column of silicic acid; elution with 0.5% or 1% ether in hexane gave the active material. Three further chromatographic treatments on silicic acid, followed by chromatography on a column of silicic acid impregnated with 15% silver nitrate, gave the pure attractant after elution with 5% ether in hexane. Gas chromatography of the colorless oil (6.1 mg) on a column (45 m × 0.25 mm) coated with Ucon Oil LB-550X at 160°C showed a single peak. Hydrogenation, hydrolysis, ozonolysis, and determination of the infrared, mass, and nuclear magnetic resonance spectra showed the compound to be a 9,12-tetradecadien-1-ol acetate, and synthesis of the 4 possible isomers of this compound identified the pheromone as the *cis*-9,*trans*-12 isomer [7], identical with the attractant of *Anagasta kühniella*.

The discovery by Takahashi et al. (*1165*) that only female moths of the FT strain emerged when the insects were reared at 20°C was very helpful for collecting large numbers of virgin female moths to be used in isolational work. The same structure was arrived at independently by Brady et al. (*206*) by using a very similar isolation procedure on an extract of 10,000 virgin female abdomens.

*Choristoneura fumiferana*

Findlay and Macdonald (*402*) collected the attractant by passing a stream of air through plastic bags containing virgin females and then through a system of dry-ice traps to freeze out the volatile components. Additional attractant was obtained by washing the inside of the bags with ether. The crude oil obtained in this way lost its attractiveness after storage for several weeks at deep-freeze temperature. Its major constituents were unattractive fats and fatty acids. Steam distillation of the crude attractant gave highly active material, as did preparative gas chromatography, but the pure attractant was not obtained.

Weatherston et al. (*1251a*) succeeded in isolating the pure pheromone. Since crude extracts of virgin-female tips were not active on males in the laboratory, even after column chromatography, the pheromone was collected by rinsing, with ether, containers in which large numbers of live virgin females had been confined. Column chromatography of the rinses on

Florisil gave active fractions in the eluates with petroleum ether-ether (85:15) and petroleum ether-ether (1:1). These eluates were combined and chromatographed on a second Florisil column; the activity came off in petroleum ether-ether (4:1). Subsequent column chromatography on 25% silver nitrate-impregnated silica gel gave the active fraction in eluates with the same solvent mixture at a ratio of 3:1. The active fraction was distilled at 134°C (0.8 mm), and thin-layer chromatography of the distillate on silver nitrate-impregnated silicic acid gave the pure pheromone in petroleum ether-ether (19:1). Functional group analysis of the crude ether wash had already indicated that the pheromone was an unsaturated aldehyde. Ozonolysis of the pure pheromone, followed by gas chromatography, gave a major product identified as undecanedial. The pheromone was identified as *trans*-11-tetradecenal [9a] by gas chromatography on several column packings.

$$CH_3CH_2\overset{H}{\underset{H}{C}}=C(CH_2)_9CHO$$

[9a]

*Choristoneura rosaceana*

Roelofs and Tette (*966*), subjected a methylene chloride extract of 20,000 virgin female abdominal tips to saponification, acetylation, and the processing procedure used previously (*952*) with *Argyrotaenia velutinana* and succeeded in obtaining the pure pheromone. The usual techniques of column, gas, and thin-layer chromatography, ozonolysis, and mass spectrometry showed the compound to be *cis*-11-tetradecen-1-ol acetate [8], identical with the pheromone of *A. velutinana*.

*Crambus* spp.

Collection of the pheromone from female *C. mutabilis*, *C. teterrellus*, and *C. trisectus* was accomplished by anesthetizing virgin females, squeezing out the terminal abdominal segments, crushing them in benzene, and filtering the extract (*85*).

*Cryptophlebia    leucotreta*

The activity of a crude extract (unidentified) of females was destroyed by alkaline hydrolysis but restored by acetylation; hydrogenation also destroyed the activity. Gas chromatography on polar and nonpolar column packings indicated an unsaturated $C_{12}$ straight-chain acetate structure for the attractant. Its retention time, oxidation products, and mass spectrum were identical with those of *trans*-7-dodecen-1-ol acetate [10], and synthetic [10] was attractive to males (*912*).

$$\text{CH}_3(\text{CH}_2)_3\overset{\text{H}}{\underset{\text{H}}{\text{C}}}{=}\text{C}(\text{CH}_2)_6\text{OCOCH}_3$$

[10]

*Danaus gilippus berenice*

Meinwald *et al.* (*779*) isolated 2 major components of the male's hair-pencil secretion by extracting the hair-pencils with methylene chloride and subjecting the extract to preparative thin-layer chromatography on silica gel developed with 5% methanol in methylene chloride, following removal of 1 of the components by vacuum sublimation; this latter component was identified as 2,3-dihydro-7-methyl-1*H*-pyrrolizin-1-one [11] from its infrared spectrum. The major polar component was identified by infrared, mass, and nuclear magnetic resonance spectroscopy as 3,7-dimethyl-*trans*-2,*trans*-6-decadien-1,10-diol [12]. Recoveries of crude [11] averaged 0.02–0.1 mg/male. Compounds 11 and 12 were also isolated from the hair-pencils of male *D. gilippus strigosus*.

[11]

$$\text{HO}(\text{CH}_2)_3\overset{\text{CH}_3}{\underset{\text{H}}{\text{C}}}{=}\text{C}(\text{CH}_2)_2\overset{\text{CH}_3}{\underset{\text{H}}{\text{C}}}{=}\text{CCH}_2\text{OH}$$

[12]

*Danaus plexippus*

Meinwald *et al.* (*776*) isolated from a methylene chloride extract of 6500 male hair-pencils 190 mg of highly polar, colorless oil by thin-layer chromatography on silicic acid. Its infrared spectrum indicated the presence of hydroxyl and $\alpha,\beta$-unsaturated carboxyl groups. Methylation and acetylation of this acid gave the methyl ester of an acetoxy acid. Hydrogenation, mass and nuclear magnetic resonance spectra showed that the polar constituent was 10-hydroxy-3,7-dimethyl-*trans*-2,*trans*-6-decadienoic acid [13]. By repeated thin-layer chromatography of the methylene chloride extract, Meinwald *et al.* (*777*) were also able to isolate 11.8 mg of a second component which they identified, by ultraviolet, infrared, nuclear magnetic resonance, and mass spectra as the related compound 3,7-dimethyl-*trans*-2,*trans*-6-decadiene-1,10-dioic acid [14].

$$\text{HO}(\text{CH}_2)_3\overset{\text{CH}_3}{\underset{\text{H}}{\text{C}}}{=}\text{C}(\text{CH}_2)_2\overset{\text{CH}_3}{\underset{\text{H}}{\text{C}}}{=}\text{CCO}_2\text{H}$$

[13]

$$\text{HO}_2\text{C}(\text{CH}_2)_2\overset{\text{CH}_3}{\underset{\text{H}}{\text{C}}}{=}\text{C}(\text{CH}_2)_2\overset{\text{CH}_3}{\underset{\text{H}}{\text{C}}}{=}\text{CCO}_2\text{H}$$

[14]

*Diatraea saccharalis*

Pérez (*857*) and Pérez and Long (*858*) found that extracts prepared from virgin-female abdominal tips with methylene chloride or benzene were equally attractive, whereas those prepared with acetone, petroleum ether, and 95% ethanol were much less attractive to males.

A methylene chloride extract (49.7 gm) of 17,613 virgin-female abdominal tips was subjected to crystallization in acetone at $-12°C$ and $-70°C$ by Tribble (*1197*). The active acetone filtrates were evaporated to dryness (2.3 gm) and chromatographed on a column (50 gm) of silica gel prewashed with petroleum ether, ethyl acetate, and ether; active material eluted with 10% ether in petroleum ether was rechromatographed on prewashed, neutral alumina and was eluted with 20–60% ethyl acetate in hexane. The resulting 52 mg of active oil contained 4 components in the active zone with retention times between 14.0 and 14.6 minutes by gas chromatography on a column (6 feet; diameter not given) of 10% SE-30 on Chromoport XXX (80/100 mesh) programmed at 120°–300°C (8°/minute) (helium flow rate 150 ml/minute). The crude extract was rendered inactive by alkaline saponification or bromination, but acetylation of the saponification alcohols and treatment of the brominated extract with zinc dust regenerated the activity. By comparative gas chromatography with known acetates, Tribble concluded that the attractant is the acetate of an unsaturated $C_{16}$ alcohol.

Hammond and Hensley (*506*) bioassayed Tribble's chemical fractions on males and found activity in the crude extract, a steam distillate of the extract, and the acetylated neutral fraction resulting from saponification; the saponification neutral (alcohols) and acidic fractions, as well as the methyl esters of these acids, were all inactive.

*Diparopsis castanea*

Tunstall (*1210*) collected the female's attractant by passing a stream of air through a container of virgin females and condensing it at dry-ice temperature or on a column of powdered cellulose. Ether extraction of the condensate or the cellulose gave active material. Although Tunstall was unable to obtain attractive material by macerating female abdominal tips with a solvent, Moorhouse et al. (*801*) and Nesbitt (*36, 76, 825*) reported that crude methylene chloride extracts and air-stream condensates were very attractive to males. These investigators, working together (*801, 825*), have succeeded in isolating 3 components through gas chromatography of the extract on 5 different column packings. One of these substances excites male moths into flight and attracts them into laboratory traps. The other 2 compounds produce no behavioral response in the laboratory but probably play a part in eliciting the full sexual response in the field.

*Ephestia elutella*

Brady and Nordlund (*204*) were able to isolate 2.4 μg of the sex pheromone from 1313 virgin females by extracting the abdominal tips with hexane-ether (1:1), chromatographing the extract on silica gel (elution with 4% ether in hexane), and subjecting the active fraction to preparative gas chromatography. It was identified as *cis*-9,*trans*-12-tetradecadien-1-ol acetate [7] by gas co-chromatography on several column packings with synthetic material.

*Epiphyas postvittana*

The pheromone has been subjected to gas chromatography and shown to be an unidentified acetate (*94*). The active extract obtained from virgin female abdomens was hydrolyzed to an inactive product which was then acetylated to a product as attractive to males as the crude extract. Passage through a gas chromatograph gave 2 fractions, neither of which was significantly active alone, but the combination was fully attractive (*94a*).

*Euproctis chrysorrhoea*

Females (8000) were placed in containers connected to 4-liter flasks containing a drying agent and connected to several glass tubes filled with glass rings (to give a larger surface area); the flasks were immersed in Dewar cold flasks containing liquid air. The connecting glass tubes led to an uncooled tube filled with charcoal, and finally to an oil pump providing 2.5 m³ of air/minute. The pump was operated for 16 hours/day from July 2 to July 24. The first tube contained most of the water unabsorbed by the drying tube. The middle tube contained 100 ml of a mobile liquid readily soluble in petroleum ether and probably containing some 2-hexenal. Inhoffen (*573*) concluded that the attractant must be a gas at 20°C, condensing at −5° to a stable liquid, but it is highly probable that he was working with the incorrect material.

*Euxoa ochrogaster*

The sex pheromone was extracted from the female moths by squeezing the abdomens with forceps to protrude the last 2 segments, which were then excised and macerated in a mortar with methylene chloride. The combined extracts were filtered and stored under nitrogen at −20°C (*1155*).

*Galleria mellonella*

Air passed through a container of live males was adsorbed on glass plates coated with a 2:1 mixture of rendered, cleaned, retroperitoneal tallow and mesenteric lard, containing 0.02% hydroquinone as a preservative. The fat was scraped from the plates and subjected to molecular distillation at 120°C and 0.5 mm pressure; material that volatilized was condensed on a dry ice-cooled finger. The yield of distillate from 50,000 males was 1.92 gm; it was subjected to preparative gas chromatography on a column (60 × 0.9

cm) of 15% Carbowax 20M on 60/80 mesh Chromosorb W at 146°C from which the active material emerged in 36 minutes. Rechromatography on a capillary column (450 × 0.01 cm) coated with γ-nitro-γ-methylpimelonitrile at 123°C resulted in elution of the pure pheromone (319 mg) in 9–10 minutes. The yield of pure pheromone obtained was approximately 6.4 μg/male (*972, 973*). The pheromone was identified by infrared, ultraviolet, nuclear magnetic resonance, and mass spectra as *n*-undecanal [5], identical with that obtained from *Achroia grisella*.

*Grapholitha molesta*

Ether (*962*) and benzene (*446, 960*) were used to prepare active extracts of the female abdominal tips. Saponification and acetylation of the crude methylene chloride extract resulted in a hundredfold increase in pheromone concentration. The attractant was isolated in pure form by a combination of column, gas, and thin-layer chromatography, and it was identified by ozonolysis and mass spectrometry as *cis*-8-dodecen-1-ol acetate [15] (*960*).

$$\overset{\text{H}\quad\text{H}}{CH_3(CH_2)_2C=C(CH_2)_7OCOCH_3}$$

[15]

*Heliothis virescens*

Berger *et al.* (*139*) reported that crude methylene chloride extracts of female abdominal tips were not attractive to males but that a gas chromatographic fraction of such extracts was attractive. However, Shorey *et al.* (*1093*) reported that a crude ether extract of female tips was attractive to males.

Gaston *et al.* (*441*) showed that an ether extract of female tips could be distilled in a molecular still. The active distillate, by thin-layer chromatography on silica gel plates (benzene developer), gave an $R_f$ of 0.3–0.5 for this active zone.

*Heliothis zea*

Berger *et al.* (*139*) reported that males showed no interest in filter paper impregnated with a crude methylene chloride extract of female abdominal tips, extracts prepared from mating pairs, filter papers exposed to virgin females with various solvents, or to air passed over live females. A sexual response was elicited in males by a component of the methylene chloride extract emerging from a gas chromatographic column (6 feet × ¼ inch) of 12% diethyleneglycol succinate on 60/70 mesh Anakrom AS at 155°C (flow rate 70 ml argon/minute) in 12.3 minutes. Shorey *et al.* (*1093*) reported that males responded to a crude ether extract of female tips.

McDonough *et al.* (*768*) were able to confirm the findings of Berger *et al.* (*139*). By chromatographing a methylene chloride extract of 80 virgin female moths on alumina and subjecting the active fraction eluted with pentane-ether (30:1 and 4:1) to gas chromatography, they obtained 2 active compounds which appeared to be pure when chromatographed on polar and nonpolar columns. Chemically bound pheromone was released from inactive liquid chromatographic fractions by saponification followed by acetylation. By chemical tests and comparison of the retention times of the 2 pheromones with those of known compounds, the pheromones appeared to contain $C_{14}$, monounsaturated, straight-chain alcohol moieties, 1 being a free alcohol and the other an acetate. The double bond in the ester was reported to be no further than the eighth carbon atom from the acetate group (*768*).

*Holomelina* spp.

Roelofs and Cardé (*953*) have isolated and identified a sex pheromone occurring in many sibling species of the *H. aurantiaca* complex, *H. laeta*, and *Pyrrharctia isabella*. A methylene chloride extract of the female abdominal tips was chromatographed on a column of silica gel; elution of the active component with petroleum ether indicated that it was a hydrocarbon, and this was substantiated by the failure of saponification or acetylation to deactivate the material. Activity was retained following treatment with bromine, indicating that the pheromone was not unsaturated. The attractant behaved like octadecane on several gas chromatographic columns. Mass spectrometry and comparison of retention times with authentic material identified the pheromone as 2-methylheptadecane [16].

$$(CH_3)_2CH(CH_2)_{14}CH_3$$

[16]

*Lasiocampa quercus*

The female scent was collected by passing filtered air over the females and then into a U-tube immersed in liquid oxygen (*666*). It was sealed and stored in nitrogen.

*Laspeyresia pomonella*

Butt and Hathaway (*262*) obtained active filtrates from abdominal tips or whole females using acetone, acetonitrile, benzene, chloroform, cyclohexane, ethanol, ether, hexane, methylene chloride, or petroleum ether. Extracts prepared by homogenization with acetone, cyclohexane, or hexane were inactive, probably due to masking by the additional amounts of fat obtained.

McDonough *et al.* (*767*) obtained pure sex pheromone from whole virgin female moths by a procedure consisting of extraction with methylene chloride, refluxing in methanolic sodium hydroxide to release chemically bound attractant, column chromatography on alumina, preparative gas chromatography on Carbowax 20M, and gas chromatography on Apiezon L. One-day-old and 3-day-old females contained 0.03 and 0.30 ng of free pheromone, respectively, and 0.22 and 0.25 ng of bound pheromone. Bromination, hydrogenation, attempted reduction with lithium aluminum hydride, and gas chromatography showed that the attractant was an unsaturated alcohol lacking a carbonyl group but containing at least 1 other functional group.

Without reporting the methods used, Roelofs *et al.* (*62*) announced in 1970 that they had identified the sex attractant from a total of 50 female tips as *trans*-8,*trans*-10-dodecadien-1-ol [17]. In 1971, they reported (*959*) that the abdominal tips had been extracted with methylene chloride and the extract subjected to gas chromatography on a column of 5% cyclohexanedimethanol succinate on Chromosorb Q at 170°C and on a column of 10% JXR on Chromosorb Q at 190°C. The column effluent in each case was collected at 1-minute intervals and assayed for activity on isolated male antennae by the electroantennogram (EAG) method. Comparisons with EAG's of known synthetic compounds served to identify the active compound as [17].

$$CH_3\overset{H}{\underset{H}{C}}=C-\overset{H}{\underset{H}{C}}=C(CH_2)_7OH$$

[17]

*Lycorea ceres ceres*

Meinwald *et al.* (*778, 780*) isolated 3 components of the male's hair-pencil secretion by extracting 300 hair-pencils with methylene chloride or carbon disulfide. Fractional sublimation and spectrometry showed 1 of these components, mp 74°–75°C, to be 2,3-dihydro-7-methyl-1*H*-pyrrolizin-1-one [11], having the characteristic odor of the male. The 2 remaining major components were separated by column chromatography on silicic acid (elution with methylene chloride-methanol) and preparative gas chromatography on a 5% SE 30 column (6 feet × ⅛ inch) (*780*). Hydrogenation and periodate-permanganate oxidation, combined with nuclear magnetic resonance and mass spectrometry, served to identify the components as *n*-hexadecyl acetate [18] and *cis*-11-octadecen-1-ol acetate [19] (*778*).

$$CH_3(CH_2)_{15}OCOCH_3$$

[18]

$$CH_3(CH_2)_5\overset{H}{C}=\overset{H}{C}(CH_2)_{10}OCOCH_3$$

[19]

Ten mg of [11], 4.5 mg of [18], and 8.5 mg of [19] were obtained from 100 male hair-pencils.

*Malacosoma disstria*

The last 2 abdominal segments of virgin females were macerated in a mortar and extracted with 3 aliquots of purified methylene chloride or peroxide-free ether. The extract was filtered, concentrated under reduced pressure, and stored under nitrogen at −20°C (*1154*).

*Manduca sexta*

After the last 2 or 3 abdominal segments of virgin females were demonstrated to be attractive to males, they were clipped and extracted with ether or ethanol, and only the ether extracts were attractive to males (*17*).

*Orgyia leucostigma*

Active extracts were prepared by extracting virgin-female abdominal tips with methylene chloride, filtering and drying the solution with sodium sulfate, and removing the solvent at 30°C under reduced pressure. The resulting golden-brown liquid was partially purified by column chromatography on 60/100 mesh Florisil, eluting successively with petroleum ether, 15% and 50% ether in petroleum ether, and acetone (*856*).

*Ostrinia nubilalis*

Klun (*687*) succeeded in isolating the pure sex pheromone by extracting whole virgin females with ethylene chloride, cooling in acetone and methanol solutions at −70°C, filtering off the inactive solids, evaporating the methanol filtrate to dryness, and chromatographing it successively on a column of silicic acid (elution with 5% methanol in ethylene chloride) and 2 further columns of silicic acid (elution with 0.5% methanol in benzene). Preparative gas chromatography on a column (1.83 m × 4 mm) of 5% SE 30 on 60/80 mesh Chromosorb G at 200°C (helium flow rate 120 ml/minute) showed 2 unresolved active peaks that were separable by gas chromatography on a polar column (1.83 m × 4 mm) of 5% diethylene-glycol succinate on acid-washed, dimethylchlorosilane-treated 60/80 mesh Chromosorb W at 185°C (helium flow rate 120 ml/minute); the single active peak had a retention time of 2 minutes. The pheromone was reported to be a strong electron captor, unsaponifiable, soluble in polar organic solvents, and highly stable (for at least a year) at room temperature in an open dropper.

Klun and Brindley (*54, 56, 688*), and Klun and Robinson (*690a*) subsequently reported that the pheromone is *cis*-11-tetradecen-1-ol acetate

[8], identical with that obtained from *Argyrotaenia velutinana*, but that a secondary substance may be necessary with it to attract in the field (*775*).

*Pectinophora gossypiella*

Berger *et al.* (*140*) obtained approximately 30 mg of crude attractant by extracting 1000 female abdominal tips with methylene chloride. Attempts to purify a whole-moth extract by dissolving it in acetone and freezing the impurities at −15°C were unsuccessful, although large amounts of sterols and triglycerides could be removed in this manner. Active preparations could also be obtained by steam distillation, although some attractant was lost by this procedure. Some purification of the extract was obtained by chromatography on a column of silicic acid, followed by thin-layer chromatography on the same adsorbent or on aluminum oxide. Gas chromatography of material thus purified showed several components to be present, only 1 of which showed activity on males when it emerged from the instrument; this component was reported as possibly a $C_{18}$ ester, according to comparisons of its retention time with those of several known esters. The extract was completely inactivated by boiling with acid or alkali or by treatment with bromine in carbon tetrachloride, but activity was retained after treatment with bisulfite solution.

Jones *et al.* (*632*) succeeded in isolating the pure sex pheromone by extracting 850,000 whole virgin females with methylene chloride, cooling in acetone at −20°C, filtering off the inactive solid, evaporating the filtrate to dryness, dissolving the residue in methanol, again subjecting to acetone crystallization at −20°C, and chromatographing successively on 2 Florisil columns (active components eluted with 3% ether in hexane) and 2 columns of silicic acid impregnated with 25% silver nitrate (active components eluted with 50% ether in hexane). Preparative gas chromatography on a column (24 m × 0.63 cm) of 5% SE 30 on Chromosorb W at 185°C (helium flow rate 33.3 ml/minute) showed 4 peaks, only 1 of which (retention time 11.5 minute), obtained as 1.6 mg of colorless oil, was active in laboratory bioassays on males. Hydrogenolytic gas chromatography and infrared, mass, and nuclear magnetic resonance spectroscopy served to identify the pheromone as 10-propyl-*trans*-5,9-tridecadien-1-ol acetate [20] (*37*).

$$(CH_3CH_2CH_2)_2C{=}CH(CH_2)_2\overset{H}{\underset{H}{C}}{=}C(CH_2)_4OCOCH_3$$

[20]

Although compound [20], designated "propylure," was highly active on males in the laboratory, it was unattractive when used in field traps. Jones and Jacobson (631) isolated a compound from female extract which, when combined with propylure, acted as an activator, resulting in field attraction. The portion of extract soluble in acetone at $-20°C$ was freed of propylure by being shaken repeatedly at room temperature with portions of methanol; the insoluble oil was chromatographed on a column of silica gel impregnated with 25% silver nitrate. The fractions eluted with 25% and 50% ether in hexane (combined weight 608 mg) comprised the activator, proved to be pure by gas chromatography. The activator was identified by combined infrared, mass, and nuclear magnetic resonance spectroscopy as $N$,$N$-diethyl-$m$-toluamide ("Deet") [21].

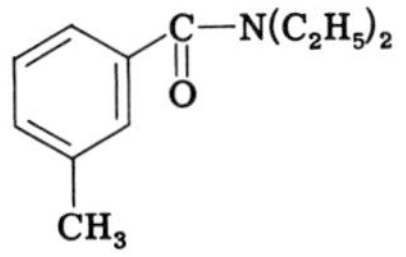

[21]

Compound [21] was found in large amounts in adult females from 3 geographic locations and in smaller amounts in female pupae; it was not found in female larvae or in males at any stage of development (631).

*Philosamia cynthia ricini*

Tomida and Ishii (1190) extracted the sex attractant from virgin-female abdominal tips with methylene chloride. Activity was destroyed by treatment with reagents specific for aldehydes and ketones, but the biological activity was recovered when the carbonyl group was regenerated. Inactivation also resulted from hydrogenation, bromination, and treatment with maleic anhydride, lithium aluminum hydride, and sodium borohydride, but not by hydrolysis. The results suggest that the attractant is an unsaturated aldehyde or ketone.

*Phthorimaea operculella*

The pheromone was extracted from whole female abdomens with methylene chloride. Crude extract from newly emerged females was 10–20 times less potent than that from 2-day-old virgins (564).

*Plodia interpunctella*

Barth (101) concluded that the sex attractant must be highly volatile, since the production gland turned black on treatment with osmium-containing reagent, and must have the nature of an ether-oil. It has no odor perceptible to humans.

Kuwahara *et al.* (706) extracted the whole bodies of 670,000 males and females (equal numbers) with methylene chloride, subjected the extract to crystallization from methanol at $-20°C$ for 24 hours, evaporated the filtrate to dryness, and hydrolyzed the residual oil with 5% methanolic potassium hydroxide. The neutral fraction was acetylated with acetic anhydride and chromatographed twice on a column of silicic acid, the active material being eluted with 0.5% ether in hexane. The final purification of the pheromone was accomplished on a column of silicic acid impregnated with 10% silver nitrate, giving 500 μg of colorless oil. The structure was established as *cis*-9,*trans*-12-tetradecadien-1-ol acetate [7] by ozonolysis and gas chromatography.

The same structure was arrived at independently by Brady *et al.* (*63*, *206*), who collected the attractant by extracting, with hexane-ether (1:1),

*Fig. X.2.* Abdominal tips of virgin female gypsy moths are clipped and dropped into benzene for extraction. [By permission of the U.S. Department of Agriculture.]

filter papers previously used to line glass jars containing 19,000 live virgin females. Chromatography of the extract on a column of silicic acid, eluting the active material with 4% ether in hexane, followed by gas chromatography on several types of column packings, gave 200 µg of pure pheromone.

Dahm *et al.* (*326*) isolated the sex pheromone in pure form by chromatographing an ether extract of 2000 female abdominal tips on a column of Sephadex LH 20, and subjecting the active fraction to successive thin-layer (silica gel) and gas chromatography. Saponification of an ether extract of whole females followed by acetylation gave much larger amounts of pheromone. About 1 ng of pure pheromone was obtained from each abdominal tip and 20 µg from 13,000 whole females. The pheromone was identified by combined hydrogenation, ozonization, and mass spectrometry as 9,12-tetradecadien-1-ol acetate, but the geometric isomerism about the double bonds was not determined.

*Porthetria dispar*

According to Collins and Potts (*304*), chemical studies on extracts of the female abdominal tips (Fig. X.2) were initiated in 1925 by Bloor, who concluded that the attractant was a relatively stable, unsaponifiable substance soluble in fat solvents and slightly soluble in water. Fiske, in 1926, suspected that the substance was an aldehyde but failed to isolate it, and the following year Souther concluded that the attractant was a saturated fat, protein, or an ester that was destroyed by acid or base and continuously generated by the female through hydrolysis of a more complex compound.

In 1957, Prüffer (*895*) determined that 85% ethyl alcohol was the best solvent for extraction of the attractant from the females or their abdomens. Allowing the females to steep in the solvent for up to 12 hours gave a very active extract, whereas steeping for 24 hours gave a less potent extract, possibly due to extracting out inactive, masking substances. Extracts stored in glass-stoppered containers in the dark remained attractive for at least 3 years, whereas activity was lost after 1 year's storage in the light, the extract turning from a yellow to a gray-brown color.

Extracts prepared by von Zehmen (*1332*) in 1942 using petroleum ether, 95–100% ethyl alcohol, chloroform, acetone, and ether were all active, but a carbon tetrachloride extract was unattractive to males. Males were attracted to loosely corked flasks containing the petroleum ether extract, as well as to the pants' pockets of a forester who had carried such flasks in his pockets. Ether extracts were pale green in color, whereas extracts prepared with other solvents were colorless. Removal of the solvents from all extracts left a substance with a cucumberlike odor.

In 1942, Haller *et al.* (*502*) reported that benzene was probably the solvent of choice for obtaining attractive extracts from the female abdominal tips. Attractiveness of the extract was markedly increased by hydrogenation (7). The attractant, residing in the neutral fraction of the extract, reacted with phthalic anhydride, from which it could be recovered by saponification. Attractant obtained in this way was designated "gyptol" by Acree (4), who obtained considerable purification of this material and its ester derivatives by successive column chromatography on magnesium carbonate and magnesium oxide (5, 6). Acree concluded that the activity of the gypsy moth sex attractant could be attributed to 2, or possibly 3, fatty esters derived from at least 2 different alcohols.

Stefanović (1137, 1138), working with a hydrogenated benzene extract of 500,000 female tips, obtained, by steam distillation, a highly attractive, fluorescent, yellow oil with a characteristic odor. This oil was separated into 246 fractions, 19 of which were attractive in field tests. These 19 fractions were combined to give 100 mg of yellow-red oil.

In 1960, Jacobson et al. (603) reported that they had succeeded in isolating and identifying the sex attractant. Their procedure involved clipping the last 2 abdominal segments of many virgin female moths, separating the neutral fraction from a benzene extract of the tips, and either chromatographing on adsorbent columns or dissolving the neutral fraction in acetone, precipitating out the inactive solids at low temperature, and subjecting the attractive yellow oil to paper chromatography. Of the 5 spots obtained, only 1 was attractive to males, and this was separated into a highly attractive colorless liquid (the major attractant) and a solid of much lower activity. A total of 20 mg of major attractant and 3.4 mg of minor attractant was isolated from 500,000 females. On the basis of degradation and other chemical studies, as well as by synthesis, the structure d-10-acetoxy-cis-7-hexadecen-1-ol [22] was proposed for the major attractant (603, 604).

$$CH_3(CH_2)_5\overset{}{\underset{\overset{|}{OCOCH_3}}{C}}HCH_2\overset{H}{C}=\overset{H}{C}(CH_2)_6OH$$

[22]

Inconsistencies in the attractiveness of synthetic laboratory and commercial samples of the major attractant (gyptol), as evidenced by erratic activity in the field and laboratory bioassays, prompted a reinvestigation of its physical, chemical, and biological properties. In 1970, Jacobson et al. (612) reported that gyptol rigorously prepared by 3 different methods to exclude any trans isomer was not attractive to male gypsy moths, although its chemical and physical properties were identical with gyptol previously reported (604). It was then determined that carefully conducted basic hydrolysis, ozonolysis, and catalytic hydrogenation did not affect the activity of a crude extract of virgin females, whereas activity was destroyed by acidic hydrolysis or reduction with lithium aluminum hydride. The activity appeared to survive pyrolytic conditions that completely destroyed synthetic gyptol. It was concluded that, while synthetic gyptol itself was unattractive, it sometimes contained, as a contaminant, a small amount of extremely active compound.

Later in 1970, Bierl et al. (59, 61, 158) were able to determine the true structure of the fantastically potent sex attractant of this insect. A benzene extract of 78,000 virgin-female abdominal tips was saponified with ethanolic potassium hydroxide and the active neutral fraction was chromatographed on a column of Florisil; the active material was eluted with 2–6% ether in hexane. Chromatography on Florisil was again carried out, followed by thin-layer chromatography on silica gel, giving a partially purified material that appeared to be, on the basis of chemical and pyrolytic tests, a $C_{18}$-$C_{20}$ alkyl epoxide. Although the amount of this material obtained was insufficient for complete characterization, the investigators speculated that the insect might contain an olefin precursor from which the attractive epoxide could be formed, and that this olefin could be used to generate additional amounts of the sex attractant. This proved to be true, since treatment of the original neutral and hydrocarbon fractions with $m$-chloroperbenzoic acid caused a 10-fold enhancement of the activity. Chromatography on a column of silica gel impregnated with silver nitrate, and then by preparative gas chromatography on a column (0.9 m × 0.9 cm) of 10% OV-17 on 70/80 mesh Anakrom ABS at 180°C, as well as on a column (1.2 m × 0.45 cm) of 5% diethyleneglycol succinate on 60/80 mesh Chromosorb W at 110°C, gave the pure attractant. Ozonolysis and mass spectrometry established the structure as cis-7,8-epoxy-2-methyloctadecane [23]. It has been designated "disparlure" (61).

$$CH_3(CH_2)_{10}\overset{H}{\underset{\diagdown O \diagup}{C}}\!\!-\!\!\overset{H}{\underset{}{C}}(CH_2)_4CH(CH_3)_2$$

[23]

Enhancement of the activity of a crude extract of female tips by treatment with $m$-chloroperbenzoic acid to generate additional amounts of the attractive epoxide has been discussed in detail by Bierl et al. (159).

*Porthetria monacha*

Virgin female extracts were prepared with petroleum ether, ether, acetone, chloroform, and 95–100% ethyl alcohol. The odor of the solvents did not mask the attractiveness of the extracts in the field (1332). Extracts of the females and their abdomens with 85% ethanol or xylene became inactive during 1 year's storage in the dark at room temperature, whereas alcohol extracts lost none of their activity during this period (21).

*Samia cynthia*

A crude extract prepared by homogenizing virgin female abdomens with hexane failed to evoke a sexual response in caged males, but column chromatography of the extract on Florisil gave an active fraction on elution with 5% ether in hexane (614).

*Sitotroga cerealella*

Benzene and ether extracts prepared by homogenizing the whole females were more attractive to males than methylene chloride and acetone extracts (667). Homogenization of female abdominal tips with absolute ethanol also gave an active extract, whereas such an extract prepared from females without abdominal tips was completely inactive (201).

*Spodoptera eridania*

Jones (628) and Jacobson et al. (611) succeeded in isolating and identifying 2 sex pheromones from virgin females, each of which sexually excites males in the laboratory. An extract of 309,000 virgin female abdomens prepared with methylene chloride, hexane, and ethanol was subjected to low temperature crystallization at $-10°C$; the filtrate was subjected to sweep codistillation in hexane, and the distillate was chromatographed successively on a column of Florisil (active material removed by elution with 3% ether in hexane) and a column of silica gel impregnated with 25% silver nitrate (active material eluted with 10 and 25% ether in hexane). The active material was then subjected to preparative gas chromatography, first on a column (3.05 m $\times$ 0.63 cm) of 5% OV-1 on 60/80 mesh Gas Chrome Q at 200°C (helium flow rate 60 ml/minute) and then on a column (3.66 m $\times$ 0.63 cm) of 5% stabilized diethyleneglycol succinate on 60/80 mesh Gas Chrome Q at 175°C (helium flow rate 32 ml/minute). The active compounds were identified by a combination of ozonolysis and infrared, mass, and nuclear magnetic resonance spectra as *cis*-9-tetradecen-1-ol acetate [24] and *cis*-9,*trans*-12-tetradecadien-1-ol acetate [7]; the yields obtained were 4 mg of [24] (designated prodenialure A) and 0.8 mg of [7] (designated prodenialure B) (60, 65, 914). In field trials, a combination of both pheromones was necessary to attract male moths into traps.

$$CH_3(CH_2)_3\overset{H}{C}{=}\overset{H}{C}(CH_2)_8OCOCH_3$$

[24]

Weatherston (*1249a*, *1251b*) has also identified 9-hexadecen-1-ol acetate and 9-octadecen-1-ol acetate as defense substances in the millipede, *Blaniulus guttulatus*.

*Spodoptera frugiperda*

Sekul and Cox (*1062*, *1063*) were able to achieve considerable purification of the pheromone by extracting 12,000 virgin-female abdominal tips with peroxide-free ether, chromatographing on a column of silica gel (active material eluted with 3 and 6% ether in petroleum ether), subjecting to crystallization from acetone at −20°C, and then conducting preparative thin-layer chromatography on silica gel. Of 6 spots observed in ultraviolet light following development with hexane-ether-acetic acid (90:5:1), only 1 elicited a sexual response in males in the laboratory.

In 1967, Sekul and Sparks (*1064*) succeeded in isolating and identifying the sex pheromone. An extract of 135,000 virgin female tips was partially purified as before and then chromatographed on a column of silica gel impregnated with 25% silver nitrate (active material eluted with 0.75 and 1% ether in pentane) followed by thick-layer chromatography on acetylated cellulose (developed with methanol-water (8:1)) and then by preparative gas chromatography on a column (1.5 m × 6 mm) of 10% SE 30 on Chromosorb P at 187°C (helium flow rate 45 ml/minute). Of the 5 components obtained, only 1 (retention time 9 minutes) was active. A combination of saponification, hydrogenation, ozonolysis, ultraviolet, and infrared spectroscopy was used to identify the pheromone (900 μg from 135,000 females) as *cis*-9-tetradecen-1-ol acetate [24].

*Spodoptera littoralis*

Flaschenträger (*407*, *408*) found that isolated abdominal segments of virgin females lost their attractiveness in a vacuum but again became attractive to males within a short time after removal from an evacuated atmosphere. The attractant could be collected by extraction of the abdominal segments with purified ether, or by freezing it out in a stream of air followed by steam distillation and extraction with various solvents.

In 1963, Zayed et al. (*1331*) homogenized and lyophilized the abdominal tips from 23,000 females and then extracted them with ethanol-ether (3:1). The unsaponifiable neutral fraction of this extract was freed of carbonyl-containing compounds and sterols and treated with succinic anhydride; the resulting half ester was saponified to give an alcohol that was treated with 4′-nitroazobenzenecarboxylic acid chloride. The resulting ester was extracted with acetone, and saponification of the 100 mg of acetone-soluble portion gave a substance attractive to males at $10^{-5}$ μg/ml. The investigators concluded that the attractant was probably an alcohol. However, in the following year, they decided (*429*), as the result of a chromatographic analysis of the insect fat, that the sex attractant was either an alcohol, aldehyde, or a ketone.

In 1965, Zayed and Hussein (*1329*) reported that only carbonyl- and hydroxyl-containing fractions of the female extract were attractive to males. The unsaponifiable fraction, which was active at $10^{-2}$ $\mu$g/ml, was separated into a carbonyl fraction (active at $10^{-2}$ $\mu$g/ml) by treatment with 2,4-dinitrophenylhydrazine, and an alcoholic fraction (active at $10^{-3}$ $\mu$g/ml). The latter was treated with 4'-nitroazobenzenecarboxylic acid chloride and the resulting ester was saponified to give the free alcohols, active at $10^{-5}$ $\mu$g/ml. The active compounds were not identified.

*Synanthedon pictipes*

The attractive substance could be removed from the female abdominal segments by gently wiping them with a wad of cotton, or by extracting them with ethyl alcohol but not with benzene, methylene chloride, ether, hexane, or distilled water. Abdominal segments wiped with cotton were no longer attractive, whereas the cotton was highly attractive to males (*302*).

*Tineola bisselliella*

A crude petroleum ether extract of filter papers over which females had crawled remained attractive to males during refrigeration for 1 year. Active extracts could also be prepared from shark-skin filter paper sheets or combed, unspun wool exposed to virgin females in mason jars; the filtered extract, concentrated at 70°C, remained attractive for more than 1 year. A petroleum ether extract of 1000 females was freed of solvent to give a thick yellow oil which was then extracted with ether; a dilution of the ether extract to 1:500,000 still induced males to court (*986*).

*Trichoplusia ni*

In 1933, Ignoffo et al. (*567*) prepared a methylene chloride extract of virgin female abdominal tips and subjected it to gas chromatography. A strong peak, having a retention time midway between those of methyl laurate and methyl myristate, was highly attractive to males, indicating that the attractant was fairly volatile with a relatively low molecular weight. No corresponding peak could be found on gas chromatograms of extracts of male abdominal tips.

Gaston et al. (*441*), in 1966, tried a number of solvents for extracting female tips. Ether, methylene chloride, benzene, carbon disulfide, methyl formate, and hexane were all equally satisfactory. However, the hexane extract lost 40% of its activity during drying with sodium sulfate, whereas the other extracts retained full activity. An ether extract of 15,100 tips gave 17.2 gm of oil containing 4.5 mg of pheromone. Distillation of this oil in a falling film still yielded 1.153 gm of oil containing 3.9 mg of pheromone. Preparative thin-layer chromatography of the oil on silica gel plates, using benzene developer, showed the pheromone to have an $R_f$ in the region 0.30–0.54. Subsequent preparative gas chromatography on 4 different column packings indicated that the pheromone was probably an acetate with approximately the same molecular weight as lauryl (saturated $C_{12}$) acetate. (The investigators gave excellent information on thin-layer and preparative gas chromatography.)

In the same year (1966), Berger (135) reported the isolation and identification of the sex attractant. Approximately 200 mg of yellow oil containing the attractant were obtained by grinding 2500 female tips in methylene chloride; it was freed of solvent and chromatographed on a column of silicic acid, giving 8–10 mg of highly active oil eluted with 10% ether in petroleum ether. Subsequent preparative gas chromatography on a column (6 feet × 4 mm) of 12% diethyleneglycol succinate on 60/70 mesh Anakrom ABS at 172°C (nitrogen flow rate 50 ml/minute) gave pure attractant (approximately 2 $\mu$g/female) having a retention time of 7.8 minutes. It was identified by saponifying to the alcohol and characterizing the latter by acetylation, hydrogenation, infrared spectroscopy, and ozonolysis. The structure is cis-7-dodecen-1-ol acetate [25].

$$CH_3(CH_2)_3\overset{H}{C}=\overset{H}{C}(CH_2)_6OCOCH_3$$

[25]

### Zeiraphera diniana

Gas chromatography of a crude benzene extract of female abdominal tips gave an active fraction with a retention time similar to trans-11-tetradecen-1-ol acetate. Male antennal responses to a series of $C_{12}$ and $C_{14}$ alcohols, aldehydes, and acetates gave a pattern of responses indicating an extremely close relationship between trans-11-tetradecen-1-ol acetate and the main component of the insect sex pheromone. Field tests with this compound showed that it was very attractive to males and not to females (953a).

## COLEOPTERA

### Acanthoscelides obtectus

A substance presumed to be a sex attractant has been isolated and identified by Horler (560) from a hexane wash of the male beetles. The wash from 57,000 males was chromatographed on a column of silica gel and the benzene eluate was passed through a column of acid-washed alumina. The benzene eluate was chromatographed on a column of silica gel, and then on a column of Sephadex LH20. The ester obtained was identified by hydrogenation, hydrolysis, permanganate-periodate oxidation, infrared and nuclear magnetic resonance spectroscopy as (−)-methyl *trans*-2,4,5-tetradecatrienoate [26]. It is believed to be the first example of an allenic compound occurring in the animal kingdom.

$$CH_3(CH_2)_7\overset{H}{C}=C=\overset{H}{CH}C=CCO_2CH_3$$
$$\underset{H}{}$$

[26]

The compound, produced in amounts of 10–20 μg/male, does not survive gas chromatography on Apiezon L or neopentyl glycol adipate columns, probably isomerizing to methyl 2,4,6-tetradecatrienoate.

*Agriotes ferrugineipennis*

Paper chromatography of crude ether extracts of the female abdomens with ethanol-ammonia (95:5) as solvent gave an active spot at $R_f$ 0.93 (*733*).

*Anthonomus grandis*

Air drawn continuously over males in a cage was passed through charcoal, which was then extracted with chloroform; removal of solvent from the extract gave an active substance with a musty, faintly minty odor (*655*).

Tumlinson *et al.* (*1205, 1209*) studied various techniques for collecting the attractant. Condensate collected by aerating live males was passed through a column of Carbowax 20M-coated silica gel, and the latter was eluted with 10% ether in pentane to give active material. It was less active than extracts prepared from males or their feces with methylene chloride and then passing through Florisil (elution with 5% ether in pentane), Carbowax 20M-coated silica gel (elution with 10% ether in pentane), or steam-distilling.

The procedure used to collect and isolate the attractant has been described by Tumlinson *et al.* (*1205–1208*). Males (67,000) and mixed weevils of both sexes (4,500,000) or male fecal material (54.7 kg) were extracted with methylene chloride, and the extracts were concentrated under a vacuum and steam-distilled. The distillate was extracted with the same solvent and chromatographed on a column of Carbowax 20M-coated silica gel, eluting successively with pentane and increasing amounts of ether in pentane. The active fractions, eluted with 10% and 50% ether in pentane, were further chromatographed separately on columns of silica gel impregnated with 25% silver nitrate; in each case, the active material was eluted with pentane-ether (1:1). Gas chromatography of the combined active compounds on Carbowax 4000 and SE 30 resulted in the isolation of 4 pure compounds which together make up the attractant, known as "grandlure."

A combination of micro-ozonolysis, infrared, mass, and nuclear magnetic resonance spectrometry was used to identify each of the components of grandlure (*1206–1208*). They are (+)-*cis*-2-isopropenyl-1-methylcyclo-butaneethanol [27], comprising 0.76 ppm in the feces; *cis*-3,3-dimethyl-$\Delta^{1,\beta}$-cyclohexaneethanol [28], 0.57 ppm; *cis*-3,3-dimethyl-$\Delta^{1,\alpha}$-cyclo-hexaneacetaldehyde [29], 0.06 ppm; and *trans*-3,3-dimethyl-$\Delta^{1,\alpha}$-cyclo-hexaneacetaldehyde [30], 0.06 ppm. The laboratory response of females to males was best elicited by a mixture containing 0.99 µg of [27], 0.07 µg of [28], and 0.12 µg each of [29] and [30] (*50, 58*).

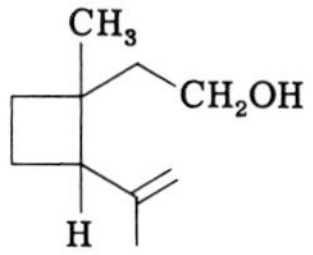

[27] 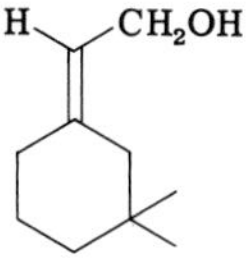          [28]

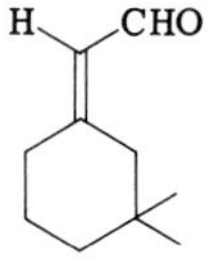

[29] 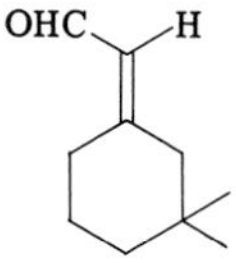          [30]

*Attagenus megatoma*

Attractive material was obtained by rinsing, with ether, filter papers and flasks that had been exposed to virgin females, or by passing a stream of air through containers of females (*35, 224*). The attractant was obtained in pure form by Silverstein *et al.* (*1104*), who extracted 30,000 virgin females with benzene, distilled the extract in a short-path still at 100°C and 0.01 mm pressure, and extracted with ice-cold 0.1 $N$ sodium hydroxide an ether solution of the distillate. The acidic fraction (1.7 gm) obtained after acidification with dilute acid was chromatographed on a column of silica gel, and the active fraction (0.8 gm from 17,000 females) was chroma-tographed on an anion-exchange column (AG1-X4, chloride form); active material (0.6 gm) eluted with methanol-water (3:1) did not survive gas chromatography and was therefore converted to the methyl ester with diazomethane. Preparative gas chromatography of the ester on a column (1 m × 4 mm) of 5% SE 30 on Gas Pack F at 150°–200°C (helium flow rate 100 ml/minute) and then on a column (2 m × 4 mm) of 4% Carbowax 20M on Chromosorb G at 160°C (helium flow rate 51 ml/minute) gave pure methyl ester of the attractant (4 mg from 8000 beetles) (1099).

Identification was obtained by infrared, ultraviolet, mass, and nuclear magnetic resonance spectroscopy of the methyl ester. The structure of the attractant is trans-3,cis-5-tetradecadienoic acid [31] (1100, 1104).

$$CH_3(CH_2)_7\overset{H}{C}=\overset{H}{C}-\overset{H}{\underset{H}{C}}=CCH_2CO_2H$$

[31]

*Costelytra zealandica*

Henzell (540) and Henzell and Lowe (541) isolated the attractant by washing the abdomens of 1500 virgin female beetles with ether, subliming the concentrate for 3 hours at 30°C and 0.0001 mm pressure, and either chromatographing the sublimate on silica gel plates (elution with chloroform or 10% ether in hexane), on Whatman No. 1 filter paper (elution with water, ethanol, or ether-ethanol [1:1]), or on preparative gas chromatographic columns of 2% Atlox G-1292 on 40/70 mesh Silocel. The 130 μg of pure attractant was identified as phenol [32] by mass spectrometry and comparison of chromatographic retention times with those of authentic phenol.

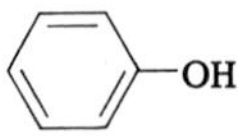

[32]

*Ctenicera destructor*
*Ctenicera sylvatica*

Paper chromatography of crude ether extracts of female abdomens with ethanol-ammonia (95:5) as solvent gave active spots at $R_f$ 0.86 and 0.88, respectively (733).

*Dendroctonus brevicomis*

In 1967, Renwick (921) isolated, by gas chromatography of an ether extract of female hindguts, a compound identified as trans-verbenol [33]. He also isolated, in like manner, verbenone [34] from an ether extract of male hindguts. Both compounds were identified by means of mass, ultraviolet, infrared, and nuclear magnetic resonance spectrometry, but their function in the insect was not understood.

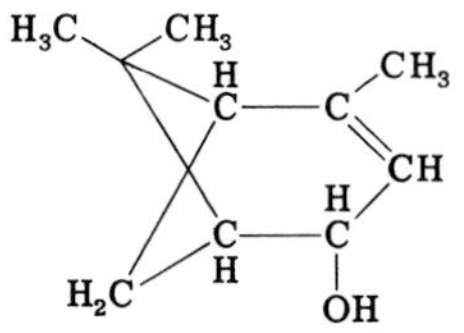

[33]

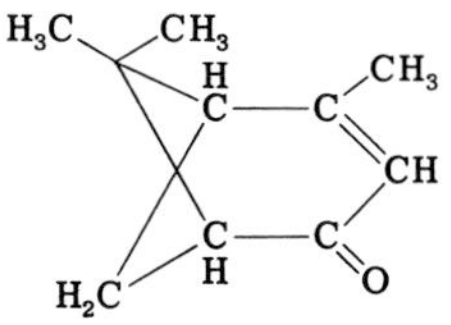

[34]

In 1968, Silverstein *et al.* (*1101*) isolated the sex aggregant produced by virgin females boring into ponderosa pine by extracting 1.6 kg of frass with benzene, under nitrogen, at 70°C, concentrating the filtrate, and distilling the concentrate in a short-path still at 100°C and 0.03 mm pressure. Chromatography of the distillate on a silica gel column followed by preparative gas chromatography on a glass column (91 cm × 11 mm) of 14% SE 30 on 45/60 mesh Chromosorb G at programmed temperature of 100°–200°C (helium flow rate 50 ml/minute) gave 2 mg of pure attractant. A combination of hydrogenolysis and mass spectrometry was used to identify the attractant as *exo*-7-ethyl-5-methyl-6,8-dioxabicyclo[3.2.1]-octane [35], designated "brevicomin" (*1100*).

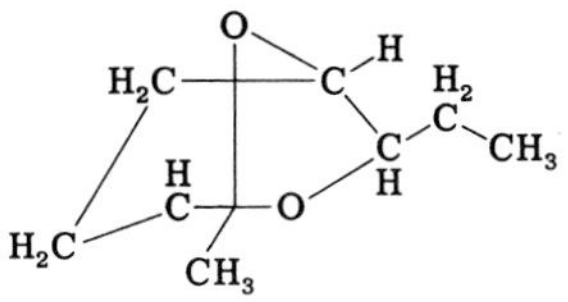

[35]

Brevicomin was subsequently isolated by Pitman *et al.* (*876*) from the hindguts of feeding females. The chemistry has been reviewed by Renwick (*922*).

From the hindgut contents of males, Kinzer *et al.* (*671*) and Renwick and Vité (*924*) isolated a prominent component to which they ascribed no biological significance, but which has been shown to possess biological activity (*1100*). It was identified as 1,5-dimethyl-6,8-dioxabicyclo[3.2.1]-octane [36].

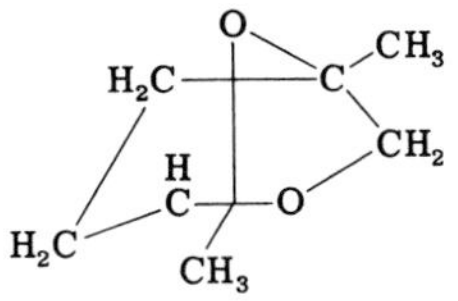

[36]

*Dendroctonus frontalis*

Renwick (*921*) and Pitman *et al.* (*876*) isolated *trans*-verbenol [33] from an ether extract of female hindguts and verbenone [34] from an extract of male hindguts. Verbenone, but not verbenol, has been shown by Vité and Crozier (*1221*) to cause some aggregation of this insect.

In 1968, Renwick and Vité (*924*) obtained, by gas chromatography of the volatile substances from a pentane extract of female hindguts, a compound which caused aggregation of both sexes in the field. Although they determined by mass spectrometry that its molecular weight was 142, they were unable to identify the compound. Identification was successfully accomplished by Kinzer *et al.* (*671*), who had previously isolated it from male *D. brevicomis*, as compound [36]. Despite the fact that Kinzer *et al.* had isolated it from another species and it had not yet been shown to be identical with the compound isolated by Renwick and Vité from *D. frontalis*, these investigators named the compound "frontalin" (*1100, 1291*).

Both frontalin and *trans*-verbenol were subsequently isolated by Coster (*312*) through gas chromatography of the hindguts of reemerged and virgin female *D. frontalis*. Virgin females contained 4.1 times as much frontalin and 1.9 times as much *trans*-verbenol as did reemerged beetles.

The chemistry of frontalin has been reviewed by Renwick (*922*), and Vité and Renwick (*1232*) have proposed the use of comparative gas chromatographic analyses and field bioassays to detect compounds active in bark beetle (*D. brevicomis* and *D. frontalis*) aggregation.

*Dendroctonus ponderosa*

*trans*-Verbenol [33] was isolated from the hindguts of female beetles by extraction with water, extracting the aqueous fractions with ether, and chromatographing the solution on thin layers of silica gel G, using 15% ether in chloroform as developer. The compound appeared as a single spot, $R_f$ 0.45, and was purified by gas chromatography on a column (10 feet $\times$ $\frac{1}{4}$ inch) of 10% Carbowax 20M on 60/80 mesh Chromosorb W at 140°C. Identification was made through infrared, mass, and nuclear magnetic resonance spectroscopy (*875, 1226*). Pitman *et al.* (*876*) obtained indications of the presence of brevicomin [35], as well as *trans*-verbenol, in the hindguts after 18 hours of feeding.

*Dendroctonus pseudotsugae*

Frass produced by virgin females boring into fresh Douglas-fir logs was extracted with benzene, and the concentrated extract was distilled in a short-path still at 100°C and 0.025 mm pressure. The concentrated distillate was chromatographed on a silica gel column to give a highly active fraction that was eluted with a mixture of benzene and ether, but the attractive component was not identified (*185*).

In the same year (1968), Renwick and Vité (*924*) obtained a compound from the volatile substances of female hindguts with a molecular weight of 142, as determined by mass spectrometry. In 1970, Pitman and Vité (*873*) reported the identification of this component as 1,5-dimethyl-6,8-dioxabicyclo[3.2.1]octane [36]; it is therefore identical with frontalin, obtained previously from *D. brevicomis* and *D. frontalis*. The compound was isolated by extracting female hindguts with carbon disulfide and subjecting the extract to preparative gas chromatography; retention times were compared with those of authentic frontalin. Hindguts of emergent females contained more frontalin than females that fed in phloem tissue for 6–48 hours. Camphene, also isolated from the hindguts, synergized the aggregating effect of frontalin.

In 1971, Kinzer *et al.* (*670*) reported the isolation from female hindguts of another component of the pheromone complex, and identified it as 3-methyl-2-cyclohexenone [37]. Female hindguts were extracted with ether, the extract was concentrated, and part of the concentrate was subjected to gas chromatography on a column (10 feet × $\frac{1}{4}$ inch) of 10% Carbowax 20M TPA on 60/80 mesh Chromosorb W at 140°C. The remainder of the concentrate was fractionated by thin-layer chromatography on silica gel, using 10% ether in chloroform as developer, and the active spot ($R_f$ 0.4) was eluted and gas chromatographed. A total of 60 $\mu$g of compound [37] was obtained from 14,500 hindguts.

[37]

A 1:1 mixture of compounds [36] and [37] was more attractive to both sexes than either compound alone.

*Diabrotica balteata*

Hexane extracts of virgin-female abdominal tips were subjected to crystallization in acetone at −10°C; the acetone-soluble fraction was subjected to sweep codistillation in hexane, and the distillate was chromatographed on a column of Florisil. The active material, removed with 7.5% ether in hexane, was gas chromatographed on a column (1.83 m × 0.64 cm) of 5% DC 110 on 60/80 mesh base-washed Chromosorb W at 180°C (helium flow rate 60 ml/minute). The active material (retention time 13–15 minutes) showed a single peak after reinjection on a nonpolar column (SE 30) but it showed 1 major peak and 2 minor peaks on a polar column (diethyleneglycol succinate). The total active material obtained from 37,500 tips was approximately 5 μg; it was unsaponifiable with alkali (not an ester) but the activity was reduced considerably on heating with dilute acid. Activity survived hydrogenation and ozonolysis (no unsaturation), but not treatment with lithium aluminum hydride. These tests and a mass spectrum indicate that the attractant is a carbonyl compound (probably a ketone) with possibly an epoxy group (*1052*).

*Ips confusus*

By means of a vacuum pump, air was drawn (at 20 liters/minute) through a box containing male-infested ponderosa pine log sections; it was then passed through a series of 2 freezing traps, the first at −4°C and the second at −70°C. The thawed contents of both cold traps were highly attractive to the insects in field olfactometers. The attractive substance could be extracted from its aqueous emulsion by the use of petroleum ether and concentrated by distillation of the solvent at 80°C; the colorless oil obtained did not lose activity during storage under refrigeration for several weeks, but it became inactive through prolonged exposure to temperatures above 85°C (*1224*).

Renwick *et al.* (*923*) drew into a gas chromatograph the air from vials holding the hindguts of males that had previously fed on fresh ponderosa pine phloem. A material emerging from a column (5 feet × ⅛ inch) of 5% Apiezon L on Chromosorb W (acid-washed) with a retention time of 6 minutes at 110°C (nitrogen flow rate 25 ml/minute) was highly attractive to walking beetles.

In 1966, Wood *et al.* (*1289*) described methods for mass producing pheromone-laden male frass in kilogram quantities. Individual males were found to produce 9.4 mg of frass per day for about 15 days. Isolation of the attractant was initiated by soaking and pulverizing the frass in warm benzene, removing the solvent, and distilling the concentrate over a short path at 85°–90°C and 0.01 mm pressure. The distillate was chromatographed on a column of silica gel, and the active fraction (effective at $10^{-6}$ gm) eluted with benzene-ether was subjected to preparative gas chromatography on a column (3 feet × ⅜ inch) of 8% SE 30 on Chromosorb G with programming from 80° to 200°C (flow rate 45 ml/minute). The active fraction, 0.5 gm from 2.5 kg of frass, emerged at 14–23 minutes, between 2 other peaks that were identified as nonanal and geranyl acetate (*1102, 1289*). Further chromatography on a polar column (8% Carbowax 20M on Chromosorb G) at 120°C gave 14 mg of material active at $5 \times 10^{-5}$ gm as a number of partially resolved minor peaks. The 3 components jointly responsible for the activity were identified by a combination of infrared, mass, and nuclear magnetic resonance spectrometry as

(−)-2-methyl-6-methylene-7-octen-4-ol [38] (*1107*), *cis*-verbenol [39], and (+)-2-methyl-6-methylene-2,7-octadien-4-ol [40] (*1105*). The methodology for isolation and identification of these components has been reviewed by Silverstein *et al.* (*1099, 1100, 1106*).

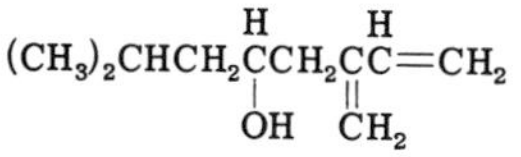

$$(CH_3)_2CHCH_2\underset{\underset{OH}{|}}{C}H CH_2\underset{\underset{CH_2}{||}}{C}C=CH_2$$

[38]

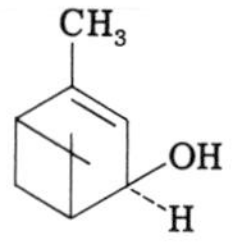

[39]

$$(CH_3)_2C=\underset{\underset{OH}{|}}{C}\overset{H}{C}CH_2\underset{\underset{CH_2}{||}}{C}\overset{H}{C}=CH_2$$

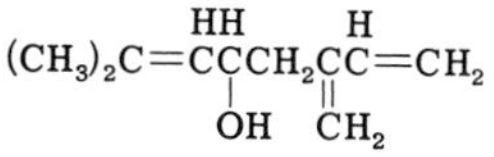

[40]

Also in 1966, Pitman *et al.* (*870*) reported the gas-chromatographic isolation, from the hindguts, malpighian tubes, and excrement of males who had fed for 2 days in small ponderosa pine logs, of an unidentified fraction which elicited a response from walking beetles. Although they believed this material to contain a single pheromone, it was actually a complex mixture of substances (*1100*).

Although Vité (*1219*) denied the validity of using walking beetles to follow the steps in a chemical isolation, as had been done by Wood *et al.* (*1289*), the latter investigators (*1288, 1293*) demonstrated that the chemoklinotaxis exhibited by pedestrian beetles in the laboratory olfactometer reflected the flight response of the insects to the sex pheromone.

*Limonius californicus*

Paper chromatography of a crude ethanol extract of the female abdomens with 0.1 N ammonium hydroxide or ethanol-ammonia (95:5) as solvents gave spots at $R_f$ 0.9 and 0.85 (or 0.82), respectively, that were attractive to males in laboratory tests. The attractant was apparently quite stable at room temperature (*732, 733*).

Jacobson *et al.* (*609*) succeeded in isolating and identifying the attractant. An ether extract (50 mg) prepared from 18 virgin-female abdomens was chromatographed on a column of silicic acid and the active fraction, eluted with 25% ether in hexane, was separated into neutral (14.2 mg) and acidic (13.0 mg) portions by shaking with ice-cold, dilute potassium hydroxide. Only the acidic portion was active, and it was separated into solid and liquid acid fractions by acidifying the combined alkaline fraction at 5°C, filtering off the solid acid (palmitic acid), and extracting the filtrate with ether. The highly active liquid acid fraction (2.4 mg) possessed a strong odor similar to that of valeric acid; it was distilled to give a colorless liquid identified by paper chromatography and gas chromatography of the methyl ester as valeric acid [41]. The large amount (more than 100 $\mu$g) of valeric acid obtained from each female suggests that the pheromone is stored in the female body in bound form, and released in the free form as needed. This hypothesis is supported by the fact that a strong odor of the acid was detected only after the insect fraction was treated with alkali and subsequently acidified.

$$CH_3(CH_2)_3CO_2H$$

[41]

*Limonius canus*

The pheromone was extracted from virgin females with methylene chloride or ether. Shaking the ether solution with dilute sodium hydroxide, followed by acidification of the alkaline solution, showed that the pheromone is an acid, but it has not yet been identified (*833*).

*Popillia japonica*

Although it has been shown that the attractant is produced in female abdomens, extracts prepared from virgin female abdomens with acetone, benzene, ethyl acetate, methylene chloride, and ethanol were all unattractive to males in field traps. However, vessels which had contained attractive females were highly attractive to males for 30 minutes after females were removed (*708, 709*).

*Tenebrio molitor*

Happ and Wheeler (*512*) have collected the pheromone by passing air through desiccators containing thousands of live virgin females, condensing the volatile substances in cold traps, and dissolving them in ether. Although acetone extracts of both males and females elicited male response in the laboratory, female extracts were 5–10 times more potent than male extracts (live females but not live males excited males in test chambers). Preliminary purification (700-fold) of the pheromone was obtained by column chromatography on silicic acid; active material was eluted with 4–6% ether in pentane; thin-layer chromatography showed that the active material obtained was a mixture of substances. Infrared and nuclear magnetic resonance spectra indicated the presence of a long-chain ester, but mass spectra were inconclusive. Attempts at further purification by thin-layer, column, and gas chromatography led to loss of activity.

Tschinkel (*1203, 1204*) obtained from female extracts 2 synergistically active compounds, 1 of which (A) is extractable in pentane, unstable above 60°C, and not recoverable after exposure to chromatography on alumina. The other compound (B) is extractable in more polar solvents and stable to at least 80°C; it is soluble in dilute base, recoverable on acidification of the alkaline solution, inactivated by bromination, and forms a methyl ester with diazomethane. Although compound B thus appears to be an unsaturated acid, neither it nor its methyl ester could be satisfactorily purified by column chromatography on silica gel or Florisil, thin-layer chromatography on silica gel, gel filtration, ion-exchange chromatography, or gas chromatography.

*Trogoderma glabrum*

A pheromone attractive to males was obtained by washing, with ether, absorbent paper discs, filter paper, or glass jars that had been in contact with the virgin females, or by passing a stream of air through flasks containing virgin females (*35, 223, 224*).

*Trogoderma granarium*

Levinson and Bar Ilan (*725, 726*) obtained an assembling pheromone by extracting whole females with ether. The extract was readily soluble in ether or chloroform, less soluble in ethanol, and sparingly soluble in water. The active material was distillable at 80°–90°C under atmospheric pressure and contained unsaponifiable and nonsterol lipids. Body extracts of males were less active as aggregants.

The assembling scent has been found to be a mixture of fatty acid esters by Ikan *et al.* (*568, 570*) and Yinon *et al.* (*1326*) working as a team. An extract of 21,300 virgin females was separated into acidic, basic, and neutral fractions, and the latter was subjected to gas chromatography on a column (2 m × 6 mm) of 15% diethyleneglycol succinate on 80/100 mesh Chromosorb P at 185°C. The esters, identified by comparing their retention times with pure authentic samples and by coinjection, are ethyl oleate [42] 44.2%, ethyl palmitate [43] 34.8%, ethyl linoleate [44] 14.6%, ethyl stearate [45] 6.0%, and methyl oleate [46] 0.4%. The free acids produced by saponification were identified by reversed-phase thin-layer chromatography, infrared spectra, and mass spectra. Each ester is attractive alone, as is the whole mixture.

$$CH_3(CH_2)_7\overset{H}{C}{=}\overset{H}{C}(CH_2)_7CO_2C_2H_5 \qquad CH_3(CH_2)_{14}CO_2C_2H_5$$

$$[42] \qquad\qquad\qquad [43]$$

$$CH_3(CH_2)_4\overset{H}{C}{=}\overset{H}{C}CH_2\overset{H}{C}{=}\overset{H}{C}(CH_2)_7CO_2C_2H_5$$

$$[44]$$

$$CH_3(CH_2)_{16}CO_2C_2H_5 \qquad CH_3(CH_2)_7\overset{H}{C}{=}\overset{H}{C}(CH_2)_7CO_2CH_3$$

$$[45] \qquad\qquad\qquad [46]$$

*Trogoderma inclusum*

A pheromone attractive to males only was obtained by washing, with ether, absorbent paper discs, filter paper, or glass jars that had been in contact with the virgin females, or by passing a stream of air through flasks containing virgin females (*35, 223, 224*).

Two components of the sex attractant were isolated and identified by Rodin *et al.* (*949*). A benzene extract of 250,000 virgin females was freed of solvent and the concentrate was distilled over a short path at 65°C under 0.01 mm pressure. A solution of the distillate was extracted with ice-cold 0.1 *N* sodium hydroxide and chromatographed on a column of silica gel; the ether eluate was fractionated by gas chromatography on a column (0.6 m × 7.5 mm) of 5% SE 30 on 60/80 mesh Gas Chrome Q at 160°–180°C (helium flow rate 50 ml/minute). The fraction eluting between 14 and 27 minutes was further fractionated by gas chromatography on a column (1.7 m × 4 mm) of 5% Carbowax 20M on 60/80 mesh Gas Chrom Q at 110°–200°C (helium flow rate 32 ml/minute). The fraction eluting between 30 and 32 minutes (1 mg from 100,000 females) was identified by hydrogenolysis, ozonolysis, optical rotation, infrared, mass, and nuclear magnetic resonance spectra as (−)-14-methyl-*cis*-8-hexadecen-1-ol [47]. The fraction eluting between 21 and 27 minutes (3 mg from 100,000 females) was purified by further gas chromatography, and the active compound (0.2 mg) was identified by ozonolysis, optical rotation, infrared, mass, and nuclear magnetic resonance spectra as (−)-methyl 14-methyl-*cis*-8-hexadecenoate [48]. Each of the 2 identified compounds is attractive to males.

$$CH_3$$
$$\underset{[47]}{CH_3CH_2\overset{|}{C}H(CH_2)_4\overset{H}{C}=\overset{H}{C}(CH_2)_7OH}$$

$$CH_3$$
$$\underset{[48]}{CH_3CH_2\overset{|}{C}H(CH_2)_4\overset{H}{C}=\overset{H}{C}(CH_2)_6CO_2CH_3}$$

*Trypodendron lineatum*

Castek *et al.* (*287*) succeeded in purifying a volatile substance, attractive to both sexes, produced by feeding, mated females. The material was collected by condensation of air passed through heated frass or by extraction of the boring dust with chloroform, carbon disulfide, acetone, or 95% ethanol, but not with benzene. The volatile substances collected in the condensate were fractionated by gas chromatography on a column (6 feet × $\frac{3}{16}$ inch) of 20% Dow-710 on 60/80 mesh Chromosorb W at 145°C (helium flow rate 70 ml/minute). The active peak, having a retention time of 6 minutes, was shown to be pure by gas chromatography; its mass spectrum was very similar to that of β-ocimene, but authentic β-ocimene was not attractive to either males or females.

The substance in logs responsible for attracting the female beetles initially, so that they may feed and produce the aggregant, has been identified as ethanol by Moeck (794).

## HYMENOPTERA

### Apis mellifera

By means of gas chromatography of ether and ethanol extracts of queen-bee heads, Callow et al. (277) and Shearer et al. (1074) were able to identify 9-oxo-trans-2-decenoic acid (queen substance) [49] as the sex attractant. Queen substance had previously been isolated and identified by Pain et al. (842) and shown to be attractive (258) to drones. It has also been isolated from the heads of queens of A. cerana (253, 1017, 1074), A. dorsata (1017, 1074), and A. florea (253, 1017). Butler (248) has obtained evidence that queen substance, after attracting drones, will act as an aphrodisiac eliciting mounting behavior in drones.

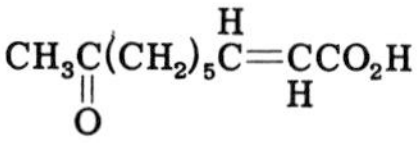

$$CH_3\underset{\underset{O}{\|}}{C}(CH_2)_5\overset{H}{C}{=}\underset{H}{C}CO_2H$$

[49]

### Bombus spp.

Stein (1142, 1143) extracted the heads of male B. terrestris with pentane and subjected the extract to thin-layer chromatography on silica gel with petroleum ether-ether (1:1) as solvent. Spraying the plates with antimony pentachloride solution showed the swarming attractant as a brown spot ($R_f$ 0.45) with an ultraviolet maximum at 260 m$\mu$. On the basis of infrared spectroscopy, the attractant was identified as farnesol [50]. However, Bergström et al. (141) were unable to detect farnesol by mass spectrometry of a distillate from dry, male mandibular glands of this species; they did, however, isolate and identify 2,3-dihydrofarnesol [51] as the main component. Ställberg-Stenhagen and Stenhagen (1131) have concluded that

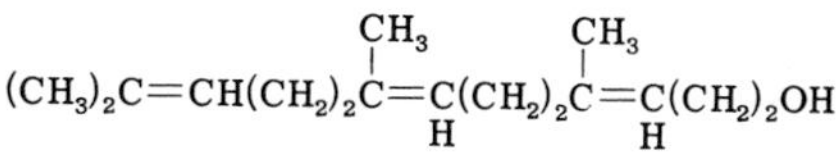

$$(CH_3)_2C{=}CH(CH_2)_2\overset{\overset{CH_3}{|}}{C}{=}\underset{H}{C}(CH_2)_2\overset{\overset{CH_3}{|}}{C}{=}\underset{H}{C}(CH_2)_2OH$$

[50]

$$(CH_3)_2C{=}CH(CH_2)_2\overset{\overset{CH_3}{|}}{C}{=}\underset{H}{C}(CH_2)_2\overset{\overset{CH_3}{|}}{C}H(CH_2)_3OH$$

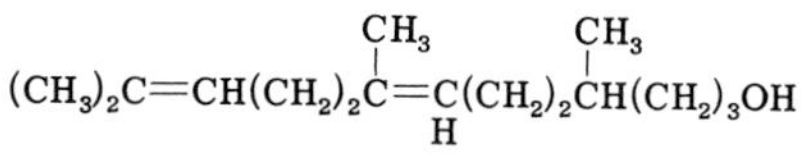

[51]

the chief component of the pheromone is not yet completely characterized, although a mixture of dihydrofarnesols of molecular weight 224 are indicated as the active constituents.

Calam (*266*) has identified the major components in the heads of males of 5 other species of *Bombus* through gas-chromatographic examination of acetone-ether extracts. The compounds identified were *cis*-7-hexadecen-1-ol [52] in *B. agrorum* and *B. lapidarius*, ethyl *cis*-9-tetradecenoate (ethyl myristoleate) [53] in *B. lucorum*, *n*-tricosene, *n*-tricosane, *n*-pentacosene, and *n*-pentacosane in *B. derhamellus*, and a $C_{15}$ terpene alcohol (probably farnesol) in *B. pratorum*.

$$CH_3(CH_2)_7\overset{H}{C}=\overset{H}{C}(CH_2)_6OH \qquad\qquad CH_3(CH_2)_3\overset{H}{C}=\overset{H}{C}(CH_2)_7CO_2C_2H_5$$

[52]                                                    [53]

*Diprion similis*

In 1960, Coppel *et al.* (*310*) reported that an attempt to trap the attractant by passing air rapidly over virgin females and then through various solvents was unsuccessful. The crude attractant was obtained by extracting crushed whole females with acetone or benzene and by rinsing, with ether, glassware that had contained the live or dead females. Ether extracts of filter paper that had been exposed to virgin females were also attractive to males, although the activity of several extracts was masked by unknown materials until these were removed by column chromatography (*286*). Considerable purification of the attractant was obtained on columns of Florisil or silicic acid, but not on alumina. An aliquot of material thus purified, weighing 0.02 μg, attracted males within 30 seconds in the field; within 5 minutes, 500–1000 males were attracted from distances up to 100–200 feet. The attractant appears to be a saturated ester that may contain a free hydroxyl group.

Additional sex attractant was obtained by Jones *et al.* (*627*), who trapped the volatile fraction from virgin females obtained with an air stream. Although amounts collected in this way were small, attractant so collected was less contaminated than that collected from extracts.

*Halictus albipes*
*Halictus calceatus*

The characteristic scent of females of these species is possibly due to a sex pheromone (*32, 700*). Andersson *et al.* (*32*) collected the odorous material by heating the bees with 6% silicone high vacuum grease on Chromosorb W (60/80 mesh) and driving out the adsorbed compounds in a molecular still (enfleurage method). Gas chromatography on polar and nonpolar columns and mass spectrometry of the fractions showed that the odor of females of these species is due to a mixture of the omega lactones of 16-hydroxyhexadecanoic acid (dihydroambrettolide), 18-hydroxyocta-decanoic acid, and an 18-hydroxyoctadecenoic acid (position of double bond not determined).

## DIPTERA

### Ceratitis capitata

Chemical substances produced by males to attract and sexually excite females have been collected, fractionated, and partially characterized (610).

### Cochliomyia hominivorax

An active volatile substance was collected from 24-hour-old virgin males by drawing air through polyethylene-covered cages into a cold trap. Condensate collected from 77,000 males over a period of $5\frac{1}{2}$ months was extracted with ether or hexane (414).

### Dacus tryoni

The oil from the hindgut reservoir of the male was examined by combined gas chromatography-mass spectrometry. The major components were identified as amides derived from isopentylamine, but their separate or combined activity on females did not compare with that of the oil itself. An active fraction was collected from the gas chromatograph (132a).

### Drosophila melanogaster

A lipid found exclusively in the ejaculatory bulb of adult males has been isolated and identified as cis-vaccenyl acetate [19], previously isolated from male Lycorea ceres ceres. The males were extracted with chloroform-methanol (2:1), the extract was chromatographed on a column of silica gel, and the active fraction was eluted with hexane-ether-acetic acid (85:15:1) and subjected to thin-layer chromatography on silica gel. The pure lipid was obtained by preparative gas chromatography and identified by oxidation, infrared, mass, and nuclear magnetic resonance spectroscopy. Although the physiological function of compound 19 has not yet been determined in this insect, its function may involve an aspect of reproduction since it is transferred to females during mating (207).

### Musca domestica

In 1964, Rogoff et al. (969) reported that a male sex excitant could be extracted from female flies with benzene, water, or ethanol. It could be removed from aqueous solutions with benzene, but water did not remove it from benzene. Mayer and Thaggard (762) and Mayer (760) subsequently reported that they had demonstrated an olfactory attractant for males in the feces of both males and females, and that the active material was localized in the nonpolar lipid fraction. In 1971, Mayer and James (761) described the isolation of crude attractant from benzene extracts of virgin females and from benzene or chloroform-methanol extracts of fecally contaminated, gauze cage covers. The neutral lipids obtained from a silicic acid slurry were active. These investigators were of the opinion that the sex pheromone reported by Rogoff *et al.* might not be the same as theirs, and that the insect probably uses both an attractant and an excitant in courtship and mating.

Silhacek *et al.* (*1097a, 1097b*) showed that nonpolar lipids from housefly excrement and cuticular hydrocarbons from the flies attracted sexually mature males. Cuticular hydrocarbons from sexually mature females, both mated and unmated, were attractive, whereas hydrocarbons from unmated males were not. Cuticular hydrocarbons from *Periplaneta americana*, adult lipid extracts from *Attagenus megatoma*, and pupal lipid extracts from *Trichoplusia ni* were all unattractive to male houseflies.

A sex attractant was isolated from the cuticle and feces of sexually mature females by Carlson *et al.* (*69, 283*) in 1971; they used hexane or ether washes of the insect cuticle. The concentrate was chromatographed on a column of silicic acid and the active material, eluted with hexane, was subjected to thin-layer chromatography on silica gel impregnated with silver nitrate. Development of the plates with 1% ether in hexane gave 4 zones, only one of which was active; it appeared to be a long-chain mono-olefin. Column chromatography on silver nitrate-impregnated silica gel gave larger amounts of this fraction, which was subjected to preparative gas chromatography. The pure attractant was identified by a combination of hydrogenation, ozonolysis, mass spectrometry, and nuclear magnetic resonance spectroscopy as *cis*-9-tricosene [54]; it has been designated "muscalure."

$$CH_3(CH_2)_{12}\overset{H}{C}{=}\overset{H}{C}(CH_2)_7CH_3$$

[54]

In addition to being attracted to muscalure, male flies appear to be sexually stimulated by it and attempt to mate.

# CHAPTER XI

# SYNTHESIS OF THE SEX PHEROMONES

The numbers assigned to the sex pheromone structures are those assigned to these compounds (in Arabic numerals) in Chapter X.

*Periplaneta americana*, American cockroach

After the structure 2,2-dimethyl-3-isopropylidenecyclopropyl propionate [1] was proposed for the sex attractant of this insect, numerous attempts to synthesize it were made by several groups. Foremost among these groups was the Pesticide Chemicals Research Branch of the U.S. Department of Agriculture, several of whose members formed the team that isolated the attractant *(605)*. The structure was finally independently synthesized by several groups and found to be unattractive to male American cockroaches; it is therefore not the sex pheromone, whose structure remains to be determined.

Structure 1 was successfully synthesized by Day and Whiting *(335, 336)*, Wakabayashi *(602, 1239)*, and Matsui and Liau *(754)*; attempts by Meinwald *et al.* *(781)*, Chapman *(295)*, and Singh *(1109)* were unsuccessful.

*Lethocerus indicus*, Indian water bug

Synthetic *trans*-2-hexen-1-ol acetate [2] (bp 165°–166°C), identical in all respects with the substance produced by the male insect, may serve to excite the female immediately before or during mating *(242)*, and the corresponding synthetic butyrate, identical with the natural material

209

isolated from the gland liquid ("Meng Da"), may serve the same purpose (*348*).

The acetate has been synthesized by the following procedure (*242*).

$$CH_3(CH_2)_2CHO \ + \ HO_2CCH_2CO_2C_2H_5 \ \xrightarrow{C_5H_5N}$$

$$CH_3(CH_2)_2CH{=}\overset{t}{CH}CO_2C_2H_5 \ \xrightarrow[\text{2. } CH_3COCl]{\text{1. } LiAlH_4} \ [2]$$

*Adoxophyes orana*, lesser tea tortrix

The 2 sex pheromones identified in this insect are *cis*-9-tetradecen-1-ol acetate [24] and *cis*-11-tetradecen-1-ol acetate [8] (*944*). See pages 229–230 and 210–211, respectively, for the procedures used to synthesize these compounds.

A mixture of compounds [8] and [24] is necessary to attract males in the laboratory; the individual compounds are ineffective (*944*).

*Anagasta kühniella*, Mediterranean flour moth

The pheromone is *cis*-9,*trans*-12-tetradecadien-1-ol acetate [7] (*704*). It has been synthesized by Jones (*628*) and Jacobson *et al.* (*611*) (see page 228) as well as by Kuwahara *et al.* (*706*).

*Argyrotaenia velutinana*, red-banded leaf roller

An economical synthesis of this pheromone, *cis*-11-tetradecen-1-ol acetate [8] (riblure), was achieved by means of a novel, one-pot reaction sequence, involving the condensation of 11-bromoundecyl acetate and propionaldehyde by a Wittig reaction, using dimethylformamide as solvent and sodium methoxide as catalyst (*952*). The multiple-step synthesis shown in Scheme XI.1 was also successful in preparing the cis and trans isomers (*951, 952*).

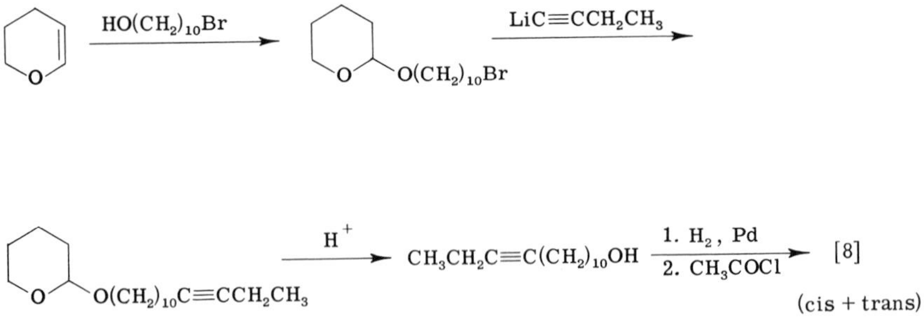

Scheme XI.1

Tamaki *et al.* (*1169a*) synthesized compound [8] using Scheme XI.1a.

$$CH_3(CH_2)_2Br \xrightarrow[\text{2. } OHC(CH_2)_9CO_2Me]{\text{1. } Ph_3P} CH_3CH_2CH{=}CH(CH_2)_9CO_2Me \xrightarrow[\text{2. } Ac_2O]{\text{1. } LiAlH_4} [8]$$

(cis + trans)

Scheme XI. 1a

Mixtures of the cis and trans isomers were separated by thin-layer chromatography on silver nitrate-impregnated silica gel, using benzene-petroleum ether (80:20) as the solvent; the isomers showed $R_f$ values of 0.28 and 0.48, respectively (*952*).

The cis isomer at 0.1 $\mu$g elicited maximum responses from males in the laboratory, whereas the trans isomer was inactive up to 100 $\mu$g (*952*).

Of numerous analogs tested, the activity of [8] was best synergized by undecyl acetate, dodecyl acetate, 10-propoxydecan-1-ol acetate, 11-methoxyundecanol, 11-methoxyundecan-1-ol acetate, 11-(ethylthio)-undecan-1-ol acetate, *cis*-5 dodecen-1-ol acetate, *cis*-7-dodecen-1-ol acetate, and 10-undecen-1-ol acetate. Potent inhibitors of [8] in field traps were *trans*-11-tetradecen-1-ol acetate, *cis*-11-tetradecen-1-ol, *cis*-11-tetradecen-1-ol formate, and *trans*-9-tetradecen-1-ol acetate (*958*).

Field traps baited with compound [8] attracted males of *Choristoneura rosaceana*, *Phalonia* sp., and *Sparganothis sulfureana*, in addition to *A. velutinana*. The corresponding alcohol (*cis*-11-tetradecen-1-ol) attracted *C. fructivittana* Clemens, *Nedra ramosula* Guenée, and *S. groteana* Fernald (*957*).

*Bombyx mori*, silkworm moth

During their investigations into the nature of the sex attractant of the female silkworm moth, Butenandt and his co-workers (*229*) had ascertained the fact that the attractant was an alcohol containing 2 conjugated double bonds. As a consequence, they bioassayed several hundred synthetic substances for attractiveness, including unsaturated aliphatic and alicyclic alcohols (*229, 528*). It had already been shown (*229*) that a steam distillate of mulberry leaves, the natural food of silkworm larvae known to contain 2-hexenal, evoked a weak sexual response in the adult male following its reduction to 2-hexenol with lithium aluminum hydride. A synthetic sample of 2-hexenol was attractive at 4 mg/ml. Increasing the carbon chain and the number of double bonds increased attractiveness, 2,4-hexadienol (sorbyl alcohol) and 2,4,6-octatrienol being active at 0.1 and 0.001 mg/ml, respectively (*229*); cycloheptanone was active at 0.1 mg/ml. However, these three compounds caused wing flutter in females as well as males (*528*). The four possible geometric (cis-trans) isomers of sorbyl alcohol were synthesized (*240*) and found to show no observable difference in attractiveness (*528*).

The natural sex attractant, identified in 1959 (*234, 530*) as 10,12-hexadecadien-1-ol, was at first thought to have the *cis*-10,*trans*-12 configuration, but this became highly improbable when this isomer was prepared in the same year (*243*) and showed a minimum activity level of $10^{-2}$ μg/ml as contrasted with $10^{-10}$ μg/ml for the natural product. Nevertheless, the natural and synthetic preparations showed the same physical constants. Mixtures of other cis, trans isomers were also prepared synthetically (*243*).

Subsequent synthesis of the pure geometric isomers of [9] showed that the natural isomer (bombykol) possessed the *trans*-10,*cis*-12 form. These pure isomers were prepared simultaneously through independent procedures by teams of chemists at the Max-Planck-Institute of Biochemistry in Munich (*237–239, 531, 693*) and at Farbenfabriken Bayer AG in Leverkusen (*1198*). The comparative attractancy of bombykol and its synthetic isomers to male silkworm moths, as reported by these teams, is shown in Table XI.1. There seems to be no doubt from these results that the natural attractant is mainly the *trans*-10,*cis*-12 form, although the fact that its potency is 1000 times lower than that of the synthetically prepared material having this structure casts doubt on the absolute purity of the natural form. Investigators from both teams combined talents to prepare and patent (*1200, 1201*) the following compounds, whose attractiveness (μg/ml) to males is given: *trans*-10,*cis*-12-tetradecadien-1-ol, 100; *trans*-10,*cis*-12-hexadecadien-1-ol,  $10^{-13}$;  *trans*-10,*cis*-12-octadecadien-1-ol, $10^{-6}$.

Munich synthesis of bombykol [9] is shown in Scheme XI.2.

$$CH_3(CH_2)_2Br \xrightarrow{NaC\equiv CH} CH_3(CH_2)_2C\equiv CH \xrightarrow[HCHO]{Mg} CH_3(CH_2)_2C\equiv CCH_2OH \xrightarrow{PBr_3}$$

$$CH_3(CH_2)_2C\equiv CCH_2Br \xrightarrow{Ph_3P} CH_3(CH_2)_2C\equiv CCH_2P^+Ph_3Br^- \xrightarrow[EtONa/EtOH]{OHC(CH_2)_8CO_2Et}$$

$$CH_3(CH_2)_2C\equiv CCH=CH(CH_2)_8CO_2Et \xrightarrow{urea} CH_3(CH_2)_2C\equiv CCH\overset{t}{=}CH(CH_2)_8CO_2Et \xrightarrow{H_2,\ Pd}$$
(cis + trans)

$$CH_3(CH_2)_2CH\overset{c}{=}CHCH\overset{t}{=}CH(CH_2)_8CO_2Et \xrightarrow{LiAlH_4} [9]$$

Scheme XI.2

Leverkusen synthesis of bombykol is shown in Scheme XI.3. The product showed bp 130°–133°C (0.005 mm), $n_D^{20}$ 1.4835.

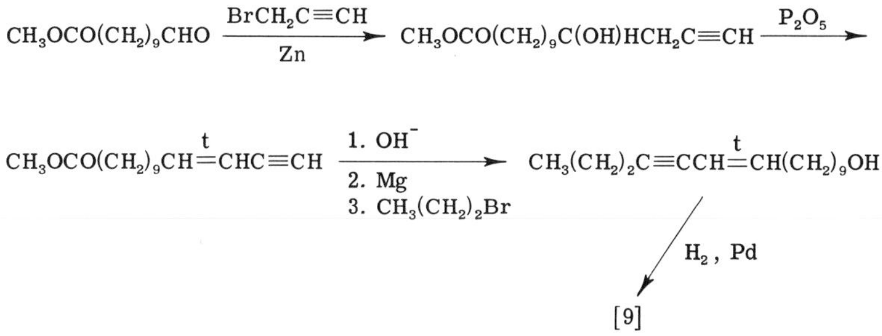

$$CH_3OCO(CH_2)_9CHO \xrightarrow[\text{Zn}]{\text{BrCH}_2C\equiv CH} CH_3OCO(CH_2)_9C(OH)HCH_2C\equiv CH \xrightarrow{P_2O_5}$$

$$CH_3OCO(CH_2)_9\overset{t}{CH}=CHC\equiv CH \xrightarrow[\substack{2.\ Mg \\ 3.\ CH_3(CH_2)_2Br}]{1.\ OH^-} CH_3(CH_2)_2C\equiv CCH\overset{t}{=}CH(CH_2)_9OH$$

$$\Big\downarrow H_2,\ Pd$$

[9]

Scheme XI.3

By reducing *trans*-10-hexadecen-12-yn-1-ol with tritium, Kasang (*651*) obtained labeled bombykol, useful in studying the olfactory properties of bombykol.

Butenandt *et al.* (*235*) and Hoffmann-LaRoche (*553*) have patented the use, alone or mixed with insecticide, of the following unsaturated alcohols as insect attractants: 6,8-tetradecadien-1-ol, 5,7-tetradecadien-1-ol, 6,8-pentadecadien-1-ol, 10,12-pentadecadien-1-ol, 9,11-hexadecadien-1-ol, 10,12-hexadecadien-1-ol, 11,13-hexadecadien-1-ol.

Guex *et al.* (*493*) have patented the preparation of the cis,cis forms of 5,8-tetradecadien-1-ol, 10,13-hexadecadien-1-ol, and 9,12-hexadecadien-1-ol. Alkaline isomerization of these compounds caused conjugation to give mixtures of *trans*-6,*cis*-8- and *cis*-5,*trans*-7-tetradecadien-1-ol, *cis*-10,*trans*-12- and *trans*-11,*cis*-13-hexadecadien-1-ol, and *cis*-9,*trans*-11- and *trans*-10,*cis*-12-hexadecadien-1-ol, respectively, as colorless oils useful as insect attractants.

Truscheit and Eiter (*1199*) have patented the preparation of the following alcohols useful as insect attractants: *trans*-10,*trans*-12-tetradecadien-1-ol, *cis*-10,*trans*-12-tetradecadien-1-ol, 10,12,14-hexadecatrien-1-ol, 11-(2,6,6-

TABLE XI.1

COMPARATIVE ATTRACTANCY OF BOMBYKOL AND ITS
SYNTHETIC ISOMERS TO MALE *Bombyx mori*

| Isomer | Attractancy ($\mu$g/ml) | | |
|---|---|---|---|
| | Ref. *1198* | Ref. *237* | Ref. *239* |
| *cis-10,cis-12* | — | 1 | 1 |
| *cis-10,trans-12* | $10^{-5}$ | $10^{-3}$ | $10^{-2}$ |
| *trans-10,cis-12* | $10^{-13}$ | $10^{-12}$ | $10^{-12}$ |
| *trans*-10,*trans*-12 | 100 | 10 | 100 |
| Bombykol (natural) | $10^{-10}$ | $10^{-10}$ | $10^{-10}$ |

trimethyl-1-cyclohexen-1-yl)-10-undecen-1-ol,      13,17-dimethyl-10,12,16-octadecatrien-1-ol,    and    a    mixture    of    10,12-tetradecadien-1-ol    and 11-methyl-10,12-tridecadien-1-ol. Farbenfabriken Bayer AG (393) has patented the preparation of various 10,12-hexadecadiene derivatives.

Butenandt et al. (241) have reported that 1,10-octadecanediol (prepared by lithium aluminum hydride reduction of 10-ketostearic acid), a mixture of palmityl and stearyl alcohols, and oleyl, linoleyl, and linolenyl alcohols (prepared by reduction of the corresponding acids) are all unattractive to male Bombyx, showing an attractancy level of at least 10 μg/ml.

It is of interest to note that a mixture of the trans, trans and cis, trans forms of 10,12-hexadecadien-1-ol is reported (897) to show strong anti-catabolic activity in rats. Oral administration of the trans, cis and especially the cis,trans forms caused a weight increase in rats kept on a protein-deficient diet. The cis,trans form, at 1–1000 μg/kg, showed anticatabolic activity which is only caused by anabolic steroids in much higher doses. No androgenic or estrogenic effect was caused by the isomers at doses above 1 mg/kg. The compound (isomer not given) inhibits ovarian atrophy of old female rats, and it is nontoxic to these animals; the $LD_{50}$ (intra-peritoneal) is 6.5 gm/kg.

The attractive power of 10,12-hexadecadien-1-ol prompted Buchta and Fuchs (216) to synthesize 10,13-diketo- and 7,10,13-triketohexadecanoic acids, 10,13-dihydroxyhexadecanoate, and 1,10,13-hexadecanetriol. None of these compounds were attractive to male Bombyx mori, Porthetria dispar, or Spodoptera eridania in laboratory bioassays (217).

Riemschneider and Kasang (940) synthesized cis-10,cis-12- and trans-10,trans-12-heptadecadien-1-ol, which attracted male silkworm moths at concentrations of 1 and 10 μg/ml, respectively. Attempts to isomerize one of the cis double bonds to the trans form, in order to increase attractiveness, were unsuccessful. The cis,cis isomer lost none of its activity on storage for 2 years as long as it was protected from air and light. 10,12-Docosadien-1,22-diol and other unidentified primary and secondary conjugated diols were unattractive to males.

Kasang (650) and Riemschneider et al. (941) prepared a large number of acetoxy dienols in an effort to obtain compounds attractive to both B. mori and Porthetria dispar. The compounds were tested electrophysiologi-cally on male P. dispar and behaviorally on B. mori, Antheraea pernyi, Drepana binaria Hufn., Euproctis chrysorrhoea, Lasiocampa trifolii Esp., Notodonta dromedarius L., Orgyia antiqua, and Phoesia tremula Cl. 13-Acetoxy-cis-8,cis-10-heptadecadien-1-ol was reported to elicit an EAG in male gypsy moths at a concentration of $10^{-9}$ μg/ml (941). However, Jacobson et al. (615) repeated the synthesis of this compound and found it to be a mixture of cis and trans isomers showing no activity on males electrophysiologically, behaviorally, or in the field.

*Cadra cautella*, almond moth

The pheromone, *cis*-9,*trans*-12-tetradecadien-1-ol acetate [7] was synthesized by Kuwahara *et al.* (*706*), who also synthesized the other 3 isomers of this structure. The procedure used for the syntheses was not described.

Compound [7] is also the sex pheromone of *Anagasta kühniella* (*704*), *Ephestia elutella* (*204*), *Plodia interpunctella* (*63, 206, 706*), and *Spodoptera eridania* (*611, 628*). See scheme XI.19, p. 228 used by Jones to synthesize [7].

Brady *et al.* (*206*) have reported that an unidentified compound isolated from female abdominal tips of *C. cautella* elicits no behavioral response in males but serves to synergize compound [7] in this insect.

A series of 36 synthetic acetate analogs of compound [7] was tested for sexual excitation of male *C. cautella* in the laboratory (*1164*). Strong activity was shown by *cis*-7-tridecen-1-ol acetate and *cis*-9-tetradecen-1-ol acetate [24]; lesser activity was shown by *cis*-8-tetradecen-1-ol acetate, *cis*-8-dodecen-1-ol acetate, *cis*-7-dodecen-1-ol acetate, *cis*-9-tridecen-1-ol acetate, and *cis*-9-pentadecen-1-ol acetate.

*Choristoneura fumiferana*, spruce budworm

*trans*-11-Tetradecenal [9a], the sex pheromone, was synthesized from 11-tetradecyn-1-ol, an intermediate in the preparation of *cis*-11-tetradecen-1-ol acetate (compound [8]) (see page 210) (*1251a*).

$$CH_3CH_2C{\equiv}C(CH_2)_{10}OH \xrightarrow[NH_3]{Na} CH_3CH_2CH{=}^{t}CH(CH_2)_{10}OH \xrightarrow[2.\ DMSO;\ 160°]{1.\ CH_3SO_2Cl} [9a]$$

Findlay and Macdonald (*403*) found that traps baited with commercial, technical-grade palmitic acid captured small numbers of males in the field. The active component (about 0.000013% of the preparation) was isolated and identified by Findlay *et al.* (*404*) as *n*-octadecanenitrile. A few traps baited with 1 mg of this compound were able to compete successfully in the field with traps containing a live virgin female. However, Findlay (*401a*) subsequently found that *n*-octadecanenitrile is not an efficient attractant for this insect. Oleic, linoleic, and palmitic acids were not attractive in the field to males of this species (*403*).

*Choristoneura rosaceana*, oblique-banded leaf roller

The pheromone of this species was identified as *cis*-11-tetradecen-1-ol acetate [8] (*57*), which has been synthesized by Roelofs (*951, 952*) (see page 210).

Strangely, although [8] is a good attractant for males of this species,

compounds which synergized [8] for male *Argyrotaenia velutinana* acted as
its inhibitors for this insect. *trans*-11-Tetradecen-1-ol acetate, which was an
excellent inhibitor for the attractant with *A. velutinana*, was also an
inhibitor with *C. rosaceana* (*958, 966*).

*Cryptophlebia leucotreta*, false codling moth

*trans*-7-Dodecen-1-ol acetate [10], the sex pheromone, has been synthesized by the following 2 schemes [XI.4, Berger and Canerday (*137*);
XI.5, Henderson and Warren (*533*)].

Henderson and Warren (*533*) also synthesized the propionate and
butyrate esters by acylation with the corresponding acid chlorides, respectively. The boiling points shown by the esters are as follows: acetate,
83°–84°C (0.1 mm); propionate, 113°–115°C (2 mm); butyrate, 117°–120°C
(2 mm). The acetate was the only one that stimulated male behavioral

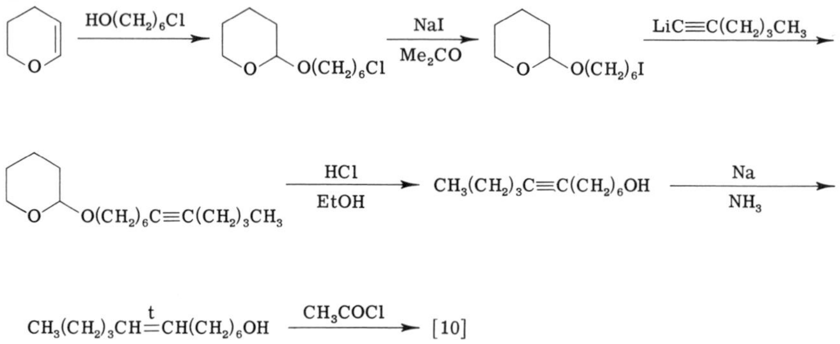

Scheme XI.4

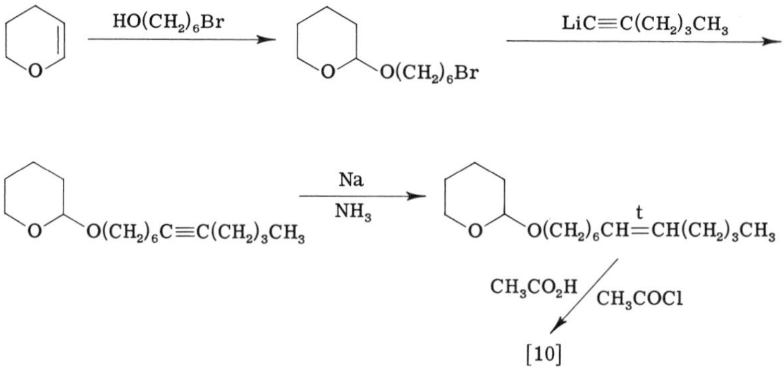

Scheme XI.5

response; the others stimulated some flight response. No behavioral
response was stimulated in males by synthetic *cis*-7-dodecen-1-ol acetate,

7-dodecyn-1-ol and its acetate, or *trans*-7-dodecen-1-ol [Scheme XI.6, Green *et al.* (*480*)]. The product showed bp 78°–82°C (0.05 mm), $n_D^{25}$ 1.4410.

*Danaus gilippus berenice*, queen butterfly

Of the 2 major components of the hair-pencil secretion of males of this species, 2,3-dihydro-7-methyl-1*H*-pyrrolizin-1-one [11] was synthesized in 1966 by Meinwald and Meinwald (*778*) following its isolation from the secretion of male *Lycorea ceres ceres* (see page 220). The other component, 3,7-dimethyl-*trans*-2,*trans*-6-decadien-1,10-diol [12], was synthesized in 1969 by Meinwald *et al.* (*779*) from *trans*-farnesol according to Scheme XI.7.

*Danaus plexippus*, monarch butterfly

The synthesis of 3,7-dimethyl-10-hydroxy-*trans*-2,*trans*-6-decadienoic acid [13] (*776*) is shown in Scheme XI.8.

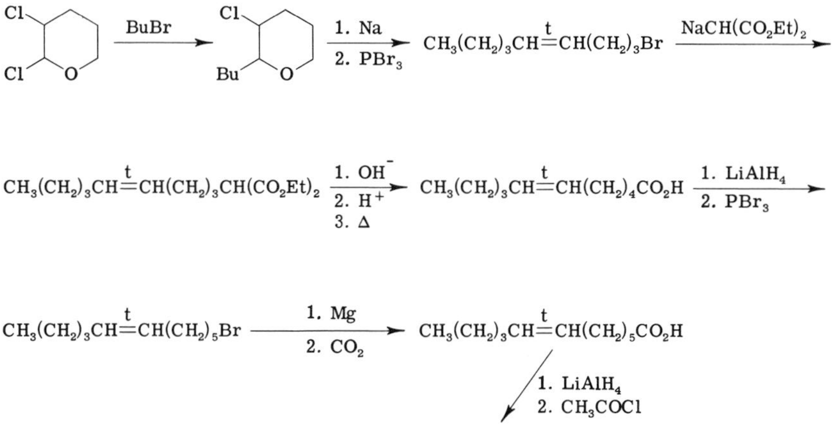

Scheme XI.6

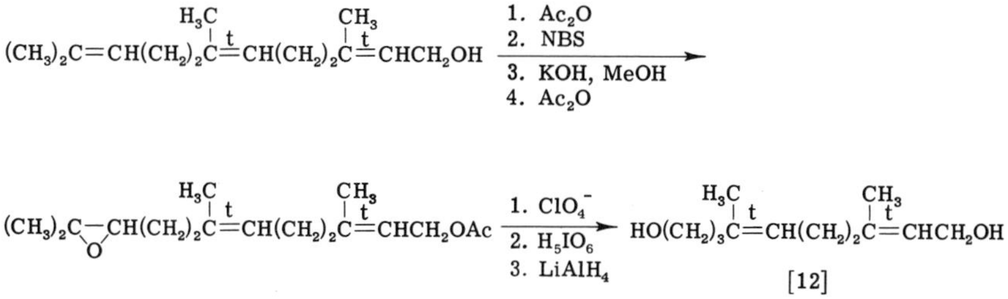

Scheme XI.7

*Diatraea saccharalis*, sugarcane borer

Although he was unable to identify the sex pheromone of this insect, Tribble (*1197*) determined that it was an unsaturated acetate. He synthesized and tested, by laboratory bioassay, the following acetates: trans-2-decen-1-ol, *trans*-2-dodecen-1-ol, *cis*-9-tetradecen-1-ol, *cis*-9-hexadecen-1-ol, *cis*-9-octadecen-1-ol, *cis*-9,*cis*-12-octadecadien-1-ol, decyl, undecyl, dodecyl, tetradecyl, hexadecyl, and octadecyl. None of these compounds was active on males.

*Ephestia elutella*, tobacco moth

The major component of the sex pheromone, *cis*-9,*trans*-12-tetradecadien-1-ol acetate [7], has been synthesized (see page 228). It is also the sex pheromone of *Anagasta kühniella* (*704*), *Cadra cautella* (*206, 706*),

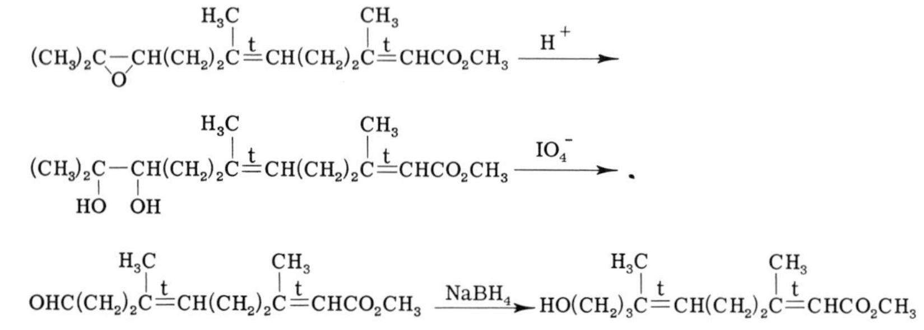

[13]

Scheme XI.8

*Plodia interpunctella* (*63, 206, 706*), and *Spodoptera eridania* (*611, 628*). Evidence indicates that female *E. elutella* also produce a second, as yet unidentified, component, since compound [7] alone is only a weak attractant for males whereas crude female extracts attract males strongly (*1207*).

*Grapholithia molesta*, oriental fruit moth

Roelofs et al. (*960*) reported synthesizing this pheromone (*cis*-8-dodecen-1-ol acetate) [15] by 2 independent routes which were not described. The compound is also attractive to male *G. prunivora* Walsh, and the corresponding trans compound is attractive to male *G. packardi* Zeller (*957*). Male *G. prunivora* are not attracted to *G. molesta* females, however, suggesting the possible role of secondary chemicals (*960*). Large numbers of male *Hemiarcha thermochroa* (Low), another lepidoptera, were attracted to field traps baited with compound [15] (*991*).

*Heliothis zea*, corn earworm

Although the natural sex pheromones of this species have not yet been identified, a number of synthetic model compounds with gas chromatographic properties close to those of the natural materials were tested by laboratory bioassay (*768*). The minimum concentrations (gm/ml) necessary to elicit male response were *cis*-5-tetradecen-1-ol acetate, $10^{-6}$; *trans*-5-tetradecen-1-ol acetate, $\leq 10^{-8}$; *cis*-7-tetradecen-1-ol acetate, $10^{-8}$; *trans*-7-tetradecen-1-ol acetate, $10^{-18}$; *cis*-9-tetradecen-1-ol acetate, $10^{-15}$; and *trans*-9-tetradecen-1-ol acetate, $\leq 10^{-3}$. *cis*-11-Tetradecen-1-ol acetate was inactive at $10^{-3}$ gm/ml.

*Laspeyresia pomonella*, codling moth

The compound *trans*-8,*trans*-10-dodecadien-1-ol [17], reported to be the natural sex attractant of this insect by Roelofs (*62*), was synthesized by Roelofs and co-workers (*959*) by the procedure shown in Scheme XI.9.

Isomerization of the resulting mixed trans,trans (75%) and cis,trans

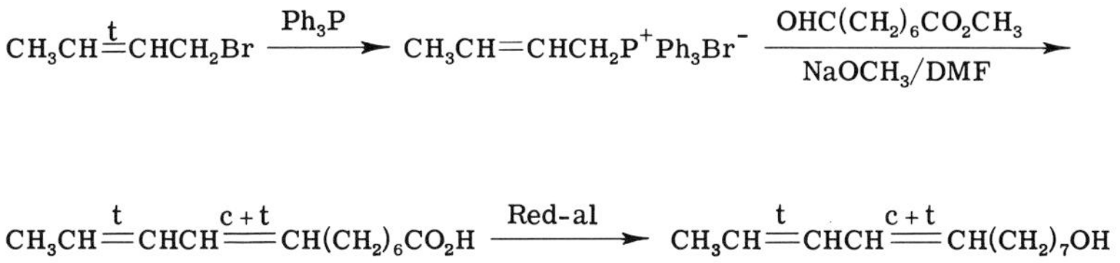

$$CH_3CH\overset{t}{=}CHCH_2Br \xrightarrow{Ph_3P} CH_3CH=CHCH_2P^+Ph_3Br^- \xrightarrow[\text{NaOCH}_3/\text{DMF}]{OHC(CH_2)_6CO_2CH_3}$$

$$CH_3CH\overset{t}{=}CHCH\overset{c+t}{=}CH(CH_2)_6CO_2H \xrightarrow{\text{Red-al}} CH_3CH\overset{t}{=}CHCH\overset{c+t}{=}CH(CH_2)_7OH$$

Scheme XI.9

(25%) isomers with iodine and sunlight gave the all-trans compound [17]. Although this compound is an effective attractant for male codling moths in the field (1 μg was attractive for more than a month), Roelofs *et al.* (*959*) are not positive that it is the natural sex pheromone of this insect.

The trans,cis and cis,cis isomers were also synthesized but were unattractive. The 75:25 mixture of trans,trans and cis,trans isomers was quite attractive (*959*), so the latter isomer apparently is not an inhibitor.

Butt *et al.* (*261*) tested a large series of synthetic aldehydes and nitriles in the laboratory and field with male codling moths. Compounds excitatory in the laboratory were citral and 19 nitriles, but only 5,7,7-trimethyl-2-octenenitrile, heptanenitrile, and phenoxyacetonitrile attracted small numbers of males to field traps.

During studies of the chemical structure of the natural sex attractant, McDonough *et al.* (*766*) obtained evidence indicating that it might contain an epoxy group. They therefore synthesized a series of 21 cis and trans epoxides and bioassayed them for activity in the laboratory. Although many of these compounds elicited excitation in males, the most potent was *cis*-9,10-epoxyoctadecan-1-ol, which evoked a response at $10^{-9}$ gm/ml. None of these epoxides was attractive to males when tested in the field at various concentrations (*766a*).

2,3-Dihydro-7-methyl-1*H*-pyrrolizin-1-one

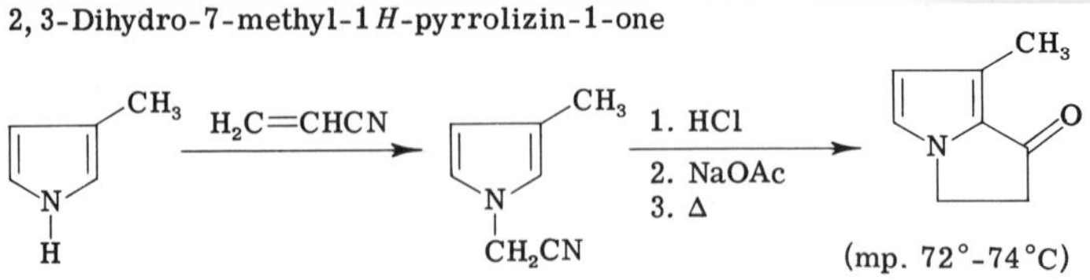

$$\text{(pyrrole with CH}_3\text{)} \xrightarrow{H_2C=CHCN} \text{(N-CH}_2\text{CN pyrrole with CH}_3\text{)} \xrightarrow[\substack{\text{2. NaOAc}\\\text{3. }\Delta}]{\text{1. HCl}} \text{(bicyclic ketone with CH}_3\text{)}$$

(mp. 72°-74 °C)

[11]

Hexadecyl acetate

$$CH_3(CH_2)_{15}OH \xrightarrow{Ac_2O} CH_3(CH_2)_{15}OCOCH_3$$

[18]

*cis* -11-Octadecen-1-ol (vaccenyl) acetate

$$CH_3(CH_2)_5CH\overset{c}{=}CH(CH_2)_9CO_2CH_3 \xrightarrow[\text{2. Ac}_2O]{\text{1. LiAlH}_4} CH_3(CH_2)_5CH\overset{c}{=}CH(CH_2)_{10}OCOCH_3$$

[19]

Scheme XI.10

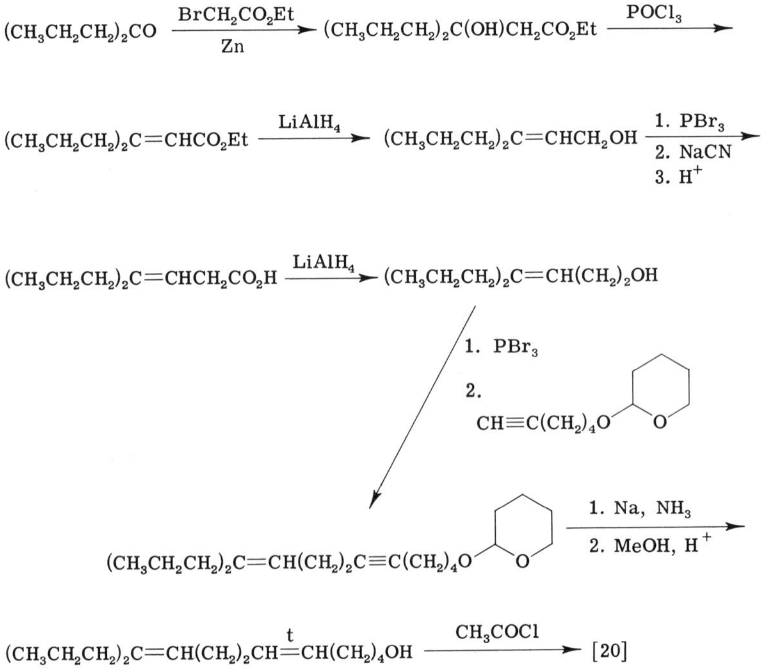

$$(CH_3CH_2CH_2)_2CO \xrightarrow[\text{Zn}]{BrCH_2CO_2Et} (CH_3CH_2CH_2)_2C(OH)CH_2CO_2Et \xrightarrow{POCl_3}$$

$$(CH_3CH_2CH_2)_2C=CHCO_2Et \xrightarrow{LiAlH_4} (CH_3CH_2CH_2)_2C=CHCH_2OH \xrightarrow[\substack{\text{2. NaCN}\\\text{3. H}^+}]{\text{1. PBr}_3}$$

$$(CH_3CH_2CH_2)_2C=CHCH_2CO_2H \xrightarrow{LiAlH_4} (CH_3CH_2CH_2)_2C=CH(CH_2)_2OH$$

1. PBr₃

2.
$$CH{\equiv}C(CH_2)_4O\text{(tetrahydropyranyl)}$$

$$(CH_3CH_2CH_2)_2C=CH(CH_2)_2C{\equiv}C(CH_2)_4O\text{(tetrahydropyranyl)} \xrightarrow[\text{2. MeOH, H}^+]{\text{1. Na, NH}_3}$$

$$(CH_3CH_2CH_2)_2C=CH(CH_2)_2CH\overset{t}{=}CH(CH_2)_4OH \xrightarrow{CH_3COCl} \text{[20]}$$

Scheme XI.11

*Lycorea ceres ceres*

The 3 components of the male hair-pencils of this species were synthesized in 1966 by Meinwald *et al.* (*778, 780*) (Scheme XI.10).

*Ostrinia nubilalis*, European corn borer

This pheromone, *cis*-11-tetradecen-1-ol acetate [8], is identical with that of the red-banded leaf roller, and was synthesized by Klun and Brindley (*54, 56, 688*) according to the method described by Roelofs (*951, 952*) (see page 210).

Klun and Brindley (*688*) also synthesized *cis*-10-, *cis*-12-, and 13-tetradecen-1-ol acetates by modifications of the above method (*951*). These isomers, as well as the propionate and butyrate of *cis*-11-tetradecen-1-ol, were inactive in laboratory bioassays (*688*).

*Pectinophora gossypiella*, pink bollworm moth

10-Propyl-*trans*-5,9-tridecadien-1-ol acetate (propylure) [20], the sex pheromone of the insect (*37*), has been synthesized by the procedures shown in Schemes XI.11–16 [XI.11, Jones *et al.* (*632*); XI.12, Eiter *et al.* (*376*)]. The product obtained in Scheme XI.11 in 0.2% overall yield showed bp 135°C (0.1 mm), $n_D^{25}$ 1.4635.

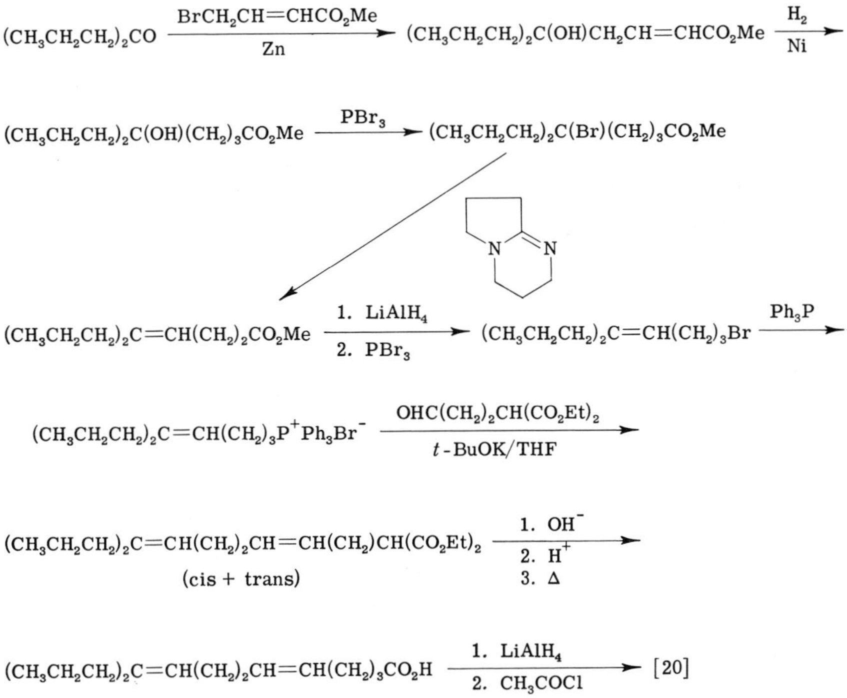

$$(CH_3CH_2CH_2)_2CO \xrightarrow[Zn]{BrCH_2CH=CHCO_2Me} (CH_3CH_2CH_2)_2C(OH)CH_2CH=CHCO_2Me \xrightarrow[Ni]{H_2}$$

$$(CH_3CH_2CH_2)_2C(OH)(CH_2)_3CO_2Me \xrightarrow{PBr_3} (CH_3CH_2CH_2)_2C(Br)(CH_2)_3CO_2Me$$

$$(CH_3CH_2CH_2)_2C=CH(CH_2)_2CO_2Me \xrightarrow[\text{2. } PBr_3]{\text{1. } LiAlH_4} (CH_3CH_2CH_2)_2C=CH(CH_2)_3Br \xrightarrow{Ph_3P}$$

$$(CH_3CH_2CH_2)_2C=CH(CH_2)_3P^+Ph_3Br^- \xrightarrow[t\text{-BuOK/THF}]{OHC(CH_2)_2CH(CO_2Et)_2}$$

$$(CH_3CH_2CH_2)_2C=CH(CH_2)_2CH=CH(CH_2)CH(CO_2Et)_2 \xrightarrow[\substack{\text{2. } H^+ \\ \text{3. } \Delta}]{\text{1. } OH^-}$$
(cis + trans)

$$(CH_3CH_2CH_2)_2C=CH(CH_2)_2CH=CH(CH_2)_3CO_2H \xrightarrow[\text{2. } CH_3COCl]{\text{1. } LiAlH_4} [20]$$

Scheme XI.12

The product obtained by the procedure in Scheme XI.12, bp 90°–100°C (0.05 mm) (376), was also obtained by Jacobson (597) following the same procedure; Jacobson's preparation showed bp 130°–135°C (0.1 mm). Jacobson showed that propylure prepared by this procedure consisted of 60% of the desired trans form and 40% of the cis form. Since as little as 15% of the cis isomer will completely nullify the activity of the trans isomer (597), this accounts for the fact that Eiter's preparation was not attractive to male pink bollworm moths (376).

Scheme XI.13 was devised by Pattenden (848); the product, bp 120°C (0.4 mm), $n_D^{24}$ 1.4610, was a single isomer and was obtained in 18% overall yield. Stoll and Flament (1149) followed the procedure shown in Scheme XI.14; the product was reported to consist of a single isomer (trans). Stowell (1150) obtained a product, bp 120°C (0.1 mm), that contained about 5% of the cis isomer; it was separable by chromatography on silver nitrate-impregnated silica gel, and the overall yield was 7% (Scheme XI.15). Scheme XI.16 [Shamshurin et al. (1067)] gave a product that showed bp 100°–110°C (0.05 mm), $n_D^{20}$ 1.4630 after thin-layer chromatography on alumina.

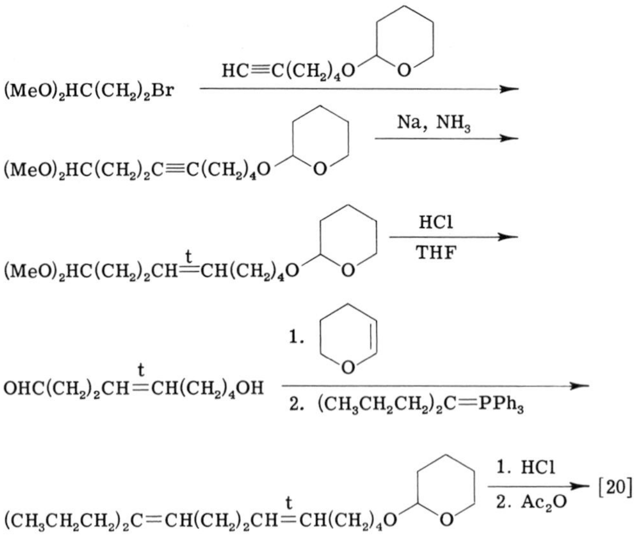

Scheme XI.13

Since propylure alone does not attract male pink bollworm moths in the

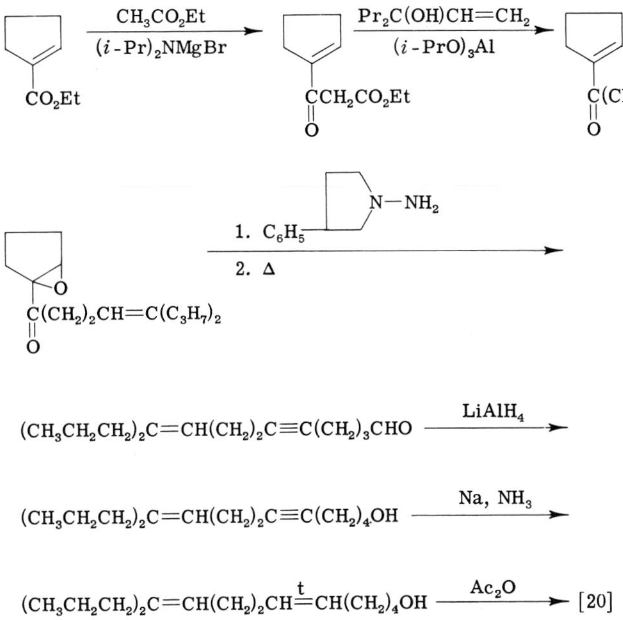

$$\text{(cyclopentene-CO}_2\text{Et)} \xrightarrow[(i\text{-Pr})_2\text{NMgBr}]{\text{CH}_3\text{CO}_2\text{Et}} \text{(cyclopentene-CCH}_2\text{CO}_2\text{Et, O)} \xrightarrow[(i\text{-PrO})_3\text{Al}]{\text{Pr}_2\text{C(OH)CH=CH}_2} \text{(cyclopentene-C(CH}_2)_2\text{CH=C(C}_3\text{H}_7)_2\text{, O)} \xrightarrow{\text{H}_2\text{O}_2}$$

$$\text{(cyclopentane-epoxide-C(CH}_2)_2\text{CH=C(C}_3\text{H}_7)_2\text{, O)} \xrightarrow[\text{2. }\Delta]{\text{1. C}_6\text{H}_5\text{-N-NH}_2}$$

$$(\text{CH}_3\text{CH}_2\text{CH}_2)_2\text{C=CH(CH}_2)_2\text{C}\equiv\text{C(CH}_2)_3\text{CHO} \xrightarrow{\text{LiAlH}_4}$$

$$(\text{CH}_3\text{CH}_2\text{CH}_2)_2\text{C=CH(CH}_2)_2\text{C}\equiv\text{C(CH}_2)_4\text{OH} \xrightarrow{\text{Na, NH}_3}$$

$$(\text{CH}_3\text{CH}_2\text{CH}_2)_2\text{C=CH(CH}_2)_2\text{CH}\overset{t}{=}\text{CH(CH}_2)_4\text{OH} \xrightarrow{\text{Ac}_2\text{O}} [20]$$

Scheme XI.14

field, a number of compounds with related structures were synthesized in an effort to discover one that might be attractive. The following compounds were all unattractive in laboratory bioassays (606): 10-isopropyl-*trans*-5,9-tridecadien-1-ol acetate, 10-isopropyl-12-methyl-*trans*-5,9-tridecadien-1-ol acetate, 12-ethyl-*trans*-7,11-tetradecadien-1-ol acetate, 10-propyl-*trans*-5,9-tridecadien-1-ol, 10-propyl-*trans*-6,9-tridecadien-1-ol, the 4 geometric isomers of 5,9-tridecadien-1-ol acetate (1244), and the acetates of *trans*-3-hexen-1-ol, 7-octen-1-ol, *cis*-7-decen-1-ol, 10-undecen-1-ol, *cis*-3-, *cis*-5-, *trans*-5-, *cis*-6-, *trans*-6-, *cis*-7-, *trans*-7-, *cis*-9-, and *trans*-9-dodecen-1-ol, *cis*-5-, *trans*-5-, *trans*-7-, *cis*-9-, and *trans*-9-tetradecen-1-ol, *trans*-2-, *trans*-7-, *cis*-8-, *trans*-8-, *cis*-9-, and *cis*-11-hexadecen-1-ol, and *cis*-9-

$$(\text{CH}_3\text{CH}_2\text{CH}_2)_2\text{CO} \xrightarrow[\text{OHC(CH}_2)_4\text{OCOCH}_3]{\text{Ph}_3\text{P=CH(CH}_2)_2\text{CH=PPh}_3}$$

$$(\text{CH}_3\text{CH}_2\text{CH}_2)_2\text{C=CH(CH}_2)_2\text{CH=CH(CH}_2)_4\text{OCOCH}_3 \xrightarrow[\Delta]{\text{Se}} [20]$$
$$(\text{cis + trans})$$

Scheme XI.15

$(CH_3CH_2CH_2)_2CO$ $\xrightarrow{(EtO)_2P(O)CH_2CO_2Et}$ $(CH_3CH_2CH_2)_2C{=}CHCO_2Et$ $\xrightarrow[\text{2. PBr}_3]{\text{1. LiAlH}_4}$

$(CH_3CH_2)_2C{=}CHCH_2Br$ $\xrightarrow[\substack{\text{2. OH}^- \\ \text{3. }\Delta}]{\text{1. CH}_2(CO_2Et)_2}$ $(CH_3CH_2CH_2)_2C{=}CH(CH_2)_2CO_2H$ $\xrightarrow[\text{2. PBr}_3]{\text{1. LiAlH}_4}$

$(CH_3CH_2CH_2)_2C{=}CH(CH_2)_3Br$ $\xrightarrow[\text{2. OHC(CH}_2)_4OCOCH_3]{\text{1. Ph}_3P}$ [20]

Scheme XI.16

octadecen-1-ol. Propylure did not excite male *Sitotroga cerealella* in the laboratory (*201*).

Unexpectedly, the acetate of *cis*-7-hexadecen-1-ol proved to be highly attractive both in laboratory and field bioassays with male pink bollworm moths. This compound, named "hexalure," has been synthesized by Green *et al.* (*481*, *482*) according to the method shown in Scheme XI.17, which is being used commercially to prepare material for survey.

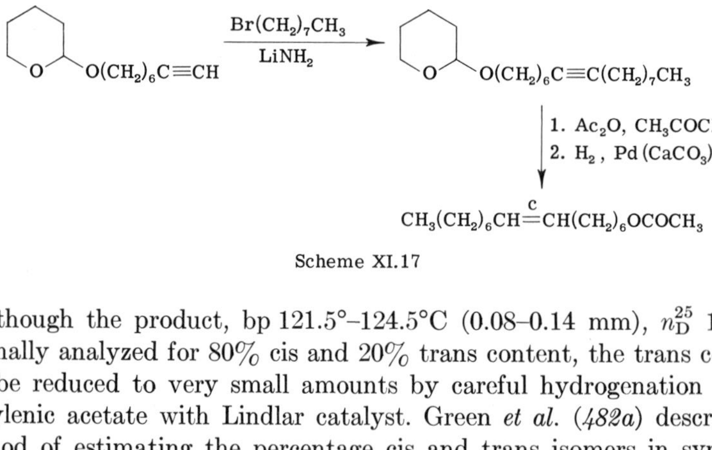

Scheme XI.17

Although the product, bp 121.5°–124.5°C (0.08–0.14 mm), $n_D^{25}$ 1.4484, originally analyzed for 80% cis and 20% trans content, the trans content can be reduced to very small amounts by careful hydrogenation of the acetylenic acetate with Lindlar catalyst. Green *et al.* (*482a*) described a method of estimating the percentage cis and trans isomers in synthetic hexalure from the height of the infrared band at 10.3 μm. The method is also applicable to cis-trans mixtures of other alkenol acetates. The trans isomer of hexalure and a mixture of 54% trans + 46% hexalure were virtually unattractive (*482a*).

Beroza *et al.* (*155*) found that a combination of 25 mg of hexalure with 10 mg of 1-tetradecyl acetate placed in a field trap captured much fewer males than did hexalure alone; at the same concentration, 1-dodecyl acetate, 1-hexadecyl acetate, and 1-octadecyl acetate inhibited the response to hexalure to a lesser extent.

*Plodia interpunctella*, Indian meal moth

The pheromone, *cis*-9,*trans*-12-tetradecadien-1-ol acetate [7], has been synthesized (see page 228). It is also the sex pheromone of *Anagasta kühniella* (704), *Cadra cautella* (206, 706), *Spodoptera eridania* (611, 628), and *Ephestia elutella* (204).

The laboratory response of male *P. interpunctella* to extracts of virgin female abdomens (containing [7]) was strongly inhibited by *cis*-7-tetradecen-1-ol acetate, *cis*-9-tetradecen-1-ol acetate, and *cis*-11-tetradecen-1-ol acetate (202). Lesser inhibition was shown by decyl acetate, 10-undecen-1-ol acetate, dodecyl acetate, tetradecyl acetate, methyl myristate, and tetradecyl aldehyde.

*Porthetria dispar*, gypsy moth

The synthesis of the potent attractant, *cis*-7,8-epoxy-2-methyloctadecane [23], named "disparlure," was carried out by Bierl *et al.* (59, 158) according to Scheme XI.18.

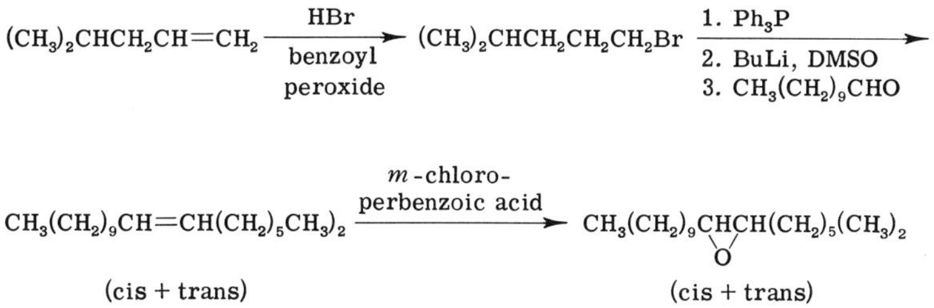

$$(CH_3)_2CHCH_2CH=CH_2 \xrightarrow[\substack{benzoyl \\ peroxide}]{HBr} (CH_3)_2CHCH_2CH_2CH_2Br \xrightarrow[\substack{2.\ BuLi,\ DMSO \\ 3.\ CH_3(CH_2)_9CHO}]{1.\ Ph_3P}$$

$$CH_3(CH_2)_9CH=CH(CH_2)_5CH_3)_2 \xrightarrow[\substack{m\text{-chloro-} \\ perbenzoic\ acid}]{} CH_3(CH_2)_9\underset{O}{CHCH}(CH_2)_5(CH_3)_2$$

(cis + trans)                                    (cis + trans)

Scheme XI.18

The product, consisting of 85% [23] and 15% of the trans epoxide, was separable by column chromatography on silica gel-silver nitrate. A number of epoxides with related structures were also prepared (158), none of which came close to disparlure in attractiveness to male gypsy moths (61).

For a number of years before the isolation and identification of the natural sex attractant, a total of several thousand organic substances were tested in the field for attractiveness to males. One of these, a commercially available formulation consisting of 11% 1,2-epoxytetradecane, 60% 1,2-epoxyhexadecane, 22% 1,2-epoxyoctadecane, 2% 1,2-epoxyeicosane, and 5% hydroxylic compounds (including a small amount of 1,2-hexadecanediol and one of its monoacetates) was particularly attractive, 0.25 gm luring a total of 50 males in a 3-week period during which 12 female abdominal tips lured 94 males (582). This formulation was also attractive to field-collected males in the laboratory, but its activity was uncertain when tested on laboratory-reared moths because their sexual responses were very erratic (166). After fractionation of the formulation, only 1,2-epoxyhexadecane and the hydroxylic fraction were attractive in both laboratory and field tests. Although the natural extract placed in the field began to attract males promptly, the synthetic formulation was attractive only after an induction period of several days. Further investigation showed that while 1,2-epoxyhexadecane is, per se, a fair attractant for the male moth it slowly hydrolyzes with moisture during the induction period to produce 1,2-hexadecanediol, which is strongly attractive. The corresponding $C_{14}$ and $C_{18}$ diols were unattractive. The use of 1,2-epoxyhexadecane (prepared in 60% yield by the peracetic acid oxidation of 1-hexadecene) and 1,2-hexadecanediol (prepared in 95% yield by the performic acid oxidation of 1-hexadecene) as male gypsy moth attractants has been patented (582).

As a result of an intensive chemical investigation of the natural gypsy moth sex attractant, Jacobson et al. (603, 604) had proposed that the structure was d-10-acetoxy-cis-7-hexadecen-1-ol [22] and designated it as "gyptol." They prepared the dl form in an overall yield of 0.2% (10 steps) and resolved it into the d- and l-forms by treating its acid succinate with L-brucine, separating the brucine salts by fractional crystallization from acetone, and saponifying the acid succinates with ethanolic alkali (586). All forms of compound [22] were equally attractive to male gypsy moths when tested in the field, and hydrogenation to give 10-acetoxyhexadecanol caused a drop in activity (167, 604).

Truscheit et al. (1202) patented the preparation of 10-hydroxy-7-hexadecyn-1-oic acid as an intermediate in preparing gyptol.

Characterization of gyptol resulted in the synthesis of a homolog, d-12-acetoxy-cis-9-octadecen-1-ol, which was designated "gyplure" (583, 608). This compound was prepared in high yield from ricinoleyl alcohol (available commercially from the reduction of ricinoleic acid, the major ingredient of castor oil) by acetylating both hydroxyl groups and then selectively saponifying the primary acetyl group with refluxing ethanolic potassium hydroxide. Male gypsy moths were lured to field traps containing as little as $10^{-5}$ μg of gyplure per trap (1135), and the compound was attractive at $10^{-12}$ μg in laboratory bioassay tests. Adlung (12) reported that 25 and 50 μg of the gyplure placed in field traps did not lure males until the traps had been out for 11 days. The preparation and use of gyplure as a gypsy moth attractant were patented (584, 585).

The trans form of gyplure, prepared by elaidinization of the cis isomer with nitrous acid (608, 630), was inactive in both laboratory and field tests. It was shown that a concentration of 20% or more crude *trans*-gyplure in formulations of *cis*-gyplure is sufficient to cause complete inactivation (*587*); the mechanism of this inactivation is not known. The propyl and butyl analogs of *cis*-gyplure were completely devoid of activity in field tests (*608*).

A number of laboratory and commercial samples of gyplure showed little or no activity in laboratory and field tests. By means of column and gas chromatographic methods, developed by Jones and Jacobson (*629*), it was shown that these samples contained only 30–39% *cis*-gyplure, being contaminated with varying amounts of stearyl, oleyl, and ricinoleyl alcohols.

Stefanović *et al.* (*1136, 1139–1141*) prepared gyplure by several methods; the properties of the product agreed closely with those given by Jacobson, but none of the samples was attractive to male gypsy moths in the field at concentrations below $10^{-1}$ to $2 \times 10^{-2}$ gm. Gyplure synthesized by Eiter *et al.* (*376*) was likewise inactive in laboratory bioassays. Gyptol prepared by Stefanović *et al.* (*1140*) and by Eiter *et al.* (*376*) was also reported to be unattractive to males in the field and laboratory, respectively. Pure gyptol and gyplure prepared by Beroza *et al.* (*150a*) were likewise unattractive to males.

Inconsistencies in the attractiveness of synthetic samples of gyptol and gyplure prompted Jacobson *et al.* (*612*) to reinvestigate their physical, chemical, and biological properties. They found that pure gyptol, although produced by the female gypsy moth, is sexually inert toward the male moth, as is pure gyplure. They concluded that a highly active contaminant of synthetic gyptol and gyplure was responsible for their original attractiveness to males.

The evaporation rate of gyplure in the laboratory has been experimentally determined by McGovern *et al.* (*770*). After 75 days, the weight loss was only about 2 mg.

Of 10 insect lures, gyplure, 1,2-epoxyhexadecane, and 1,2-hexadecanediol were the least toxic to the heterothallic fungus, *Choanephora trispora* (*1328*). Sporangia formation of both the plus and minus mating types was stimulated by the diol. Only gyplure showed positive chemotactic activity (attraction) for the hyphae. Accelerated growth was shown for both mating types toward cellulose disks containing gyplure at 0.3, 0.03, and 0.003% concentrations.

Attempting to obtain a compound attractive to both gypsy moths and silkworm moths, Kasang (*650*) and Riemschneider *et al.* (*941*) synthesized a number of acetoxy dienols combining the structural features of gyptol and bombykol. Although they reported that 13-acetoxy-*cis*-8,*cis*-10-heptadecadien-1-ol elicited electroantennograms from male gypsy moths at $10^{-9}$ μg/ml, Jacobson *et al.* *(615)* repeated the published synthesis of this compound and also prepared it by a second method and found the product (a mixture of isomers) to be inactive on males electrophysiologically, behaviorally, and in the field.

Reduction of *d*-9-hydroxy-*cis*-12-octadecenoic acid, isolated from *Strophanthus kombe* seed oil, with lithium aluminum hydride gave *cis*-12-octadecene-1,9-diol, which was acetylated with acetyl chloride to the 1,9-diacetate; saponification with 1 mole of ethanolic potassium hydroxide gave *d*-9-acetoxy-*cis*-12-octadecen-1-ol. None of these compounds elicited a typical sexual response from male gypsy moths when tested in the laboratory *(630)*.

*Porthetria monacha*, nun moth

Trimethylamine, which has an odor like that of the female nun moth, lures no males into field traps. Trials with "muscaro," musk, gum animé, civet paste, ambrette, "bear root," "pest root," and patchouli caught no males in 1 day; trapping of 1 male with civet was assumed to be accidental since 1 female in the same area lured 368 males during the same period *(507)*.

Although gyplure lured male gypsy moths to field traps, 25 and 50 μg of this substance failed to lure male nun moths in an area infested with this insect *(12)*.

*Spodoptera eridania*, southern armyworm

Of the 2 sex pheromones isolated from female southern armyworm moths and named "prodenialure A" and "prodenialure B," the former is identical with *cis*-9-tetradecen-1-ol acetate [24] from female fall armyworm moths *(60)*. See pages 229-230 for the synthesis of this compound.

Prodenialure B, *cis*-9,*trans*-12-tetradecadien-1-ol acetate [7] was synthesized by Jones *(628)* and Jacobson *et al.* *(611)* according to Scheme XI.19.

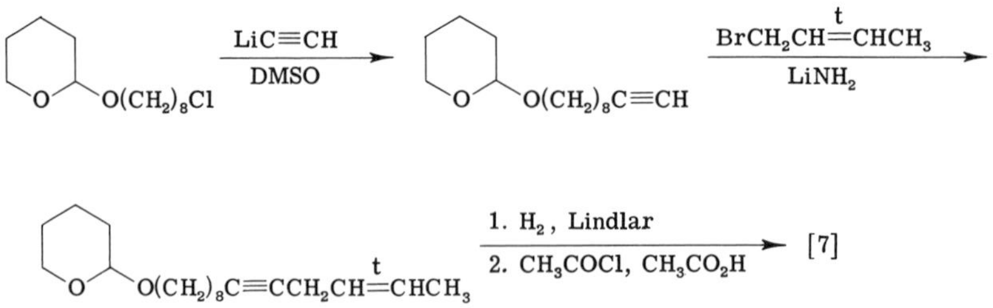

Scheme XI.19

The product showed bp 110°–112°C (0.04 mm), $n_D^{18}$ 1.4572. This compound has also been found to be the natural pheromone produced by female *Plodia interpunctella* and *Cadra cautella* (*63, 206, 706*), *Anagasta kühniella* (*704*), and *Ephestia elutella* (*204*).

*Spodoptera frugiperda*, fall armyworm

The pheromone of this species, *cis*-9-tetradecen-1-ol acetate [24], was first synthesized in 1967 by its discoverers, Sekul and Sparks (*43, 1064*), who reduced the extremely rare methyl myristoleate with lithium aluminum hydride and acetylated the resulting *cis*-9-tetradecen-1-ol. A considerable improvement in yield and cost was developed in 1968 by Warthen (*1242*) through the procedure shown in Scheme XI.20. The product showed bp 89°–95°C (0.06 mm), $n_D^{23}$ 1.4450.

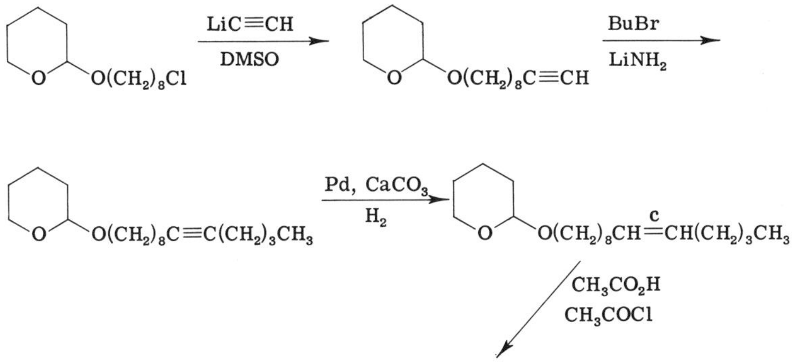

Scheme XI.20

The pheromone was also synthesized in the same year by Jacobson and Harding (*607*), who used the sex attractant of the cabbage looper moth as the starting material according to Scheme XI.21. The product showed bp 125°–126°C (0.03 mm), $n_D^{19}$ 1.4462.

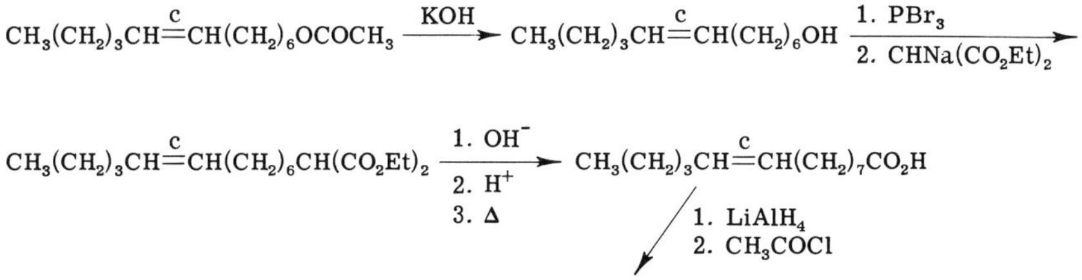

Scheme XI.21

Bestmann *et al.* (*157*) have synthesized the pheromone by the method shown in Scheme XI.22.

$$C_2H_5O_2C(CH_2)_7CO_2H \xrightarrow{\ SOCl_2\ } C_2H_5O_2C(CH_2)_7COCl \xrightarrow[\text{2. NaOH}]{\text{1. NaBH}_4}$$

$$HO_2C(CH_2)_7CH_2OH \xrightarrow{\ CH_3COCl\ } HO_2C(CH_2)_7CH_2OCOCH_3 \xrightarrow{\ SOCl_2\ }$$

$$ClOC(CH_2)_7CH_2OCOCH_3 \xrightarrow{\ NaSC_2H_5\ } C_2H_5SCO(CH_2)_7CH_2OCOCH_3 \xrightarrow[\substack{\text{2. 1, 2-Dian-}\\ \text{ilinoethane}\\ \text{3. H}_2\text{O, H}^+}]{\text{1. Ni}}$$

$$OHC(CH_2)_7CH_2OCOCH_3 \xrightarrow{\ CH_3(CH_2)_3CH=PPh_3\ } [24]$$

Scheme XI.22

Their product, containing a maximum of 5% of the trans isomer according to the infrared spectrum, showed bp 85°–86°C (0.04 mm). Strong electroantennograms were obtained in the laboratory with males of the noctuid genera *Dicestra*, *Mamestra*, *Apamea*, *Miana*, *Amphipyra*, and *Casadrina* on exposure to the preparation. Extracts of female abdomens of these species likewise showed cross attraction in all cases. The synthetic preparation elicited weak or no electroantennograms from males of 22 other species of noctuids (*157*).

Tamaki *et al.* (*1169a*) synthesized compound [24] as is shown in Scheme XI.22a.

$$CH_3(CH_2)_4Br \xrightarrow[\text{2. OHC(CH}_2)_6\text{CO}_2\text{Me}]{\text{1. Ph}_3\text{P}} CH_3(CH_2)_3CH=CH(CH_2)_6CO_2Me$$

$$\downarrow \substack{\text{1. LiAlH}_4 \\ \text{2. Ac}_2\text{O}}$$

[24]
(cis + trans)

Scheme XI.22a

The cis and trans isomers were separated by thin-layer chromatography on silver nitrate-impregnated silicic acid, with benzene as eluent.

Roelofs and Comeau (*956, 957*) have reported that males of a grayish form of *Bryotopha similis* respond to the pheromone, whereas a yellowish form responds to the corresponding trans isomer. These investigators claim to have shown that the compounds are the natural pheromones for *Bryotopha*, although they have not been isolated from the insects.

*cis*-9-Tetradecen-1-ol acetate failed to evoke a response from male cabbage loopers in the laboratory, nor did the trans isomer stimulate male fall armyworms or cabbage loopers (*606*).

Another species whose males are attracted to field traps baited with compound [24] is *Apamea interoceanica* Smith. The trans isomer attracts *Loxostege neobliteralis* Capps and *Polia grandis* Boisduval (*957*)

*Trichoplusia ni*, cabbage looper

Berger (*135*) synthesized the attractant, *cis*-7-dodecen-1-ol acetate [25], in an overall yield of 22% by Scheme XI.23. The product showed bp 98°–100°C (0.5 mm).

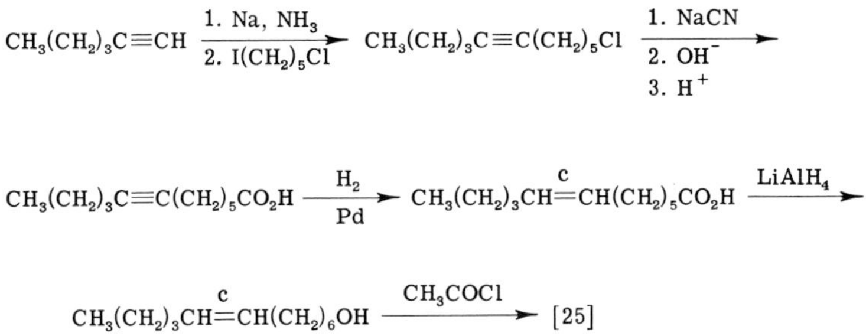

$$CH_3(CH_2)_3C\equiv CH \xrightarrow[\text{2. I(CH_2)_5Cl}]{\text{1. Na, NH_3}} CH_3(CH_2)_3C\equiv C(CH_2)_5Cl \xrightarrow[\substack{\text{2. OH}^- \\ \text{3. H}^+}]{\text{1. NaCN}}$$

$$CH_3(CH_2)_3C\equiv C(CH_2)_5CO_2H \xrightarrow[\text{Pd}]{H_2} CH_3(CH_2)_3CH\overset{c}{=}CH(CH_2)_5CO_2H \xrightarrow{LiAlH_4}$$

$$CH_3(CH_2)_3CH\overset{c}{=}CH(CH_2)_6OH \xrightarrow{CH_3COCl} [25]$$

Scheme XI.23

A more direct procedure, with an overall yield of at least 37%, was developed by Green *et al.* (*480*) (Scheme XI.24). The product showed bp 85°–90°C (0.08–0.09 mm), $n_D^{22}$ 1.4420. The procedure has been adapted to commercial production.

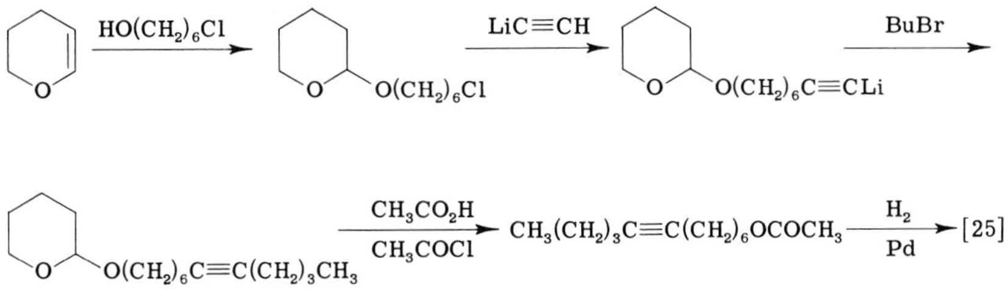

Scheme XI.24

Berger and Canerday (*137*) have described a more satisfactory modification of their original method (*135*). This procedure was identical with that used by them to synthesize *trans*-7-dodecen-1-ol acetate (see page 216), except that the 7-dodecyn-1-ol was reduced by hydrogenation over palladium catalyst to give *cis*-7-dodecen-1-ol; treatment with acetyl chloride gave the corresponding acetate. These authors also synthesized *cis*-5-, *cis*-6-, and *cis*-9-dodecen-1-ol acetates, *cis*-7-tetradecen-1-ol acetate; *cis*-7-dodecen-1-ol propionate, and *cis*-7-dodecen-1-ol butyrate, whose minimum active concentrations on males in the laboratory were $10^{-2}$, $>10$, $10^{-2}, 10^{-2}$, $4 \times 10^{-3}$, and $4 \times 10^{-2}$ gm/ml, respectively; compound [25] was active at $10^{-7}$ gm/ml and *trans*-7-dodecen-1-ol acetate at $2 \times 10^{-4}$ gm/ml.

Warthen and Jacobson (*1245*) prepared the cis and trans isomers of 5-dodecen-1-ol acetate and found them to be less attractive to male cabbage loopers than compound [25]. They were unattractive to male and female Mexican fruit flies, Mediterranean fruit flies, melon flies, oriental fruit flies, male fall armyworms, codling moths, gypsy moths, and pink bollworm moths.

Jacobson *et al.* (*616*) prepared and tested in the laboratory on male cabbage loopers *cis*-7-decen-1-ol and the acetates of *cis*-7-decen-1-ol, *cis*-7-dodecen-1-ol, *trans*-7-dodecen-1-ol, dodecanol, 7-octen-1-ol, 7-octyn-1-ol, and 7-dodecyn-1-ol. All of these compounds were much less attractive than compound [25].

Fifteen 12-carbon saturated and unsaturated alcohols and acetates related chemically to compound [25] were bioassayed in the laboratory on male *T. ni*. Some activity was shown only by acetates with unsaturation in the 6- and 7-positions. The activity of compound [25] was inhibited when it was tested simultaneously with the corresponding trans isomer (*1183*).

Species that also responded in the laboratory to compound [25] were *Rachiplusia includens* and *Autographa biloba* (*137*). Field traps baited with the compound attracted males of *A. ampla* Walker, *Anagrapha falcifera* Kirby, *Chrysaspidia contexta* Grote, *Fridatima* sp., *Phlyclaenia terrealis* Treitschke, and *Plusia aereoides* Grote (*957*).

*Anthonomus grandis*, boll weevil

[27]                      [28]                      [29]                      [30]

The 4 compounds making up the male pheromone, collectively called "grandlure," are (+)-cis-2-isopropenyl-1-methylcyclobutaneethanol [27], cis-3,3-dimethyl-Δ¹,β-cyclohexaneethanol [28], cis-3,3-dimethyl-Δ¹,α-cyclohexaneacetaldehyde [29], and trans-3,3-dimethyl-Δ¹,α-cyclohexane-acetaldehyde [30].

Compound [27] was synthesized by the following 2 procedures [Scheme XI.25, Tumlinson et al. (53, 1205, 1207, 1208) and Scheme XI.26, Zurflüh et al. (stereoselective) (53, 1336)].

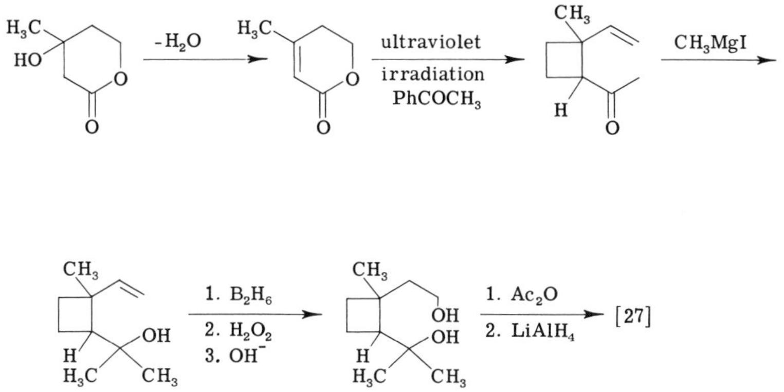

Scheme XI.25

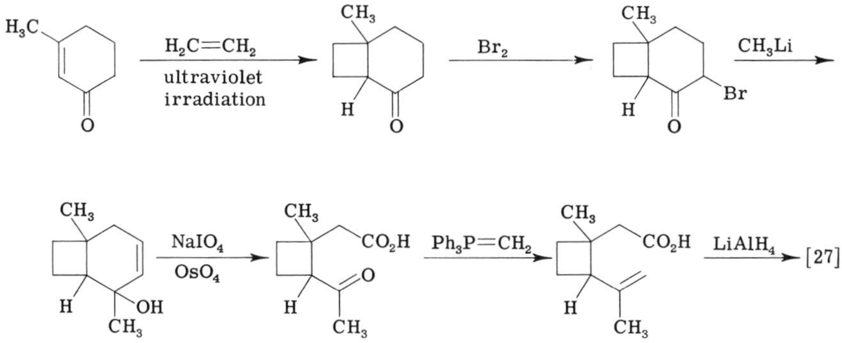

Scheme XI.26

Compounds [28], [29], and [30] were synthesized by the method shown in Scheme XI.27 (50, 1205, 1207, 1208).

Bull et al. (218) have developed a gas chromatographic method to analyze for the 4 components of grandlure.

In order to obtain attraction for female boll weevils in the laboratory, compounds [27], [28], and either [29] or [30] must be combined (1205). Field attraction requires a combination of all 4 components (1206).

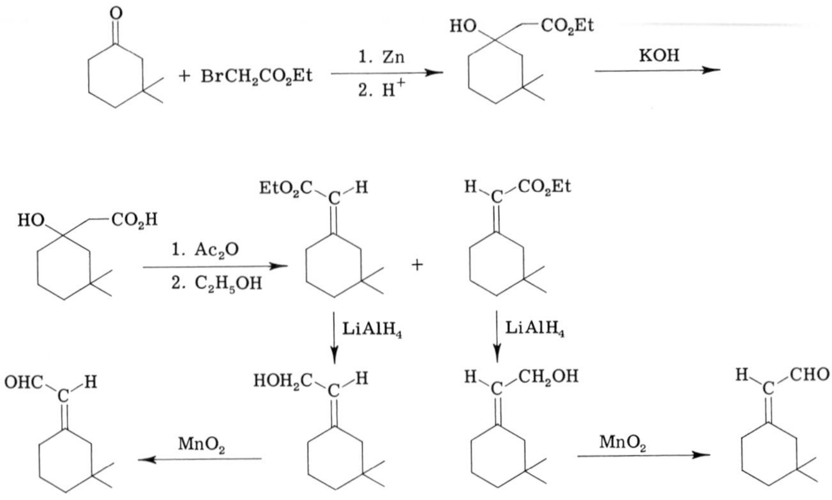

Scheme XI.27

*Attagenus megatoma*, black carpet beetle

Megatomoic acid (*trans*-3,*cis*-5-tetradecadienoic acid) was synthesized by Silverstein *et al.* (*1100, 1104*) according to the procedure shown in Scheme XI.28.

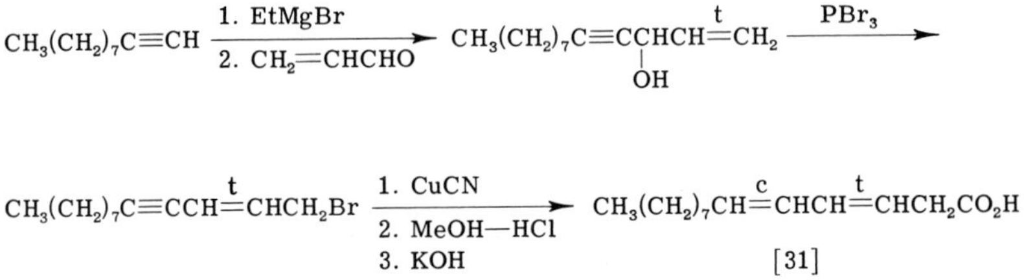

$$CH_3(CH_2)_7C{\equiv}CH \xrightarrow[\text{2. } CH_2{=}CHCHO]{\text{1. EtMgBr}} CH_3(CH_2)_7C{\equiv}C\overset{t}{C}HCH{=}CH_2 \underset{OH}{\big|} \xrightarrow{PBr_3}$$

$$CH_3(CH_2)_7\overset{t}{C}{\equiv}CCH{=}CHCH_2Br \xrightarrow[\substack{\text{2. MeOH—HCl} \\ \text{3. KOH}}]{\text{1. CuCN}} CH_3(CH_2)_7\overset{c}{CH}{=}CHCH\overset{t}{=}CHCH_2CO_2H \quad [31]$$

Scheme XI.28

The *cis*-3,*trans*-5-tetradecadienoic acid was also synthesized by Silverstein *et al.* (*225, 1104*). The cis,trans, cis,cis and trans,trans compounds were also prepared (*225, 1100, 1104*).

Laboratory tests with an olfactometer showed that the synthetic *trans*-3,*cis*-5 isomer was as effective as the natural acid in eliciting male response. Traps baited with 0.01 mg of this isomer lured males, whereas the other isomers were very much less effective (*225, 1104*).

*Costelytra zealandica*, grass grub beetle

Prior to the identification of phenol [32] as the natural pheromone of this species (*541*), it was found (*836*) that a commercial adhesive (Pliobond) and resin (Durez), both of which are synthetic products of phenol and formaldehyde, were attractive to males of this species. These products were shown to contain considerable amounts of phenol, as well as 2- and 4-hydroxybenzyl alcohols and their condensation products with formaldehyde. An unidentified constituent of the oil steam-distilled from fresh elder flowers (*Sambucus nigra*) was attractive to female *C. zealandica* (*834*).

Compounds tested and found to be unattractive to males were eugenol, geraniol, citronellol, valeric acid, caproic acid, sorbic acid, propyl sorbate, butyl sorbate, anethole, isoeugenol, limonene, and 2-methoxyethanol (*835*).

*Dendroctonus brevicomis*, western pine beetle

The major sex pheromone produced by boring females is *exo*-7-ethyl-5-methyl-6,8-dioxabicyclo[3.2.1]octane (*exo*-brevicomin) [35]. This compound has been synthesized by the procedure shown in Scheme XI.29 (*132, 1100, 1101*).

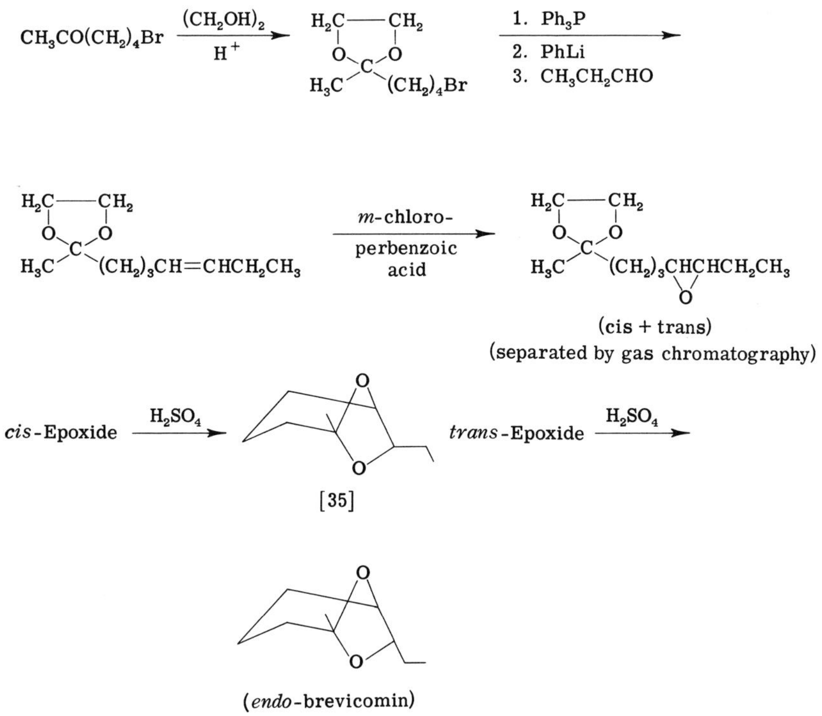

$$CH_3CO(CH_2)_4Br \xrightarrow[\text{H}^+]{(CH_2OH)_2}$$

H₂C——CH₂
|      |
O   O
 \  /
  C
 /  \
H₃C   (CH₂)₄Br

1. Ph₃P
2. PhLi
3. CH₃CH₂CHO

H₂C——CH₂
|      |
O   O
 \  /
  C
 /  \
H₃C   (CH₂)₃CH=CHCH₂CH₃

→ *m*-chloro-perbenzoic acid →

H₂C——CH₂
|      |
O   O
 \  /
  C
 /  \
H₃C   (CH₂)₃CHCHCH₂CH₃
          \ /
           O

(cis + trans)
(separated by gas chromatography)

*cis*-Epoxide $\xrightarrow{H_2SO_4}$ [35] $\xleftarrow{H_2SO_4}$ *trans*-Epoxide

[35]

(*endo*-brevicomin)

Scheme XI.29

*endo*-Brevicomin (epibrevicomin), also isolated from female frass, was not attractive to the insect (*1101*).

Wasserman and Barber (*1247*) synthesized compound [35] by the procedure shown in Scheme XI.30, which is the method of choice.

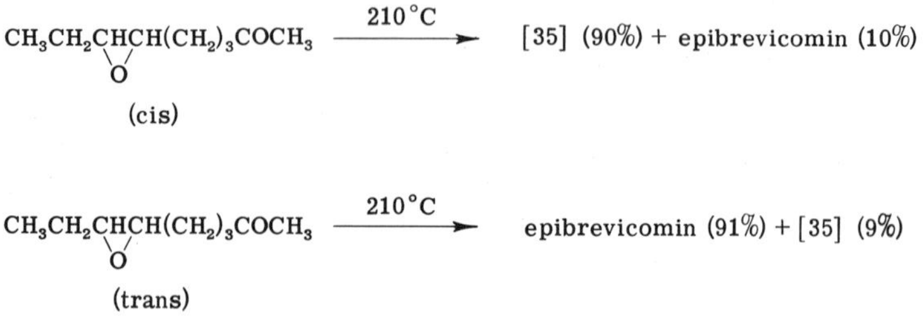

$$\text{CH}_3\text{CH}_2\underset{\underset{\displaystyle\text{O}}{\diagdown\diagup}}{\text{CH}}\text{CH}(\text{CH}_2)_3\text{COCH}_3 \xrightarrow{210\,°\text{C}} \quad [35]\ (90\%) + \text{epibrevicomin}\ (10\%)$$

(cis)

$$\text{CH}_3\text{CH}_2\underset{\underset{\displaystyle\text{O}}{\diagdown\diagup}}{\text{CH}}\text{CH}(\text{CH}_2)_3\text{COCH}_3 \xrightarrow{210\,°\text{C}} \quad \text{epibrevicomin}\ (91\%) + [35]\ (9\%)$$

(trans)

Scheme XI.30

A second active compound, 1,5-dimethyl-6,8-dioxabicyclo[3.2.1]octane [36], was isolated from male frass (*671*). For the synthesis of this compound, known as "frontalin," see below.

*trans*-Verbenol [33] and verbenone [34], have been isolated from the hindguts of female and male *D. brevicomis*, respectively (*875*, *1228*). Compound [33] has been synthesized by the oxidation of α-pinene. Compound [33] is attractive to *Blastophagus piniperda* but compound [34] is not (*641*).

*Dendroctonus frontalis*, southern pine beetle

Females boring into the host plant produce the active pheromone, 1,5-dimethyl-6,8-dioxabicyclo[3.2.1]octane [36], known as "frontalin" (*1230*). The compound has been synthesized by Kinzer *et al.* (*671*) according to Scheme XI.31.

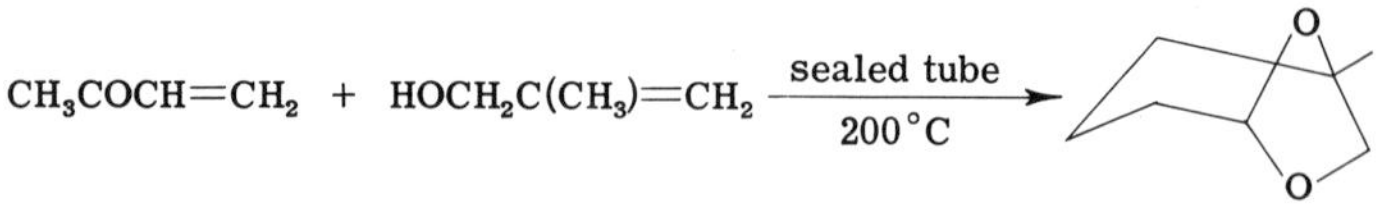

$$\text{CH}_3\text{COCH}{=}\text{CH}_2 + \text{HOCH}_2\text{C}(\text{CH}_3){=}\text{CH}_2 \xrightarrow[200\,°\text{C}]{\text{sealed tube}}$$

[36]

Scheme XI.31

Milligram quantities of synthetic frontalin combined with *exo*-brevicomin and host oleoresin effected mass aggregation of flying *D. frontalis* at bait traps (*1228*). *trans*-Verbenol [33] has also been obtained from the insect (*865*, *1230*).

A number of synthetic analogs of frontalin were tested in the field for attractiveness to *D. frontalis*, but they were all inactive. They were

1-methyl-, 5-methyl-, 1,4-dimethyl-, and 5,7-dimethyl-6,8-dioxabicyclo-[3.2.1]octanes, 2-ethyl-2,5-dimethyl-, 2-propyl-2,5-dimethyl-, 2-amyl-2-methyl-, 2-butyl-, and 2-isobutyl-2-methyl-1,3-dioxolanes (*922*).

*Dendroctonus ponderosae*, mountain pine beetle

The essential aggregation pheromone of this species is *trans*-verbenol [33], which is synthesized by oxidizing α-pinene (*872*). Optimum aggregation is obtained by combining [33] with α-pinene or myrcene (*865, 1228*).

*Dendroctonus pseudotsugae*, Douglas-fir beetle

The major pheromone of this species is 1,5-dimethyl-6,8-dioxabicyclo-[3.2.1]octane (frontalin) [36] (*873*). See page 236 for its synthesis. A second pheromone, 3-methyl-2-cyclohexenone [37], which is commercially available, has been obtained from female hindguts (*670*). A 1:1 mixture of compounds [36] and [37] was more attractive to both sexes than either compound alone (*670*).

*Ips confusus*, California five-spined ips

The pheromone complex produced by males boring into ponderosa pine bark consists of (−)-2-methyl-6-methylene-7-octen-4-ol [38], (+)-*cis*-verbenol [39], and (+)-2-methyl-6-methylene-2,7-octadien-4-ol [40].

Compound [38] was synthesized by the procedure shown in Scheme XI.32 (*40, 915, 1105*).

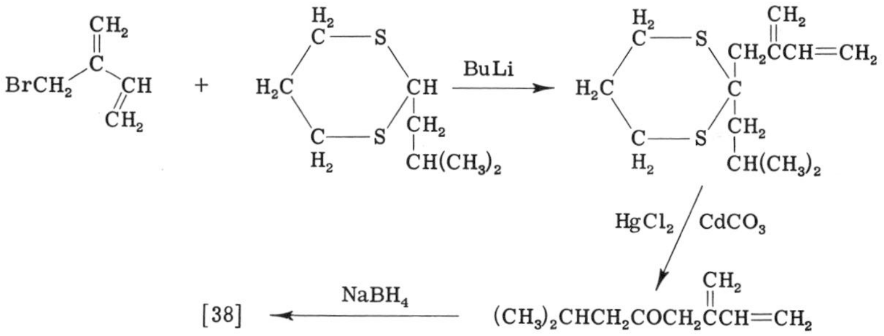

Scheme XI.32

Compound [40] was synthesized by the similar procedure shown in Scheme XI.33 (*40, 915*).

Compound [39] was prepared by the reduction of (−)-verbenone [34] with sodium borohydride (*915*).

In laboratory bioassays, a mixture of compound [38] with either compound [39] or compound [40] is necessary to evoke a response from female beetles. A mixture of all 3 compounds is needed to attract both sexes (mainly females) in the field. The individual compounds are not attractive in the laboratory or field (*40, 1105, 1294*). Female *I. latidens* was attracted

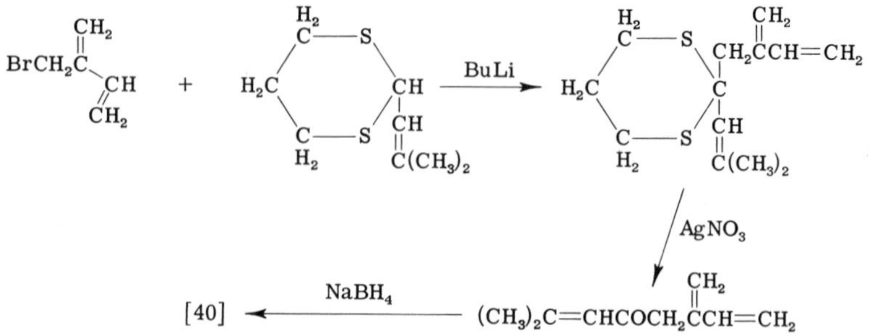

Scheme XI.33

to [38] alone and to a combination of [38] and [39], but not to [38 + 40] or to [40] alone. A combination of all 3 compounds was not attractive to *I. latidens* (*1294*).

*Limonius californicus*, sugar beet wireworm

This pheromone, *n*-valeric acid [41], is commercially available (*609*). Lehman (*719*) tested approximately 150 compounds in the laboratory for attractiveness to adult *Limonius californicus* and *L. canus*. Caproic, lactic, butyric, and valeric acids appeared to be sex attractants for males, eliciting a high degree of excitation, extension of the genitalia, and copulatory attempts. Caproic acid in field traps caught 2745 males but only 1589 female *L. canus* were attracted to 32 traps; corresponding numbers for *L. californicus* were 8 and 10, respectively. Butyric acid in the field attracted 1589 male and 156 female *L. canus* in 34 traps, and only 5 male and 1 female *L. californicus* (*720*). Caproic acid is also attractive to *Olcella parva* (Adams) (*617*).

Isomeric acids that failed to attract *L. californicus* in the field were *dl*-2-methylbutyric, isovaleric (3-methylbutyric), and pivalic (trimethylacetic). In addition, negative field tests were obtained with 14 other acids, 16 alcohols, 3 aldehydes, 4 amines, 15 esters, camphor, and cyclohexane (*609*).

*Phyllophaga lanceolata*, June beetle

Isoamylamine produces a male response similar to that elicited by crushed or sexually active females. On a day with a strong wind, isoamylamine stimulated males to activity from as far away as 50–75 feet (*1198*).

*Trogoderma granarium*, khapra beetle

The esters comprising the assembling scents produced by both sexes are methyl oleate, ethyl oleate, ethyl palmitate, ethyl linoleate, and ethyl stearate. All were prepared synthetically and recombined in their natural ratio; the mixture attracted 61.7% of females and 66.2% of males. The percentages of females and males, respectively, attracted by the natural esters were methyl oleate 90.5, 84.1; ethyl oleate 57.7, 70.2; ethyl palmitate 69.8, 77.7; ethyl linoleate 55.6, 47.0; ethyl stearate 65.8, 51.7. All compounds repelled *Tribolium castaneum* (*568*).

*Trogoderma inclusum*

The 2 sex pheromones isolated from females, $(-)$-14-methyl-*cis*-8-hexadecen-1-ol [47] and $(-)$-methyl 14-methyl-*cis*-8-hexadecenoate [48], were synthesized by Rodin *et al.* (*949*) according to Scheme XI.34.

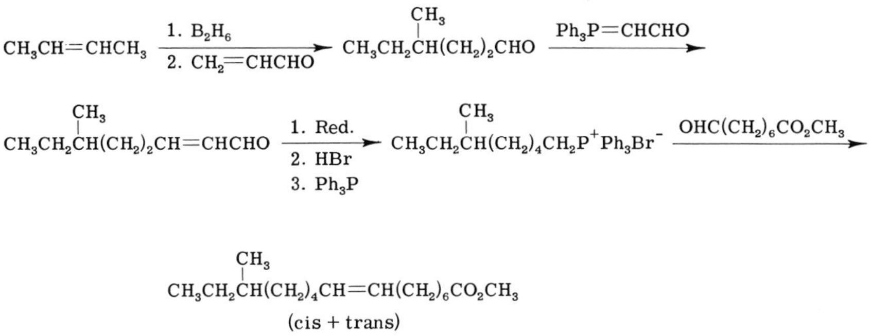

$$CH_3CH\!=\!CHCH_3 \xrightarrow[\text{2. } CH_2\!=\!CHCHO]{\text{1. } B_2H_6} CH_3CH_2\overset{\overset{\displaystyle CH_3}{|}}{C}H(CH_2)_2CHO \xrightarrow{Ph_3P\!=\!CHCHO}$$

$$CH_3CH_2\overset{\overset{\displaystyle CH_3}{|}}{C}H(CH_2)_2CH\!=\!CHCHO \xrightarrow[\substack{\text{2. HBr}\\ \text{3. } Ph_3P}]{\text{1. Red.}} CH_3CH_2\overset{\overset{\displaystyle CH_3}{|}}{C}H(CH_2)_4CH_2P^+Ph_3Br^- \xrightarrow{OHC(CH_2)_6CO_2CH_3}$$

$$CH_3CH_2\overset{\overset{\displaystyle CH_3}{|}}{C}H(CH_2)_4CH\!=\!CH(CH_2)_6CO_2CH_3$$

(cis + trans)

Scheme XI.34

The mixture of cis and trans esters was separated by thin-layer chromatography on silica gel impregnated with $AgNO_3$. Reduction of the cis isomer [48] with lithium aluminum hydride gave the corresponding alcohol [47].

The alcohol was attractive to males of this species as well as of *T. simplex* and *T. grassmani*, but not to males of *T. glabrum*, *T. parabile*, or *T. sternale*. The methyl ester was attractive only to *T. inclusum* males (*949*).

*Apis mellifera*, honeybee

Investigation of the attractiveness of queen bees for the workers showed it to be due to a mixture of volatile acids with *trans*-9-oxodec-2-enoic acid ("queen substance"). An artificial mixture of phenylacetic, phenylpropionic, *p*-hydroxybenzoic, azelaic, and sebacic acids with queen substance did not produce attractivity. Similarly, mixtures of each of the following acids with queen substance failed to attract workers: *cis*-9-oxodec-2-enoic, *trans*-8-oxonon-2-enoic, *trans*-8-oxodec-2-enoic, *trans*-9-hydroxydec-2-enoic, 6-oxoheptanoic, 7-oxooctanoic, 8-oxononanoic, 9-oxodecanoic, *trans*-2-nonene-1,9-dioic, *trans*-9-oxo-9-phenyldec-2-enoic, 9-oxo-2-decynoic (*842*).

Queen substance [49] has been synthesized by the methods in Schemes XI.35–38. Barbier *et al.* (*87*) and Tribble (*1197*) followed Scheme XI.35;

the product showed bp 163°–166°C (1.1 mm), mp 50°–54°C. Scheme XI.36 was devised by Butler *et al.* (*257*). Bestmann *et al.* (*156*) obtained a product that showed mp 52°–53°C (Scheme XI.37). Scheme XI.38 was followed by Doolittle *et al.* (*356*).

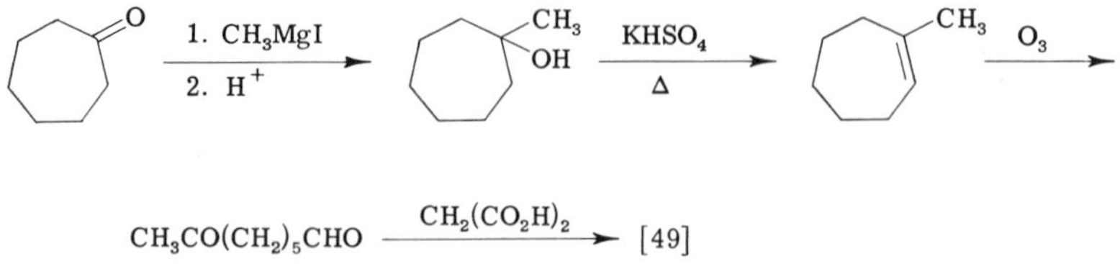

$$\text{(cycloheptanone)} \xrightarrow[\text{2. H}^+]{\text{1. CH}_3\text{MgI}} \text{(1-methylcycloheptanol)} \xrightarrow[\Delta]{\text{KHSO}_4} \text{(1-methylcycloheptene)} \xrightarrow{\text{O}_3}$$

$$\text{CH}_3\text{CO(CH}_2)_5\text{CHO} \xrightarrow{\text{CH}_2(\text{CO}_2\text{H})_2} [49]$$

Scheme XI.35

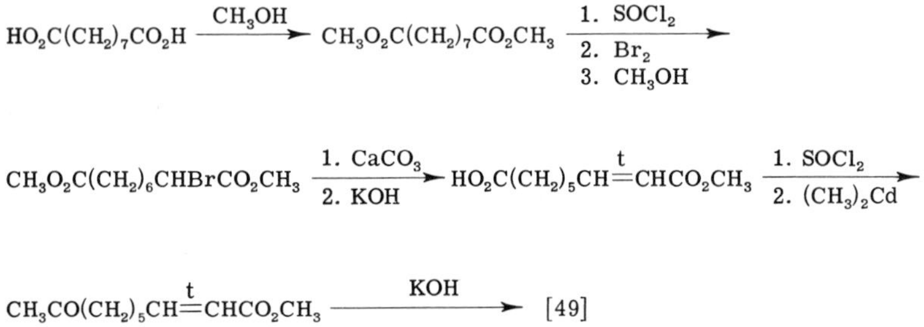

$$\text{HO}_2\text{C(CH}_2)_7\text{CO}_2\text{H} \xrightarrow{\text{CH}_3\text{OH}} \text{CH}_3\text{O}_2\text{C(CH}_2)_7\text{CO}_2\text{CH}_3 \xrightarrow[\substack{\text{2. Br}_2 \\ \text{3. CH}_3\text{OH}}]{\text{1. SOCl}_2}$$

$$\text{CH}_3\text{O}_2\text{C(CH}_2)_6\text{CHBrCO}_2\text{CH}_3 \xrightarrow[\text{2. KOH}]{\text{1. CaCO}_3} \text{HO}_2\text{C(CH}_2)_5\text{CH}\overset{t}{=}\text{CHCO}_2\text{CH}_3 \xrightarrow[\text{2. (CH}_3)_2\text{Cd}]{\text{1. SOCl}_2}$$

$$\text{CH}_3\text{CO(CH}_2)_5\text{CH}\overset{t}{=}\text{CHCO}_2\text{CH}_3 \xrightarrow{\text{KOH}} [49]$$

Scheme XI.36

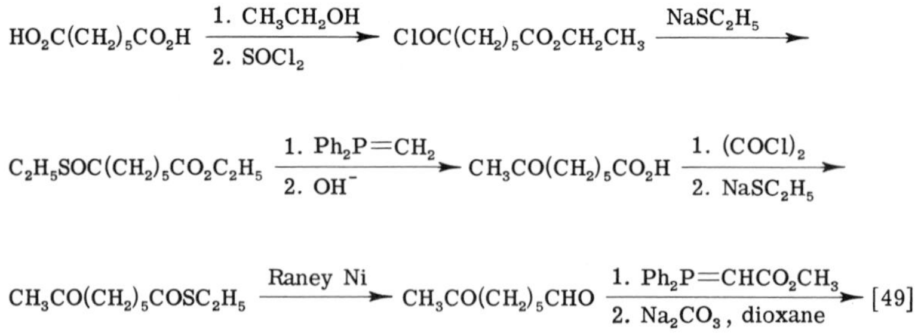

$$\text{HO}_2\text{C(CH}_2)_5\text{CO}_2\text{H} \xrightarrow[\text{2. SOCl}_2]{\text{1. CH}_3\text{CH}_2\text{OH}} \text{ClOC(CH}_2)_5\text{CO}_2\text{CH}_2\text{CH}_3 \xrightarrow{\text{NaSC}_2\text{H}_5}$$

$$\text{C}_2\text{H}_5\text{SOC(CH}_2)_5\text{CO}_2\text{C}_2\text{H}_5 \xrightarrow[\text{2. OH}^-]{\text{1. Ph}_2\text{P}=\text{CH}_2} \text{CH}_3\text{CO(CH}_2)_5\text{CO}_2\text{H} \xrightarrow[\text{2. NaSC}_2\text{H}_5]{\text{1. (COCl)}_2}$$

$$\text{CH}_3\text{CO(CH}_2)_5\text{COSC}_2\text{H}_5 \xrightarrow{\text{Raney Ni}} \text{CH}_3\text{CO(CH}_2)_5\text{CHO} \xrightarrow[\text{2. Na}_2\text{CO}_3, \text{ dioxane}]{\text{1. Ph}_2\text{P}=\text{CHCO}_2\text{CH}_3} [49]$$

Scheme XI.37

The methyl ester of [49] also attracted drones in flight (*257*). 9-Hydroxy-*trans*-2-decenoic acid was only slightly attractive to drones (*258*), but 9-chloro-*trans*-2-decenoic acid was as attractive to drones as queen substance (*1197*). 9-Oxo-*cis*-2-decenoic acid was not attractive to drones, nor

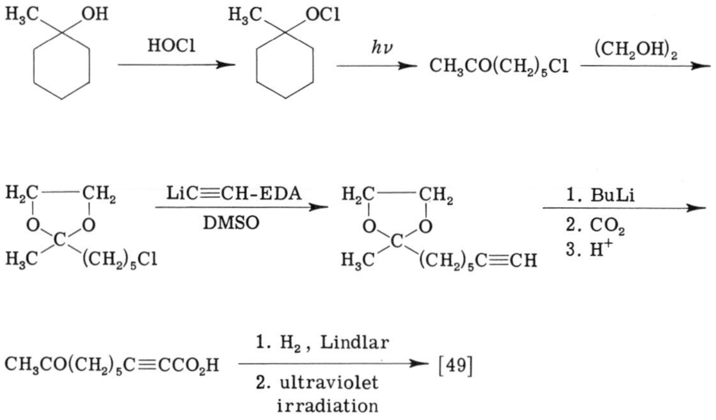

Scheme XI.38

did it mask the attractiveness of queen substance when the 2 were mixed in equal amounts and tested on flying drones (*169, 356*).

Blum *et al.* (*169*) synthesized 19 alkenoic acids closely related in structure to compound [49]. All structural modifications resulted in complete loss of attractant activity. *trans*-Alkenoic acids had no activity as masking agents for compound [49].

*Musca domestica*, housefly

The sex pheromone produced by females, identified as *cis*-9-tricosene [54] and designated "muscalure," has been synthesized by the following procedure (*69, 283*):

$$CH_3(CH_2)_{13}Br \xrightarrow[\substack{2.\ BuLi,\ DMSO \\ 3.\ CH_3(CH_2)_7CHO}]{1.\ Ph_3P} CH_3(CH_2)_{13}CH{=}CH(CH_2)_7CH_3$$

(cis + trans)

The product was separated into 85% cis isomer [54] and 15% trans isomer by thin-layer and column chromatography. A 50-$\mu$g sample of [54] attracted more flies than 200 $\mu$g of the trans isomer. No masking was observed when cis and trans isomers were mixed at ratios up to 1:3, respectively (*283*).

*Macropis labiata*, wild bee

*Bombus lapidarius*, wild bee

Kullenberg (*696*) found that male *Macropis labiata* respond to the following sex attractants, arranged in ascending order of effectiveness: methyl pelargonate, farnesal, undecalactone, nerolidol, farnesol (resembling the scent of the female). Only nerolidol and farnesol caused excitation, and the males could not distinguish with certainty between farnesol and the female scent at equal concentrations.

Farnesol and phenylethyl alcohol acted as sexual attractants and excitants for male *Bombus lapidarius*, while butyric and valeric acids repelled males (*696*).

Very extensive investigations of bumblebee sex attractants and excitants conducted by Kullenberg (*697*) showed most of the active substances to be mono- or sesquiterpene alcohols. Farnesol and nerolidol were the most effective excitants for male *Macropis labiata*; undecalactone was also attractive. The best imitation of the complex scent secreted by the female was undecalactone plus farnesol; this was very attractive to males on mating flights, even eliciting attempts at copulation.

*Crabro cribrarius*, wasps

Substances acting as sex attractants for males were allyl alcohol, benzaldehyde, rhodinol, geraniol, citronellol, nerolidol, farnesol, citronellal, hydroxycitronellal, and the scent from flowers of *Ophrys insectifera*. Citronellal, hydroxycitronellal, and the *Ophrys* scent also acted as excitants, as did farnesol and nerolidol occasionally. None of these substances evoked copulatory attempts. Allyl alcohol and benzaldehyde were not attractive at high concentrations, and capric and caprylic acids were repellent (*696*).

Citronellal was the most effective attractant and sexual excitant for *Crabro cribrarius* males, although isopulegol had some excitant effect. Butyric acid was both attractive and excitatory. Imitations of the female scent containing citronellal, nerolidol, and butyric acid, or citronellal plus nerolidol (imitation of the female abdominal scent), or nerolidol plus butyric acid (imitation of the cephalic scent) were also attractive (*697*).

# CHAPTER XII

# SEX ATTRACTANTS IN INSECT SURVEY

The high specificity of the sex attractants has made them extremely valuable for detection and estimation of insect populations. The insect, in responding to a lure-baited trap, is caught and thereby signals the presence of its species. Thus a good attractant can assure the early detection of an infestation before it can enlarge or spread. Control measures need be applied only to those areas where the insect is found and only as long as it continues to be present. A good lure can accurately delineate an infestation and locate the last few, hard-to-find insects.

*Acrolepia assectella,* leek moth

Virgin females placed in traps in a 30 × 20 m leek field effectively lured 1- to 4-day-old males released in the area during a period of 8 days (*901, 904*). A comparison was made between the numbers of released males lured by virgin females in a leek field and the numbers lured in a sugarbeet field. A total of 500 virgin males, 1 day old, were released in the center of a circular plot of sugarbeet and 1 trap containing 4 virgin females was placed in each of the 4 directions, 5 m from the point of release. A comparable parcel of leek contained 24 traps (4 females each) set in a circle 9.1 m from the point of male release. A total of 2 males was caught in the sugarbeet field, whereas 615 males were caught in the leek field (*902*).

A third trial conducted in and around infested leek plots of 400 m², free of dodder, was conducted during the summer of 1966 using 49 traps, each containing a single virgin female, spaced 5 and 10 m apart. Numerous males were trapped during the first 3 nights following flight initiation, and the numbers then slackened off. Numbers of males caught by neighboring traps were quite irregular. Traps placed inside the crop area lured many more males than did those placed on the periphery or outside the plot (*903*).

*Adoxophyes fasciata*, lesser tea tortrix

*Adoxophyes orana*, summerfruit tortrix

Field trials were conducted in a Japanese tea garden using tub-type traps baited with 50 virgin females and a 6-W fluorescent blacklight lamp; considerably more males were caught with blacklight than without it. Large numbers of males were also caught by female-baited traps without light between 2:00 AM and 4:00 AM, when mating activity of the moth is at its maximum. Cylinder-type sticky traps baited with absorbent cotton soaked with a methylene chloride extract of 100 females caught twice as many males as did traps with 50 virgin females, but the treated cotton lost 50% of its activity after 24-hours'exposure at room temperature (*1168*). Minks and Noordink (*790*) reported maximum sexual activity at around midnight. Traps containing 2 live females caught more males than those containing only 1, but more than 2 females did not increase the catch. Mated females lured much smaller numbers of males to traps.

Minks *et al.* (*790a*) were able to obtain good recapture of $^{32}$P-marked males released in an apple orchard 75 m from traps baited with 2 female equivalents of extract.

Adhesive-coated cardboard traps baited with 120 $\mu$g of a mixture of compounds [24] and [8] (2:1) caught an average of 49 male *A. fasciata* per trap per night (*1169a*). Traps containing 20 $\mu$g of a mixture of the two compounds in a 3:1 ratio attracted males of *A. orana* in an apple orchard (*1169b*).

*Anarsia lineatella*, peach twig borer

This insect is the principal pest of peaches in the state of Washington. Various types of traps containing virgin females were placed in young peach trees in the summer of 1968 and 1969 and were compared as to male catches. Eight fine-screen cylindrical tube traps containing a total of 73 females caught 119 males, 3 cylindrical plastic tube traps (with water wicks) containing a total of 30 females caught 48 males, a cotton-organdy envelope trap containing 6 females caught 10 males, and 7 unbaited check traps caught 6 male moths (*70*).

*Anthonomus grandis*, boll weevil

Virgin females released in Florida and Mississippi cotton fields in which screen cylinders holding live males had been placed were attracted by the males over more than 30 feet, especially from downwind positions (*318*).

Summer tests were conducted in several cotton-growing areas of Louisiana; stake-mounted traps were used that were composed of an adhesive-coated backboard (15 × 15 cm) placed on a plywood stage (15 × 10 cm). A 2-cm hole drilled through the center of the backboard provided a holding chamber for a caged male or female weevil. Significantly more boll weevils of both sexes were caught on traps containing males than on traps containing females; there was no significant difference in the number of each sex trapped (*200*).

Pyramid traps, adhesive (Stikem)-coated plywood (wing) traps, and cages covered with 15-cm wide strips of plywood coated with Stikem were used in the spring and fall to compare the response of overwintered boll weevils to cotton plants, males, plants plus males, and empty traps. Over-wintered and late-season weevils preferred male weevils to cotton, and high percentages of the weevils caught at these times were males. Both sexes preferred males over females, plants, or empty traps by a ratio of 15:1. The results suggested that methods of surveying for boll weevils by trap plantings of cotton in spring and examination of surface woods trash in spring and fall could be replaced or supplemented by using male weevils or the aggregant pheromone in traps (*516*).

Catches of female weevils were compared in Florida and Mississippi using 12 kinds of field traps (4 types of wing traps, a diamond trap with inserted polystyrene boxes for housing the weevils, a can trap, a cube funnel trap, a cylinder funnel trap, an oblique funnel trap, 2 modified Steiner fruit fly traps, and a cup-type gypsy moth trap) in large cotton plots and field cages. All traps were baited with live males or methylene chloride extracts of males. A small adhesive-coated plywood trap and a Plexiglas screen funnel trap proved to be the most efficient designs. Most of the captured female weevils were lured from downwind. Live males were preferred over male extracts 1.8:1 when comparisons were made on different days, and by 6.2:1 when made on the same day (*317*).

Field tests conducted in Florida, Mississippi, and Mexico in 1966–1967 showed that male weevils fed on cotton squares (flower buds) were more attractive than males fed on pellets of artificial media, and females re-sponded to males from distances of up to 82 m (*58*, *517*). Males sterilized with apholate or irradiation were as attractive as untreated males. Lack of response of recently mated females emphasized the need to capture females before they could mate with free, competitive males. The high percentage of females captured in male-baited traps in the absence of competing males, and the low percentage captured with baited male traps in an infested plot containing large numbers of competing males suggested that the pheromone might be used to suppress weevil populations in areas with low infestation (*517*). Tests conducted in 1968 showed that traps baited with 1 or 5 males fed on any diet captured more boll weevils than a trap containing a squaring cotton plant without a male. Males fed on fresh squares captured many more weevils than those fed on squaring cotton or artificial diet. Both sexes were captured during early season, midseason, or late season (*514*).

In West Texas, Stikem-coated plywood traps painted green and baited with 5 live males were effective for surveying populations of overwintered and migrating boll weevils, but in midseason, competition from males in fruiting cotton reduced their effectiveness. Larger catches were observed in August (*515*). A total of 3882 weevils was captured in $8\frac{1}{2}$ weeks in the fall on 16 male-baited wing traps in a heavily wooded area 3.2 miles from the nearest cotton field in the High and Rolling Plains of Texas (*194*).

When sheet-metal wing traps, painted yellow and baited with 4 virgin males feeding on cotton squares, were placed around small South Carolina cotton fields in a ratio of one trap per 2–3 acres, a positive correlation was found between the number of overwintered weevils captured and the number observed in the field. The estimate obtained by sampling the number of weevils hibernating in woods trash did not correlate with the number captured or observed. The location of traps in relation to over-wintering sites affected the trap captures; the time of entry of weevils into the fields was related to the beginning of fruiting by the cotton (*946*). The traps captured 85% of the emerging overwintered weevils but could not control later generations (*947*). The use of 3 × 5 foot cotton trap plots baited with males was not as feasible as the use of wing traps (*945*).

Field tests conducted in the spring and fall with grandlure, the synthetic attractant of the weevil, showed that both sexes responded, but the lure attracted mostly females in midsummer (*58*). However, grandlure had a very short residual life in the field, and attempts were made to extend its effective life by formulating it with various polymeric materials as tablets or foams for slow release. Of 13 formulations aged at room temperature for 1–8 days, 10 (at 800 $\mu$g) were more attractive then fresh standard (0.4 $\mu$g grandlure absorbed on firebrick). Only 1 formulation, 800 $\mu$g of grandlure formulated as a tablet with polyethylene glycol, was more attractive after aging at 90°F than the standard; it remained effective after being aged for 128 hours (*772*).

Synthetic grandlure in the laboratory was attractive to females at amounts as low as $5 \times 10^{-6}$ $\mu$g but was most attractive at 25–50 $\mu$g. It remained attractive longer at a given temperature when evaporation was reduced. Attractiveness of grandlure increased markedly when a cotton plant extract was added to it (*520a*).

Field tests conducted in Mississippi, Texas, and Florida showed that a wick-type formulation of grandlure containing glycerol, water, polyethylene glycol, and methanol was almost as competitive an attractant for 7 days as caged live males who were fed cotton squares (*518a*).

*Aonidiella aurantii*, California red scale

Laboratory-reared adult males, tagged by dusting the host lemons with Calco oil blue RA dye just before emergence, were released in the field and recaptured on cards that had been coated with a mixture (1:1) of Stikem and mineral oil and placed on pint ice-cream carton traps baited with a lemon holding 300 virgin females. Tagged males landing on the traps deposited dye on the cards, resulting in the formation of a distinct blue spot (*929*). This trapping and tagging technique was used in field tests conducted near Riverside, California, to obtain data on male flight habits (*69a*). Larger numbers of males were collected on traps placed on the middle and upper parts of citrus trees than on traps placed lower. During the colder fall and spring months, maximum flight occurred just before sunset, whereas flight was heaviest between sunset and dark during the summer months. Males could move at least 189 m downwind and 92 m upwind and still respond to the pheromone, although in the laboratory they were unable to fly upwind when air velocities exceeded 1 mile per hour. The average number of males attracted to a trap baited with varying numbers of females, when the traps were placed 30.5 m apart, was as follows (number of females in trap and number of males caught given): 0, 1.8; 10, 35.3; 50, 101; 100, 122.2; 200, 131.3; 400, 151.5 (*931, 1072*).

*Apis mellifera*, honeybee

Tests conducted in drone congregation areas with cylindrical tubes (5 cm high × 1.5 cm diameter) of hardware cloth containing absorbent cotton impregnated with 5 mg of synthetic queen substance showed that drones were able to discriminate between objects bearing the attractant. Large amounts of attractant produced "artificial" congregation areas. Congregation areas could not be destroyed with benzaldehyde (a known bee repellent) or with isoamyl acetate and 2-heptanone (alarm substances). A total of 25 Japanese beetle traps baited with 10 mg of sex attractant and placed in an area devoid of drones attracted 153 drones over a 2-month period; traps baited with beetle attractant caught only 1 drone (*1151*).

*Archips argyrospilus*

*Archips mortuanus*

These are very closely related species of tortricid moths. Cup traps baited with an extract of 10 female equivalents of pulverized female terminal abdominal tips of 1 species were alternated with traps containing 10 female equivalents of pulverized female tips of the other species; the traps were placed 1 m apart in apple trees during moth flight. Eighteen traps containing *A. argyrospilus* extract attracted 203 males of this species and none of the other species on the same nights that 8 traps baited with *A. mortuanus* attracted 82 males of this species and 10 males of *A. argyrospilus*. The 2 species were attracted during the same hours (*956*).

*Argyrotaenia velutinana*, red-banded leaf roller

Cup traps (12 ounces) baited with an ether extract of the terminal 2 abdominal segments of virgin female moths and hung in apple trees in the spring were very attractive to males, a single trap being capable of attracting up to 100 males in 1 day (*950*). Crude extracts baited with 10 female equivalents on dental wicks placed in cup traps were highly attractive the first day and night, but caught very few males thereafter (*456*). Live virgin females in 1-gallon cylindrical cardboard traps hung in apple trees attracted males, but their attractiveness declined with age, 1-day-old females being about twice as attractive as 9- and 10-day-old females. Laboratory-reared males marked with fluorescent powder and released in an apple orchard showed patterns of sexual activity similar to those of native males; of those released, 37% were recaptured. Methylene chloride extracts of virgin female moths were also attractive to males when used to bait traps (*1281*). The extracts attracted only *A. velutinana* males, although *Grapholitha molesta* was also in the vicinity (*962*).

Riblure, the synthetic attractant, attracted more than 3000 male moths to traps hung in apple orchards during 1968 spring flight (*952*). A comparison of various wick designs showed that a small polyethylene cap containing 10 $\mu$liters of synthetic pheromone and placed in 1-pint, round, plastic-lined ice-cream cartons or 1-quart, plastic, freezer containers with holes cut in the bottom and lid was much superior to cotton wicks or sand-filled vial wicks. Black traps were superior to white ones, and best results were obtained with traps having $1\frac{1}{2}$ inch openings in each end (*456*). In 1969, spring trials were held in an area of low infestations; mixing the synthetic lure with dodecyl acetate resulted in a 10-fold increase in catches, the traps attracting an average of 31.7 moths each (*339, 456, 958*). Other compounds that synergized the synthetic pheromone were undecyl acetate, dodecyl acetate, 10-propoxydecanyl acetate, 11-ethoxyundecanyl acetate, 11-methoxyundecanol, and 11-(ethylthio)undecanyl acetate. Inactive compounds were decyl acetate, undecyl acetate, tridecyl acetate, tetradecyl acetate, 10-methoxydecanol, 10-methoxydecanyl acetate, 10-ethoxydecanol, 10-ethoxydecanyl acetate, 10-propoxydecanol, 10-butoxydecanol, 10-

butoxydecanyl acetate, 11-methoxyundecanyl acetate, 11-ethoxyundecanol, 10-(ethylthio)decanol, 10-(ethylthio)decanyl acetate, 11-(ethylthio)-undecanol, *trans*-11-tetradecen-1-ol acetate, *trans*-9-tetradecen-1-ol acetate, *trans*-7-tetradecen-1-ol acetate, *trans*-5-tetradecen-1-ol acetate, *cis*-11-tetradecenol, *cis*-11-tetradecen-1-ol formate, and *cis*-11-tetradecen-1-ol propionate (*958*).

*Attagenus megatoma*, black carpet beetle

Comparative laboratory tests using Styrofoam traps containing filter paper impregnated with 0.01 mg (20 female equivalents) of the synthetic pheromone (*trans*-3,*cis*-5-tetradecadienoic acid) showed that an average of 7.7 males were trapped in 5 minutes, 14.3 in 15 minutes, 21.3 in 30 minutes, 29.3 in 1 hour, 34.3 in 2 hours, and 39.3 in 18 hours (*616*).

*Bryotopha similis*

*Bryotopha* sp.

These species are morphologically similar, with identical wing patterns, but the unidentified species is yellowish and *B. similis* is grayish. Field tests conducted over a period of at least 2 months showed that a total of 240 male *Bryotopha* sp. were attracted to traps containing *trans*-9-tetradecen-1-ol acetate and 45 male *B. similis* were attracted to *cis*-9-tetradecen-1-ol acetate. The attraction was highly specific, with no crossover, even though the traps were only 30 cm apart. The average hours for male activity are modally different in each case but overlap broadly (*956*).

*Choristoneura fumiferana*, spruce budworm moth

*Choristoneura pinus*, jack-pine budworm moth

Wooden boards coated with Tanglefoot were attached to tree trunks about 1.5 m above the ground in infested forests and sample vials or small screen cages containing virgin female spruce budworm moths were placed in the center of the board. Such traps attracted numerous males of this species at sunset (*402, 1015*). Identical traps baited with virgin female jack-pine budworm moths and placed in the same area with spruce budworm-baited traps caught males of *C. pinus* only (*1015*).

Specificity field tests were conducted during 3 or 4 nights with virgin females of 4 additional species of *Choristoneura* to test their attractiveness to *C. fumiferana* and *C. pinus* males. Each trap was baited with a single female. *Choristoneura occidentalis* in 17 traps attracted 181 *C. fumiferana*; *C. biennis* in 18 traps attracted 293 *C. fumiferana*; *C. orae* in 3 traps attracted 3 *C. fumiferana* and 95 *C. pinus*; *C. viridis* in 6 traps attracted no *C. fumiferana* and in 4 traps attracted 15 *C. pinus*. A total of 61 traps baited with *C. fumiferana* attracted 1163 males of this species and only 10 *C. pinus* in 3 traps. A total of 7 traps baited with *C. pinus* attracted 895 males of this species (*1016*).

Traps (Sectar) baited with 100 μg of synthetic *C. fumiferana* pheromone (*trans*-11-tetradecenal) contained in polyethylene cap stoppers were tested in infested Canadian fields. These traps were competitive with virgin female-baited traps in Alberta and Ontario, capturing numerous *C. fumiferana*. When tested in populations of *C. occidentalis* in British Columbia and Oregon, males of this species were also attracted. This pheromone may also be a major component in the pheromone system of *C. occidentalis* (*1251a*).

*Choristoneura rosaceana*, oblique-banded leaf roller moth

Traps baited with the synthetic pheromone (*cis*-11-tetradecen-1-ol acetate) caught numerous males in field tests. The trans isomer was a very potent inhibitor to male response in the field, although it had no inhibiting or masking effects in the laboratory bioassay (*966*).

*Costelytra zealandica*, grass grub beetle

The larval stage of this insect is of major economic importance in New Zealand, causing considerable pasture damage. In field tests, pure phenol mixed with water at concentrations varying from 500 to 10 ppm were attractive to the male beetle. On 1 evening 7 traps, each containing approximately 100 ppm of the mixture, caught a total of 71 males and no females. On the following evening 222 males and 19 females were caught in these traps. Control traps containing water and placed alongside the baited traps caught no beetles (*540, 541*). On one occasion about 1500 mixed male and female beetles were found in a circle (3-foot radius) around a trap containing 2 liters of a crude phenol-water mixture. Only insignificant numbers of beetles were caught with undiluted 1-naphthol, 2-naphthol, sodium 2-phenylphenolate, 3-aminophenol, 4-aminophenol, 2-cresol, 4-cresol, guaiacol, vanillin, thymol, carvacrol, catechol, resorcinol, quinol, pyrogallol, 1,2-dihydroxy-3-allylbenzene, and 1,2-dihydroxy-4-allylbenzene (*540*).

*Dendroctonus brevicomis*, western pine beetle

Field trials were conducted by Bedard *et al.* (*127*) in ponderosa pine stands using traps affixed to the boles of trees at a height of 7.5 m above the ground. Traps were either strips of hardware cloth coated with adhesive or hardware cloth cylinders. Ponderosa pine bolts infested with 50 virgin females were more attractive to both sexes than 2 mg of *exo*-brevicomin plus 20 mg of myrcene (a natural component of the pine oleoresin), which was more attractive than an extract of 2 gm of female frass; 2 mg of *exo*-brevicomin alone was least effective, but it still attracted the beetles. Myrcene alone was not attractive. Although the female-infested bolt and the frass extract attracted the beetles in a 1:1 ratio of males to females, *exo*-brevicomin alone and mixed with myrcene attracted the beetles in a ratio of 2.2 males to 1 female. The aggregation of *Temnochila virescens chlorodia*, 1 of the principal predators of *D. brevicomis* larvae and adults, on trees under mass attack is probably due to *exo*-brevicomin. Verbenone sprayed as an aqueous suspension on logs infested with *D. brevicomis* did not increase the response (*875*).

Pitman (*866*) reported that brevicomin alone was either mildly attractive or completely unattractive in field olfactometers. However, a mixture of this compound plus 3-carene was quite attractive; 3-carene was superior to myrcene and other terpenes as a synergist.

According to Vité and Pitman (*1228*), frontalin, a very close analog of brevicomin isolated from the hindgut of *D. brevicomis* males and from *D. frontalis* and *D. pseudotsugae* females, was more effective in attracting *D. brevicomis* (1 male to 2.6 females) when mixed with ponderosa pine oleoresin than *exo*-brevicomin plus oleoresin, *endo*-brevicomin plus oleoresin, several analogs plus oleoresin, or feeding *D. brevicomis* females. When the tests were conducted in tree trunk-simulating olfactometers, a mixture of frontalin, brevicomin, and oleoresin was much more attractive (1 male to 1.1 females) than brevicomin plus oleoresin (1 male to 0.4 female), which was equal in attraction to frontalin plus oleoresin (1 male to 2.6 females), frontalin plus brevicomin (1 male to 1.2 females), and feeding *D. brevicomis* females. The combination of frontalin plus brevicomin plus oleoresin was highly competitive with natural sources of attraction (*1220*). Bedard *et al.* (*125*) reported the following mean numbers of *D. brevicomis* caught over a 3-day period by various combinations of these constituents: brevicomin plus frontalin, 25.8; brevicomin plus frontalin plus myrcene, 59.7; brevicomin plus frontalin plus 3-carene, 26.0; frontalin plus 3-carene, 5.7.

Mass attack was induced in ponderosa pine forest in California by using Stikem-coated plastic tubes provided with 5 polyethylene caps each containing a mixture of 1 part of frontalin and 4 parts of 3-carene. The tubes were placed at eye level in the shade at least 5 m from ponderosa pines. The frontalin released from the caps was calculated to be about 5 mg/hour (*1229*).

The tube traps with cap wicks were used to test the effects of aggregants and oleoresin constituents on the beetles as well as their predators. Frontalin and brevicomin attracted comparable numbers of males; the female response was nearly 7 times greater to frontalin. Brevicomin attracted large numbers of the predator, *Temnochila virescens* var. *chlorodia*, which did not respond to frontalin. *Enoclerus lecontei* was not attracted by either pheromone, but it responded to myrcene, $\alpha$- and $\beta$-pinene, and camphene, whereas *T. virescens* did not (*874*). $\alpha$-Pinene is also implicated as an attractant for *Sirex* spp. wasps (*741*).

*Dendroctonus frontalis,* southern pine beetle

In field tests conducted in Texas pine forests, air was drawn from boxes containing recently infested logs and directed against test trees by means of a plastic pipe. Equal numbers of flying beetles landed in the presence of attractant material near both host and nonhost trees. The pheromone oriented the flight and stimulated subsequent activities such as landing, searching for breeding places, and entering the bark. Visual orientation supplemented the olfactory response in aggregating the population. Recently infested log sections were successfully utilized in directing the field population to attack and infest preselected trees in predetermined areas. A defined population was experimentally concentrated and decimated on resistant trees; the mortality was caused by oleoresin from the resistant trees (*436, 865*).

Emerging beetles remain in the vicinity of their brood trees and are instrumental in expanding the outbreak when emergence coincides with the availability of attractants emanating from previous attacks (*434*). The highest frequencies of flight, landing, and gallery construction are found in the immediate vicinity of the host material with olfactory stimuli dominating beetle behavior. The flight of responding beetles is terminated at wind speeds greater than $4\frac{1}{2}$ miles per hour (*313*). In forest stands, trees farther than 20–25 feet from a source of attraction are unlikely to be invaded (*435*).

Field experimental results suggest a strong influence of the host condition on the aggregation behavior of the beetle. Large trees, stumps, and log sections of *Pinus taeda* attacked became greater targets for mass aggregation than small infested trees and logs. Severing infested trees from the stump and cutting billets from highly attractive trees caused such trees and billets to rapidly lose their attractiveness to beetle flight through drying. Loss of attractiveness coincided with the beginning of extensive feeding in suitable host tissue. The physiological condition of the host material apparently determines the rate of feeding that, in turn, affects production and release of the attractant (*1221*).

Small glass capillaries containing 1 mg of synthetic frontalin placed in field olfactometers attracted small numbers of flying beetles. The response was increased by adding fresh oleoresin from *Pinus taeda* and synthetic *trans*-verbenol from the female beetle. The combination was highly competitive with an olfactometer containing oleoresin and 2500 crushed beetles (*671*).

Both males and females of the beetle predator, *Thanasimus dubius,* follow aggregations of *D. frontalis* in response to pheromones, especially frontalin released by feeding *D. frontalis* (*1233*). The sex ratio of the attracted predator was changed by adding verbenone to frontalin; this increased the ratio of responding males and reduced the total response of both sexes (*1267*). A mixture of frontalin plus α-pinene plus verbenone (1:10:0.3) was more attractive to the predator, *Medetera bistriata*, than a mixture of these components in the ratio 1:4:0.5, frontalin plus α-pinene (1:10), or frontalin plus α-pinene plus *trans*-verbenol (1:4:0.5). Billets infested with female *D. frontalis* or male *Ips grandicollis* were quite attractive to this predator (*1267*). A mixture of frontalin, verbenone, and α-pinene attracted significant numbers of the larval parasite *Heydenia unica* (*277a*).

*Dendroctonus obesus*, spruce beetle

Screen-covered logs infested with virgin females attracted mainly males initially. After first flight, approximately equal numbers of both sexes responded to logs without introduced females. Beetles continued to respond well to logs attacked by females throughout the flight period (*362*).

*Dendroctonus ponderosae*, mountain pine beetle

In Idaho field tests, billets cut from *Pinus monticola* and containing 50 female beetles were placed in field olfactometers; they attracted an average of 1.8 beetles from the natural population. Spraying petroleum ether solutions (1 mg/4 ml) of *trans*-verbenol, isolated from feeding-female hindguts, on the billets increased the number attracted to 21.2. In California field tests, the beetles responded readily to material infested with *D. brevicomis*; this cross-attraction was enhanced by applying *trans*-verbenol, which is itself unattractive when sprayed on uninfested billets (*872, 875*). Female beetles feeding on orange rinds failed to attract the flying population even while producing large amounts of *trans*-verbenol, showing that other volatile substances originating from infested host material are necessary, in addition to *trans*-verbenol, for inducing mass aggregation (*875, 1220*).

Synthetic *trans*-verbenol, the major insect pheromone, tested with oleoresin under Idaho field conditions was highly attractive to both sexes of *D. ponderosae*. Metal and canvas sleeves and various types of olfactometers were used. Emergent female *D. brevicomis* and *D. frontalis* crushed to powder at −70°C were also attractive to *D. ponderosae* (*872*).

Paperboard cylinders (8 feet × 10 inches) covered with adhesive and baited with *trans*-verbenol plus α-pinene were effective for mass-trapping *D. ponderosae*. *trans*-Verbenol plus myrcene showed some attractiveness, while limonene and 3-carene did not synergize *trans*-verbenol (*865, 868*).

*Dendroctonus pseudotsugae*, Douglas-fir beetle

Rudinsky (*993*) has shown that females are attracted by the oleoresin of fresh windthrown and cut trees, especially the α-pinene, camphene, and limonene fractions. Shortly after entering the host, the feeding beetle attracts masses of both sexes. The beetles are repelled by the resin and its fractions ($\alpha$- and $\beta$-pinene, camphene, limonene, terpineol, and geraniol) at close range in the laboratory. It has also been shown (996, 998, 999) in the field that male stridulation triggers the female to mask her own pheromone, thus rapidly stopping the mass attraction of flying beetles; the primary attraction to host oleoresin and terpenes is not affected. Masking occurs as soon as all females have been found by males. Mating is assured because the virgin females attract twice as many males as females to the invaded host. The mask prevents the aggregation of unneeded males and females. The female can quickly annul the mask if the male is destroyed by predators before she is mated.

The response of flying beetles to field olfactometers baited with logs infested with females exhibited a distinct diurnal and seasonal pattern which was influenced by several environmental factors; temperature was the most important. Response flight began in the morning, as soon as a threshold temperature of 68°F was attained, and reached its peak around noon. Later in the season, high temperatures during midday always resulted in marked depression of flight. Beetles flew toward the infested logs against the wind, and the intensity of the wind limited their flight; beetles stopped flying when the wind reached a continuous velocity of more than 5 miles per hour. The ability of baited olfactometers to compete with naturally occurring attraction centers indicates their usefulness in survey (and possibly in control) of this destructive insect. The attractive substance from the frass can be concentrated and used under field conditions. The duration of the attraction can be artificially prolonged throughout the flight season, since females under field conditions cease production of the attractive substance after mating (992).

Under field conditions, synthetic frontalin, isolated from the hindguts of emerging females, was highly competitive with natural sources of attractants when combined with camphene, a monoterpene obtained from Douglas-fir oleoresin. Also, when used in tree trunk-simulating olfactometers of a vertical and horizontal configuration, milligram amounts of frontalin plus $\alpha$-pinene added to infested billets attracted 1343 beetles (530 males and 813 females) during 1 afternoon. Frontalin with $\alpha$-pinene, limonene, or camphene attracted an occasional *D. ponderosae*. Frontalin alone attracted large numbers of male and female *Thanasimus undulatus*, a predator of *D. pseudotsugae* (873).

*Diabrotica balteata*, banded cucumber beetle

Sticky board traps each baited with an attractive female and placed in a soybean field lured males within a short time (322). Traps baited with filter papers impregnated with extracts of female abdomens were set in a

Latin square arrangement, 9–12 feet apart, in fields having high populations of the insect. During the first 24 hours, a 10-female equivalent of the extract was almost 2.5 times as attractive as a virgin female, whereas a 1-female equivalent was about one-third as attractive. Lure was lost steadily by volatilization, so that by the third day the 10-female equivalent was only half as effective as a virgin female.

*Diatraea saccharalis*, sugarcane borer

Sticky cylinder traps baited with virgin females or extracts of female abdomens were used to trap male moths in the field (*857*). The attractant was highly volatile, lasting only a few hours when exposed to the environment on filter papers. Sexual activity of moths in the field was confined mainly to the period between 1:00 AM and 4:00 AM. Since the attractant in extracts is so volatile, traps should not be baited with such extracts more than an hour or so prior to the natural mating period. More than 4 times as many moths were attracted to single virgin females as to a light trap during September and October; the females attracted only males, while the light trap caught approximately equal numbers of both sexes.

Males who were released in a lightly infested sugarcane field containing a cylindrical screen trap coated with Stikem provided with 10 virgin females were recaptured in small numbers from distances up to 320 feet north of the trap (*847*).

*Euproctis chryssorrhoea*, gold tail moth

The trapping method of Dyk (*363*) may be successfully utilized with males of this insect (*694*). A trap containing a single female lured 38 males in 1 day from a distance of 3 km and 18 males on the second day (*579*).

*Grapholitha molesta*, Oriental fruit moth

Extracts of virgin female abdomens (10 female equivalents) spaced 1 foot apart in an apple tree caught numerous males of this species only, although *Argyrotaenia velutinana* was also in the vicinity (*962*). Synthetic cis-8-dodecen-1-ol acetate, the pheromone of this species, attracted more than 1200 males in field tests; 10–200 μg of the compound absorbed on polyethylene gave the best results (*960, 990*).

*Gypsonoma haimbachiana*, cottonwood twig borer

From late June through mid-August 1970, the period of peak flight of this moth, Stikem-coated pheromone traps baited with 4 virgin female moths in a copper screen cage were kept in tree rows in a cottonwood plantation. A total of 238 males and 43 females were caught in these traps. Dead virgin female-baited traps were equivalent to unbaited traps. Male-baited traps caught very few moths. Female attractiveness decreased markedly after 1 day and was practically nil after 4 days (*854*). In Australia, survey traps baited with 1 μliter of the synthetic pheromone impregnated on polyethylene dispensers remained effective for more than 2 months (*991b*). Traps baited with $10^{-2}$ $\mu$liter of pheromone captured most males in Australian tests (*991a*).

*Harrisina brillians*, western grape leaf skeletonizer

Female moths were extracted with methyl anthranilate, a natural attractant occurring in grapes, and the extract was exposed in the field at 1.5 female equivalents per trap; $3\frac{1}{2}$ times as many males were caught as females. Males were readily attracted in the field by a benzene extract of the virgin females. A concentration of 2 female equivalents in each of a variety of traps caught the following numbers of males: (a) metal tube (Graham) trap with screen ends but a cardboard Tanglefooted liner, 19 males and no females; (b) type (a) trap with screen ends, 14 males; (c) milk carton with inside walls Tanglefooted, 3 males; (d) ice-cream carton with Kraft paper liner, 13 males. A total of 77% of the males were caught during the first 24 hours, with the limit of attractiveness being about 36 hours. The benzene extract also attracted males to water-filled fiberboard pails hung on wires strung across stream beds along which grapevines were abundant (*92*).

*Heliothis virescens*, tobacco budworm

*Heliothis zea*, bollworm

The numbers of male moths caught in 3 blacklight traps baited with virgin female *H. virescens* and in 3 similar traps baited with virgin female *H. zea* in a 4.3-acre cotton field in Texas were compared for 19 weeks. The live females were held in small cages hung below the top of the trap funnel. Although traps provided with females caught more males than light traps alone, the difference was not apparent until the data were adjusted to compensate for trap location and the effect of wind. There was interspecific attraction between male bollworm moths and female tobacco budworm moths used to bait the traps (*534*). Larger numbers of *H. virescens* males were caught in female-baited traps that were painted fluorescent green than in those painted other colors (*534a*).

*Holomelina aurantiaca* Complex, tiger moths

Field responses of males of the complex to synthetic 2-methylheptadecane, isolated from females of these species, suggested that secondary chemicals may be involved. Males of *H. aurantiaca*, *H. immaculata*, and *H. rubicundaria* were easily attracted into the traps; males of *H. lamae* approached the traps upwind but rarely entered. Lower concentrations of the attractant did not change this behavior. *H. nigricans* and *H. ferruginosa* males did not orient to the traps. Attraction of males occurred only during night hours. 2-Methylpentadecane, 2-methylhexadecane, 2-methyloctadecane, 2-methylnonadecane, 4-methylheptadecane, and *n*-octadecane were completely unattractive to male tiger moths. Although 2,15-dimethyl-hexadecane did attract some males of *H. immaculata*, it was about one-tenth as attractive as 2-methylheptadecane (*953*).

*Ips confusus*, California five-spined *Ips*

A simple field method, in which air was blown from cages containing *Ips*-attacked logs, showed that both sexes were attracted (*1222*). Large-diameter aluminum tubes were placed in vertical or horizontal positions as containers for initially infested logs known to be highly attractive. Air was drawn by a fan along the infested log and blown against a log or tree to stimulate attack, or against a board or glass for collection of the attracted insects. A modification of this procedure was used to produce large amounts of attractive material. Insects were caught in rotary net traps or window traps to which attractant was applied. Such a rotary net supplied with *I. confusus* attractant caught, in $7\frac{1}{2}$ hours, 39 *I. confusus*, 35 *I. latidens* plus *I. guildi*, 10 *Dendroctonus valens*, 72 other scolytids, and 25 ostomidae (bark-beetle predators).

Field attraction showed a definite diurnal pattern (8:00–10:00 AM and 2:00–6:00 PM). Beetles preferred the regions of least wind, but light conditions had little influence on response flights except that direct solar radiation seemed to be avoided. Beetles responded against the wind along an olfactory gradient. The distance over which populations may respond in colonizing new host material is estimated at 500–1000 m (*432*).

The 3 compounds making up the pheromone complex were tested in a treeless brush field within 1 km of naturally infested ponderosa pine logging slash in California (*1286, 1288*). Prevailing winds were moving up the valley toward the infestation. The traps consisted of hardware-cloth cylinders coated with adhesive and fastened to 1.5-m rods. Air (50 cm³/minute) was metered through a tube containing the attractants and through the traps. A mixture of 1.5 mg of compound [38] plus 1 mg of compound [39] plus 0.5 mg of compound [40], and a bolt infested with 20 males, were each highly attractive to the beetles; an extract of 2 gm of male frass was considerably less attractive. Solutions of all 3 compounds in hexane or heptane evaporated from cotton or paper wicks elicited a very weak response. Males and females of 2 predators (*Enoclerus lecontei* and *Temno-chila virescens* var. *chlorodia*) also responded to the synthetic attractants (*928, 1288*), as did an *Ips* parasite, *Tomicobia tibialis* (*1288*).

Bedard and Browne (*124*) have described a portable delivery-trapping system for evaluating insect chemical attractants, especially for *I. confusus*, in nature.

Rice (*928*) has tested in field olfactometers and Tanglefoot post traps a number of monoterpenes present in *I. confusus* host plant. α-Pinene and β-pinene were highly attractive to the *Ips* predators (*E. lecontei* and *T. virescens* var. *chlorodia*); 3-carene, myrcene, and limonene were less attractive. *Temnochila virescens*, but not *E. lecontei*, was strongly attracted to heptane.

*Lasiocampa quercus*, oak eggar moth

Using caged females and marked males released from various distances, Dufay (*359*) determined that males could be lured from a distance of about 1 km, depending upon wind velocity and direction. The sex attractant could therefore be used to determine the size of field infestations.

*Laspeyresia pomonella*, codling moth

In apple orchard tests with 4 paper-cylinder traps, each baited once with 50 female equivalents of an extract of whole virgin females, Butt and Hathaway (*262*) caught an average of 9.6 males/trap/day over 24 days from May 20 to June 13 and 18 males/trap/day from May 29 to June 13. In another test using 72–100 traps in a 12-acre apple orchard, traps baited weekly with 50 female equivalents of extract were set 65 feet apart. Each trap caught 0.86 males/day from July 22 to September 18. The difference may possibly be accounted for by masking in the summer tests, since the extract was prepared by homogenization whereas that used in the spring tests was prepared by grinding in a mortar with solvent. Orchard traps containing live females caught 3 times as many males as traps baited with female extracts.

Proverbs *et al.* (*891*) have surveyed areas for the pest by using traps made of cylindrical paper cartons (1-pint capacity), with the bottom and lid replaced by mesh screen, holding 10–15 virgin female moths. The carton was suspended within a 2-quart carton lined with adhesive-coated waxed paper, from which one-third of the lid and one-third of the bottom were removed to permit attracted males to enter. These traps could also be used with female extracts held on filter paper. This type of trap has also been used by Madsen (*742*) with 10 live virgin females replaced every 10 days during the summer; the traps were more effective when placed in the upper third of the apple trees. Traps with the females removed retained their attractiveness to males for a week or longer. Tests with up to 100 females per trap did not result in increased male attraction over that obtained with 10 females.

Catches of released, laboratory-reared males treated with doses of 25 or 40 krad of gamma irradiation were lower than those of unirradiated males in traps baited with live females (*1261*). Both Proverbs *et al.* (*891*) and Butt (*46*) have used female-baited carton traps to monitor the dispersal of sterilized male moths.

Wong *et al.* (*1282, 1283a*) used 1-gallon cylindrical cardboard cartons, coated on the inside with adhesive, to recapture released males; each trap contained 1 or 2 small screen cages baited with 1- or 3-day-old virgin females. Peak activity occurred during the hour immediately after sunset.

*trans*-8 ,*trans*-10-Dodecadien-1-ol, reported by Roelofs (*62, 959*) to be the natural sex pheromone of the female moth, successfully attracts males in the field (apple orchards). A special trap devised by Batiste (*121*) to time the flight activity of this insect utilizes a moving sticky belt as the collecting surface for males attracted to a modified carton trap holding captive live females. The belt moves at a regulated, continuous rate and indicates when males are caught by their position on the belt. Male flight was initiated well before sunset and terminated when the temperature reached 61°F. A larger modification of this trap has been described by Schoenleber *et al.* (*1049*), who reported peak catches near sunset, but a large number continued to be trapped until 1:00 AM.

Howell and Davis (*560a*) found that male codling moths captured in female-baited traps could be protected from predaceous yellow jackets, *Vespula* sp., by setting yellow-jacket traps containing heptyl butyrate nearby.

*Limonius californicus*, sugar-beet wireworm

Microscope slides moistened with a 70% ethanol female extract and placed in an infested field attracted numerous males within 10 seconds from as far away as 40 feet. Males moved rapidly toward the slides, crawling and flying excitedly, and extruded their genitals repeatedly after reaching the slides. Males were attracted in a field that had no previously recorded history of infestation, showing that the crude extract can be used as a survey tool where population density is low (*732*).

*Lobesia botrana*, grapevine moth

Trapping tests in French vineyards (*288*) showed that the numbers of males lured by females were dependent upon wind direction, the number of females used, temperature, and humidity. Six females per trap were satisfactory. Captures began about 30 minutes before sundown and ceased at nightfall. The optimum temperature was 15°–23°C. A combination of the number of males captured and the number of eggs found in the trapping area gave an idea of the size of the population in a vineyard. A total of 1173 males was caught in 1 trap in 3–4 hectares.

*Lygus hesperus*, lygus bug

Trapping experiments using live females were conducted in alfalfa hay fields in California (*1152, 1153*). The traps consisted of adhesive-coated half-gallon ice-cream cartons with both ends fitted with removable funnels made from aluminum screening; the bait insects were housed in a half-pint carton with screened ends. A hole the size of the half-pint carton was cut in the side of the large carton, and the small carton was glued in place. These traps were mounted on metal stakes and were either fixed or provided with ball bearings, and a wind vane enabled them to remain aligned with the wind at all times ("omni traps"). Traps were spaced at 12-foot intervals and examined several times daily. Omni traps caught significantly larger numbers of males than fixed traps. No males were captured at temperatures below 54°F or above 80°F, nor when the wind exceeded 12–15 miles per hour.

*Fig. XII.1.* Pheromone-blacklight trap used for attracting male tobacco hornworm and cabbage looper moths. (a) View of trap with caged live virgin females in place; (b) close-up of cage with female moths. [By permission of U.S. Department of Agriculture.]

*Fig. XII.1. (b)*

*Lygus lineolaris*, tarnished plant bug

Stand-mounted, adhesive-coated, vinyl plastic sheets supporting small cages containing virgin females were tested in fields of *Erigeron canadensis*, which is a host plant for this insect. Significant numbers of males were caught in these traps (*1020*).

*Manduca sexta*, tobacco hornworm moth

From 1 to 30 live virgin females were placed in a field with blacklight traps, as shown in Figure XII.1a, b. For each additional female, up to 10, the male catch increased by a factor equal to the male catch of the trap without virgin females. Vertical positioning of virgin females at the trap side did not influence male catches in the trap. The use of up to 10 females per trap increased the catch 10-fold. Tests were conducted from 5:00 PM to 6:00 AM (*552*). Cantelo and Skot (*281*) have described a cylindrical aluminum screen cage useful for holding live virgin females at light traps. The ends consist of clear plastic cups.

Omnidirectional funnel traps equipped with blacklight lamps and baited with 2 virgin females each were tested in the field to determine their effectiveness in capturing moths from native populations. Traps were spaced 1 or 2 miles apart in a 314-square mile circular area at 3–6 traps per square mile. The number of males captured in female-baited traps was nearly 4 times greater than the number caught in unbaited traps. Trap spacing had no effect on captures. The effectiveness of baited traps appeared to be slightly reduced when the area covered was increased from 3 to 16 square miles (769). There was no significant difference between the number of males captured by a trap containing virgin females and one blacklight lamp and by a trap containing virgin females and four blacklight lamps. Four-lamp traps collected twice as many females as the 1-lamp traps (282, 282a).

*Ostrinia nubilalis*, European corn borer

Field traps baited with 14 μg of synthetic *cis*-11-tetradecen-1-ol acetate formulated in 0.7 ml of olive oil-acetone (1:1) were as competitive for wild males in Iowa as traps baited with 4 virgin females (690a). The traps also attracted male *Argyrotaenia velutinana*.

*Pectinophora gossypiella*, pink bollworm moth

Preliminary tests by Ouye and Butt (840) and by Berger et al. (140), using 8-ounce cup-type disposable traps lined with Tanglefoot showed that a crude extract of the female moth could be used in a field survey program. Although this trap used in cotton fields captured the males satisfactorily when the Tanglefoot was fresh, the Tanglefoot tended to harden and failed to hold moths after exposure to the air for 1 or 2 days; it also interfered with fluorescent dyes used to tag moths. Tests were then made with a modification of this trap, in which a portion of the bottom of a trap was replaced with unbleached muslin; an aluminum foil container for calcium cyanide powder was taped immediately below the muslin (470). The cyanide gas released entered the trap through the muslin but kept the dead moths out of the powder. Later, a larger trap was constructed from sheet metal; its bait chamber was similar in design to the cup-type trap but a funnel connected to the bottom channelled the dead moths into a 1-quart cyanide jar. Each trap was baited with filter paper saturated with methylene chloride extract of 100 adult females at least 3 days old (470).

A study was made of the efficacy of different solvents for impregnating the methylene chloride extract on wicks (472). A methylene chloride extract was prepared of whole virgin females, freed of solvent, and made up to a concentration of 50 females per ml with acetone, benzene, hexane, methylene chloride, and ether. Each of these solutions was impregnated on cartridge-type wicks in sheet-metal field traps at 100 female equivalents per trap, and the traps were set in a one-sixth-acre cotton-field cage on October 1. Each treatment was replicated 6 times in a Latin square arrangement, a trap containing 100 female equivalents in methylene chloride recharged twice a week served as a check, and males were released in the cage daily until December 2. There was no difference in the effectiveness of lure between the solvents; all treatments lured equal numbers of males for the first 3 weeks, after which the check trap outcaught the test traps since the latter had not been recharged.

Guerra and Ouye (492) studied the catches obtained in field cages using female extract in various types of traps. Traps equipped with a lighted blacklight lamp were the most effective. A modified Frick trap (a 1-quart-ice-cream container 7 inches high and $4\frac{1}{2}$ inches in diameter with a $\frac{3}{4}$-inch diameter hole in each end and lined with Stikem) suspended horizontally was as attractive as a similar trap suspended at 45°. The sheet-metal trap and cup-type trap were less effective. Calcium cyanide was not repellent when it was used as a killing agent in baited traps.

Guerra (490) studied male catches in cotton plots using 8 modifications of the Frick trap baited with 50 female equivalents of crude extract. Modifications tried were 2 trap lengths, 2 trap diameters, and 2 diameters of hole openings in trap ends; a cup-type trap was used as the standard. All modified traps caught approximately equal numbers of males over a 10-day period, except for the trap that was $4\frac{1}{2}$ inches long and $3\frac{1}{2}$ inches in diameter with $3\frac{1}{2}$-inch openings on each end, which was less effective.

A trap baited with 6 mg of hexalure, a synthetic attractant for male pink bollworm moths, caught 51% of the male moths caught by the natural sex attractant from 25 virgin females during 7 days; over a 14-day period, the synthetic/natural catch ratio was 42%. In extensive field tests 60 mg of hexalure was initially about equal to 50 female equivalents of natural lure, but after 5–7 days the synthetic lure at all test dosages was superior to either natural lure or to live caged virgin females (481).

Extensive field testing of hexalure in Arizona and California cotton fields confirmed that hexalure was an excellent survey tool (51, 656). The traps used were made from 7-ounce polystyrene cups by using detergent and water as the trapping medium; an absorbent paper wick impregnated with the lure solution was attached to a string across the opening of the cup. One mg of hexalure per trap was equal or superior to 5 female equivalents of a methylene chloride extract of females, and 25 mg was superior to 50 female equivalents, catching males over a 10-week period while the natural lure failed much earlier. Little seasonal variation was observed in the effectiveness. The advantages of hexalure over the natural lure are chemical stability, resistance to microbial decomposition, longer test life, uniform potency, ease of synthesis, and lower cost (about 1000 times less expensive than natural lure). Hexalure is now in use by USDA's Plant Protection Division in its extensive pink bollworm survey program (481, 482).

In Phoenix, Arizona, a hexalure (25 mg) trap alone caught 189 males over an 8-month period, a light trap alone caught 374 females and 493 males, and a light plus hexalure trap caught 397 females and 556 males. In Waco, Texas, the corresponding figures were 281 males, 110,643 combined sexes, and 150,204 combined sexes. In Brownsville, Texas, the figures were 32 males, 3 females and 24 males, and 3 females and 20 males. The apparently increased efficiency of the traps at Phoenix and Brownsville was probably due to the presence of larger natural populations at Waco, with increased numbers of natural females competing with hexalure (*90*).

In the Imperial Valley of California, adhesive-coated cylindrical traps baited with 10 μliters of hexalure were more effective when placed in cotton 1.2 and 1.8 m above the soil surface than those placed at 0, 0.6, and 2.7 m above the ground. Significantly more moths were caught in traps placed in sorghum fields than in those placed in fallow ground. In August, all moths were caught between 11:00 PM and 6:00 AM, with peak activity at 2:00–3:00 AM; in late October, peak activity occurred at 9:00–10:00 PM (*1069*). Embody (*379*) has described possible methods for measuring the effective range of hexalure-baited traps for pink bollworm moths.

*Phytophaga destructor*, hessian fly

Cylindrical cages 1 inch in diameter and 3 inches tall containing 5 newly emerged virgin females were placed on the surface of the ground in a field. A larger wire cage, 3 inches in diameter and 10 inches tall, painted with Tanglefoot, was placed around each small cage. A total of 5 such cages exposed from April 2 to April 7 caught 3627 males. The males advanced against the wind; the greatest numbers were found in cages up to 3 inches above the ground. The weather was clear and warm, with a wind velocity at 6–8 miles per hour from south to southwest (*285*).

*Pleocoma minor*

Winged males of this coleopterous insect seek out flightless females in the soil. Polystyrene traps containing a live female were buried to soil level in a pear orchard and confined under a tight-fitting, wire mesh-screen baffle. An overlapping plastic container top with a 2-inch diameter entrance hole permitted males to drop into the traps. This buried trap was more effective in catching males than several other above-ground types constructed of metal funnels on stands under screen baffles with a female confined in a screen mesh cage wired to the funnel or baffle. It was not known if the females used had been inseminated or were virgin. Blacklight traps caught 10 times as many males as female-baited traps (*1337*).

*Popillia japonica*, Japanese beetle

Virgin females either tethered in open fields or confined in traps attracted males in large numbers within seconds after exposure. A single female exposed in 1 trap attracted 380 males in 1 hour and 9 females in a trap attracted almost 3000 males in the same period. Three other traps, each containing 9 females collected on successive days, attracted a total of 19,428 males and 345 females in 2 months, but 85% of these were attracted within 1 week (458, 708, 709). Traps baited with extracts prepared from females with a variety of organic solvents failed to attract males in the field (410).

*Porthetria dispar*, gypsy moth

The gypsy moth is an excellent example of an insect whose sex attractant may be used for survey and possibly for control. Males fly into the wind and pick up the scent as they approach the nonflying females (153). A very complete history of the gypsy moth problem in Europe, Asia, Africa, and the United States is given by Schedl (1021), who describes in detail the biology, life history, host plants, predators, and methods of survey and control for this insect. However, many detailed observations of male attraction by females had previously been given in 1896 by Forbush and Fernald (417), whose attempts to use traps baited with living virgin females for control proved to be unsuccessful, although many males were caught. Later the same traps were found to be efficient detectors of moth infestations.

Early attempts to find the best type of gypsy moth trap involved the use of many types of paper or cardboard traps, baited with live virgin females or their crude extracts, which were tacked to the trunks of trees in and around infested areas (304, 554). Major advances were the findings that living females could be replaced by their last 2 abdominal segments and that an extract of these tips was similarly effective. Collins and Potts (304) described in 1932 the use of such materials, from 1913 to 1931, to survey areas for infestation; their report contains many valuable detailed test results relative to type of traps, effective distance, and stability of the attractant.

In 1934, Jacentkovski (579) described tests conducted in Czechoslovakian wooded areas in which small matchboxes containing virgin female gypsy moths were tacked to tree trunks and surrounded by sticky flypaper strips. A total of 10 such traps were set out and checked twice each day for male catches. Many males were caught on the papers during daylight hours, mainly at midday, and it was estimated that such males could be lured from a distance of approximately 100 m. Empty containers that had previously held females remained effective for 2 or 3 days. This work was reviewed by Eckstein (369) in 1937. Komareck and Pfeffer (694) reported that they could verify the work of Jacentkovski (579) and that the method could be successfully used in the survey of areas for gypsy moth.

Prüffer (895) tested the attractive power of many extracts, prepared from different parts of the female's body, which were contained in cloth bags hung on tree branches in a garden at distances of 0.5–1.0 m apart; males released in the garden experienced no difficulty in locating active materials.

As a result of survey tests conducted over a period of several years with benzene extracts of virgin female gypsy moths and nun moths, Görnitz (459) reported that the gypsy moth extract lured many more male nun moths than male gypsy moths, whereas female nun moth extract lured almost exclusively male nun moths. However, Schwink (1055) reported that females of each species confined in small containers placed on glued boards lured equal numbers of males of either species in the field.

Although Görnitz (459) reported that a benzene extract of female gypsy moths lost little or no activity during storage at room temperature over a 4-year period, Haller et al. (502) found that the abdominal tip extract deteriorated rapidly, so that most of its original potency was lost when it was exposed during the flight season 1 year later. The activity of the extract could be stabilized by prompt hydrogenation, followed by storage at 4°C (7, 502). Hydrogenated lures collected 9, 4, 3, 2, and 1 years previously proved to be equally effective in survey traps. Tests to determine the effect of lure concentration on male moth captures showed that concentration (tips per trap) paralleled captures but increasing the tip strength above 16 per trap did not result in significantly greater captures. Because of this finding, traps holding natural extract and used in the field for survey work were baited with 12–16 tips each. In 1957, the charge was further reduced to 12 tips per trap because tests made each summer over a 7-year period failed to demonstrate any significant difference between traps containing charges equivalent to 12 and 16 tips (554). Bierl et al. (159) were able to increase the activity of crude benzene extract of female tips collected in Spain approximately 11-fold and of an extract of laboratory-reared female tips 2-fold by treating the crude extracts with m-chloroperbenzoic acid, thus epoxidizing the hydrocarbon fraction.

Maksimović (745, 746) set forth the results of his investigations of the relation between the number of gypsy moth egg clusters and the number of males caught in traps baited with the attractant extract. These tests, conducted in Yugoslavia, showed a ratio with a certain regularity. Traps baited with the equivalent of 12 female tips were most effective, catching males over a period of $1\frac{1}{2}$ months, so that such traps could be placed in and around infested areas immediately before the expected start of male flight. Female extract hydrogenated immediately following its preparation was somewhat more attractive than an extract hydrogenated after 3 months' storage.

In a comparison of the effectiveness in Yugoslavian forests of several types of traps baited with crude female extract, Maksimović and Marović (747) found that cylindrical metal traps (10 × 17 cm) (modified Graham traps) provided with metal funnels on the ends caught an average of 12% more males than such traps with wire mesh ends. Plastic (Steiner) traps caught 0.2–8.3% more males than the metal funnel traps. Cup-type cardboard traps were least effective.

An excellent report by Holbrook *et al.* (554) in 1960 discussed in detail trap design and color, bait dispensers, duration of bait effectiveness, bait strength requirements, trap height and placement, and other investigations related to the gypsy-moth trapping program. For many years, the U.S. Department of Agriculture made use of hydrogenated benzene extract of the abdominal tips in the metal (Graham) field traps, shown in Figure XII.2, to locate infested areas and to determine the size of the infestation by the numbers of males caught in these traps. The metal traps contained

*Fig. XII.2.* Graham field trap formerly used for luring male gypsy moths to female sex pheromone. [By permission of the U.S. Department of Agriculture.]

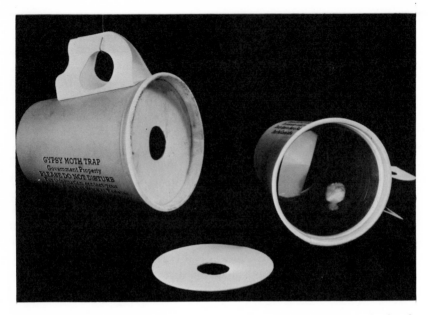

*Fig. XII.3.* Cup-type gypsy moth trap in use today. The sex attractant is placed on the cotton plug. [By permission of the U.S. Department of Agriculture.]

adhesive-coated (Tanglefoot) paper liners and rolled filter paper cartridges impregnated with lure. Males attracted to these traps flew in through a small round opening in the cardboard-cone ends and became entangled on the sticky surface of the liner. Although these metal traps were quite efficient, they had the following major disadvantages; they were (1) expensive, (2) laborious to clean after each season's use, and (3) bulky, requiring large amounts of storage space between flight seasons. The disposable, inexpensive cup-type paper trap (*826*) shown in Figure XII.3 completely replaced the Graham trap. These paper traps, containing attractant impregnated on small dental-roll wicks and coated on the inside with Tanglefoot are hung on the limbs of trees; attracted males enter through an opening in each end of the cup (Fig. XII.4) (*153*).

Although Collins and Potts (*304*) reported that under certain conditions small numbers of male gypsy moths could be lured from as far as $2\frac{5}{16}$ miles, much more recent tests have shown that the maximum effective distance of the lure is probably close to $\frac{1}{2}$ mile.

Air movement is probably the most important consideration in trap placement (*554*). Traps situated upwind from an infestation are more apt to catch insects than those downwind or across-wind. Traps placed in hollows or among dense growth do not catch well because of restricted air movement. Odors of chemicals tend to sink to the ground because they are almost always heavier than air. The male gypsy moth apparently has adapted its behavior to this property since it flies long distances close to the ground, then flies up the trunk of a tree and down, proceeds to the next tree, and repeats this behavior, apparently attempting to locate the females, which are found on tree trunks. Catches of males in traps placed between ground level and 6 feet above showed little difference but fell off at the 12-foot level.

Beroza (149) has shown that an unidentified component present in a hydrogenated, crude benzene extract of female moths may inhibit the field response of males to the attractant. This effect is nonpersistent, probably owing to its gradual disappearance through volatilization.

Sterilization of female moths by irradiation with 20,000 rad of ⁶⁰Co failed to reduce their attractiveness to males, as compared in field tests with unirradiated females and their extracts (1135).

Following the identification, by Bierl et al. (59, 158), of the gypsy moth sex attractant as cis-7,8-epoxy-2-methyloctadecane (disparlure), this com-

Fig. XII.4. Interior of gypsy moth trap, showing captured male held on the adhesive layer. [By permission of the U.S. Department of Agriculture.]

pound (85% cis) prepared synthetically was tested in the field early in May 1970. Four traps baited with 1 μg of disparlure were compared with 4 traps baited with 10 female equivalents of natural extract for attractiveness to released males. During the 10-day test period, disparlure caught a total of 110 males and the natural extract caught 3 males (*61, 150a, 158*). 9,10-Epoxy-2-methyloctadecane and 10,11-epoxy-2-methyloctadecane, each about 85% cis, were inactive at 20 μg, showing the extreme specificity of disparlure (*158*). Disparlure was active at $10^{-9}$ gm in baited traps in the field. The corresponding trans isomer was far less active (*59*). Tests with synthetic disparlure were so promising that 33,000 survey traps containing the female extract was rebaited with the synthetic in midseason 1970 by Plant Protection Division. Gypsy moths were discovered in many areas previously thought to be free of the pest.

Trioctanoin was a very satisfactory extender for disparlure; formulations containing 5 mg trioctanoin + 1 μg disparlure were highly effective in field traps for at least 3 months. Triolein degraded disparlure. Filter paper appeared to be a more effective wick than cotton for survey traps. Disparlure at concentrations up to 1 mg/trap was not repellent (*150b*).

*Porthetria monacha*, nun moth

The information given for the use of sex attractant in nun moth control (Chapter XIII) is pertinent for survey programs for this insect.

*Prionoxystus robiniae*, carpenterworm moth

This insect, a large trunk borer, causes much loss in southern hardwood forests. Preliminary field trials have shown that adult females can be used to lure males into traps (*1120*). The traps, measuring 12 × 12 × 8 inches, were constructed of wire screen and fitted with 4 cone-shaped entrances. The trap baited with a live, virgin female was a wire-screen cage (4 × 6 × 3 inches) that was stapled to the bark of a tree 5 feet above the ground and fit inside the larger cage. Marked males were released $\frac{1}{4}$, $\frac{1}{2}$, $\frac{3}{4}$, and 1 mile from the female-baited trap. Catches by females averaged 270 males, ranging from 38 to 666. Although the females continued to attract males each afternoon throughout their lives (about 9 days), their potency decreased progressively with age. About 25% of the males released $\frac{1}{4}$ and $\frac{1}{2}$ mile away, 20% of those released $\frac{3}{4}$ mile, and nearly 7% of those released 1 mile from the traps were recaptured.

*Rhyacionia buoliana*, European pine shoot moth

Cylindrical paper containers lined with Tanglefoot and baited with cotton roll wicks impregnated with 10 equivalents of a methylene chloride extract of virgin-female abdominal tips captured males released near host plants in western Washington. The duration of the bait's effectiveness was increased 3-fold by dissolving polyethylene glycol-600 distearate in the extracts. Baits could be used up to 2 weeks in a cool damp area before recharging became necessary. Males responded to these traps over distances of at least 85–90 m (*332*).

*Sitotroga cerealella*, angoumois grain moth

In a field trapping study, 40 of 120 released males were caught in 10 days in a trap containing 100 virgin females. Males were trapped only at dusk for a period of about 1 hour. A trap containing 5 virgin females captured males several hundred meters from the point of release. Two baited traps set in an uncultivated field caught 393 males, whereas 2 others set in a corn field captured 6520 males (*1148*).

*Spodoptera eridania*, southern armyworm moth

A mixture of both prodenialure A (*cis*-9-tetradecen-1-ol acetate) and prodenialure B (*cis*-9,*trans*-12-tetradecadien-1-ol acetate) was necessary to attract male moths into field traps (*60*).

*Spodoptera frugiperda*, fall armyworm moth

Of 8 types of traps tested in infested Georgian fields with live virgin females, a modification of the Steiner plastic cylinder trap proved to be most effective. The 5 females were placed in a small plastic cage which was inserted into the cylinder lined with Stikem. Traps set 3 feet above the ground were more efficient than those placed 1, 6, or 9 feet high and those with openings of $\frac{3}{4}$ inch or $1\frac{1}{2}$ inches diameter caught fewer males than those with larger openings (*1119*).

*Spodoptera littoralis*, cotton leafworm

Standard traps constructed of a wire-net cage suspended over the center of a paraffin oil bath and containing 3–5 virgin females 18–24 hours old were placed 15–25 cm above plants in an Egyptian cotton field. The greatest number of males (1200 in 84 hours) was caught by a single female in early October. Traps were attractive for 3 nights only (*1330*).

*Synanthedon pictipes*, lesser peach tree borer

Cleveland and Murdock (*302*), using sticky board traps baited with live virgin females, showed that males could be attracted from a distance of 500 feet. The traps consisted of pressed boards (6 × 8 inches), painted yellow and coated with Stikem (97% polymerized butene), bearing a centrally located hole into which was placed a 1 × 3 inch plastic-screen cage holding a virgin female. A total of 29 such traps kept in the field from May 10 to July 22 caught a total of 388 males on various days during this time. Attractive females each lured an average of 23 males.

Wong *et al.* (*1280*) used 1-gallon cylindrical cardboard cartons coated inside with Stikem and containing 1 or 2 wire-screen cages (2.5 × 7.6 cm) baited with adult virgin females. Such traps were placed near a peach tree in field cages (2.3 × 2.3 × 2.3 m) containing released males and set in an orchard. Females were attractive to males throughout their lives but attracted the greatest number of males during the first 2 days.

In 1969, 30 traps baited with 12 virgin females each were placed on 25-square-mile Washington Island, Wisconsin, in Lake Michigan. Females were replaced daily. About 4000 males were caught during the 4-month season (67). The traps had a hardboard base supporting 4 wings made of hardware cloth at right angles to one another; the wings were covered with a roof similar to the base. The females were held in small wire-screen cages secured to the angles of the wings. A larval survey estimated the borer population at 16,008; using the percentage of native moths captured versus the number of marked released males recaptured gave an estimate of 21,646 borers (1283).

*Trichoplusia ni*, cabbage looper

Shorey and Gaston (1086) observed increased male responses to a visible light source during their studies with female extracts. Field studies conducted by Henneberry et al. (535, 536) showed that blacklight traps baited with 50 2-day-old virgin female moths caught approximately 20–30 times as many males as were caught with light traps alone (see Fig. XII.1). A light trap without virgin females operated 20 feet from the baited trap caught 10–15 times as many males as did unbaited traps 1–2 miles away, indicating that the presence of virgin females in close proximity to a light source substantially influences the number of males caught. Virgin females placed as much as 40 feet from the light trap increased the number of males caught (536). A definite peak in activity occurred from 6:00 PM to midnight.

From a study using 2 blacklight traps baited with 100 virgin females each over a period of 12 months, Howland et al. (562) found that the average catch of male loopers was about 12 times higher than that of unbaited light traps. These baited traps were most effective in March and least effective in September.

Henneberry et al. (537) conducted field tests in a 690-acre experimental area in California that involved the release of marked males and their subsequent recapture in traps equipped with blacklight lamps and live virgin females. Marking was done by spraying the males with aqueous solutions of 0.1% rhodamine B or 50% red, green, yellow, or orange food coloring; all spray solutions contained 500 ppm of Tween 80. Moths were recovered as much as 10,000 feet from the release point but more than 90% of the recoveries were made within 4000 feet from the point of release. Twenty-two baited traps in the experimental area captured as much as 17% of the released males.

Henneberry et al. (539) found in laboratory tests that, although males fed with 0.5% or 1% tepa were less responsive to the female sex pheromone than untreated males, the response of males sprayed with 0.5% or 1% tepa was comparable with that of untreated males. There is every reason to believe that comparable results would be obtained in the field.

Field tests conducted with synthetic *cis*-7-dodecen-1-ol acetate (the natural sex pheromone) showed that 0.1 ml was more attractive than 100 live virgin females for a distance of about 400 feet downwind (*480*).

A California field study was conducted by Saario et al. (*1012*) with cardboard cylinder traps (lined on the inside with Stikem) containing 5 live virgin females or synthetic pheromone. The pure chemical, maintained in glass reservoirs (17 mm diameter $\times$ 4 mm deep), gave an average release rate of 0.1–0.3 $\mu$g per minute over the temperature range of 7°–20°. More males were captured in traps at an elevation of 2 m above the soil surface in a broccoli planting than in those at higher or lower elevations, but captures in traps placed on trunks of citrus or eucalyptus trees remained steady or increased through an elevation of 4 m. Captures in traps situated up to 3200 m from a cole crop were as high as captures at the crop itself. Seasonal highs and lows of males captured during a 2-year period occurred around September and January.

Catches of males in traps equipped with blacklight lamps were increased for 4 weeks by the presence of synthetic pheromone on sand substrate to reduce volatilization (*561*). A dose of 0.5 mg of pheromone replenished each night gave best results. Increasing the diameter of the container holding the mixture of sand and pheromone caused an increase in catches for the first night, but the catches were reduced on succeeding nights. Light traps baited with pheromone and mounted on barrels coated with diesel oil and containing water were twice as effective as baited, survey light traps, but pheromone suspended above the oil without the light was not effective. The male-moth catch per trap per night for survey traps plus pheromone, barrel traps plus pheromone, and unbaited survey traps was 210, 443, and 8, respectively. The use of sand as a carrier for the synthetic pheromone had been proved effective 2 years previously by Wolf et al. (*1278*), who compared the results obtained with the mixture in carton traps and light traps. Monterey sand treated with 100 $\mu$g–100 mg pheromone remained highly attractive for 12 days, whereas paraffin and Celotex remained attractive for 2 days, glass, filter paper, lanolin, Vaseline, and Celite for 7 hours, and silica gel and Sea-Sorb 43 were completely ineffective. Although carton traps baited with 10 mg and 75 mg of synthetic pheromone caught as many moths as those baited with 100 live virgin females, fewer moths were caught in these traps than in the blacklight-pheromone traps.

Studies of various bait dispensers for synthetic pheromone were conducted in Riverside, California, by Toba *et al.* (*1185*) and in Mesa, Arizona, by Debolt (*340*) in 1969 and 1970, respectively. The former investigators found that polyethylene bags were as effective as jars for dispensers with light traps and slightly more effective than wick dispensers in light traps or electric-grid traps. Debolt found that mixtures of sand and pheromone in glass jars and cloth bags and squares of carpet impregnated with the pheromone effectively increased the male catches over that with blacklight lamps alone. However, the periods of peak attractancy were short. A wick dispenser equipped with a dust shield was quite effective for longer periods.

Gaston *et al.* (*444*) found that the rate of evaporation of lauryl acetate, the saturated analog of the pheromone, on filter paper was dependent on temperature, whereas a nonadsorptive surface such as glass or stainless steel had a relatively low temperature dependence. The vapor pressure-temperature curve of lauryl acetate is almost identical to that of the pheromone. With a pheromone-baited trap, the number of males caught went through a maximum as a function of evaporation rate.

Synthetic pheromone was released in the field from several traps designed by Sharma *et al.* (*1070*). Males were caught on Stikem in an open, horizontal, cylindrical trap (A) or in a circular trap (B), and captured alive in a double cone trap (C). Optimum pheromone release rate was about 0.3 $\mu$g per minute from trap A and about 1 $\mu$g per minute from traps B and C; rates about 3 times higher than the optimum resulted in reduced captures. Traps B and C captured 4 and 26 times more males, respectively, than trap A. Electric-grid traps (*678*) baited with synthetic pheromone were more effective than blacklight traps plus pheromone, and a combination of electric grid plus blacklight plus pheromone was more effective than electric grid plus pheromone. Significant reductions in light trap catches occurred as the distances between these traps and the pheromone source were increased from 6 to 48 m; with 5 mg of pheromone, catches increased at the pheromone source as the distance from the light trap increased.

A pheromone-maze trap and a blacklight trap, each baited with synthetic pheromone, were placed about 300 yards apart in an alfalfa field in Arizona for 6 days. The maze trap caught 2229 males and 3 females; the light trap caught 558 males and 1 female. A pheromone-maze trap baited with synthetic pheromone on a wick dispenser plus a wire container holding 150 virgin female beet armyworms (*Spodoptera exigua*), placed in an alfalfa field, caught an average of 266.4 male beet armyworms and 379.8 male cabbage looper moths per trap per night (*668*).

A mixture of 25 mg of synthetic pheromone and 7 gm of sand, placed in polyethylene bags, stimulated caged males as much as 320 feet downwind when the wind velocity was 2.5–5.0 miles per hour and the temperature was

60°–63°F. When marked males were released at 350, 500, and 1000 feet downwind from the pheromone source, 30, 19, and 6%, respectively, were recovered within 15 minutes (677).

A simple, low-cost screen trap baited with 25 mg of synthetic pheromone caught as many males as a trap 0.4 km away equipped with a blacklight lamp and baited with the same amount of pheromone, but the catch per trap per night of 4 traps as much as 244 m apart was less than half that of a single trap (1186).

Synthetic pheromone exposed to outdoor conditions for 10 months at Red Rock, Arizona, was 80% less attractive in the field than fresh phero-mone. The exposed lure had a rancid odor and a yellow color, and was more viscous than fresh lure. Chemical and physical analysis indicated that autoxidation had occurred, resulting in the probable formation of a hydro-peroxide. Red Rock samples exposed for 43 days still contained a high proportion of the original pheromone (1183).

*Trypodendron lineatum*, ambrosia beetle

Chapman (292) studied the effects of attack by the female beetles on log (Douglas-fir) attractiveness using "greenhouse" cages in the field in which solar heating caused release of the odor-bearing air beneath glass-barrier traps. Attack was followed by a marked increase in attractiveness of the logs to both sexes. This secondary attraction was strong within 2–3 days after the attack, continued for many weeks if the females remained alone, and disappeared following mating. Moeck (794) isolated ethanol from the wood and bark of Douglas-fir and showed that it was mainly responsible for the initial attack by females.

*Zeiraphera diniana*, larch bud moth

Traps baited with 2 μliters of *trans*-11-tetradecen-1-ol acetate in a poly-ethylene cap did not begin to attract males until they had been in the field for 40 days; 100 μg became highly attractive in one month. Concentrations of 10 μg and 1 ng were most attractive (953a).

# CHAPTER XIII

# SEX PHEROMONES IN INSECT CONTROL

The use of sex attractants in insect control, particularly for the Lepidoptera, is a definite possibility, as has been pointed out in reviews by Götz (*468*), Knipling (*691*), Beroza and Jacobson (*153*), and Shorey (*1087*), as well as a number of other brief reviews (*75, 147, 148, 591, 824, 838, 839, 1068, 1079, 1303*). Numerous males of certain species of moths can be lured to their deaths by the use of even crude extracts of the females placed advantageously in traps with insecticide or on sticky boards (*587*), and the use of a mixture of sex attractant and a chemical sterilant merits trial. Much greater damage to the insect population can be effected by insects rendered infertile by a chemosterilant than by outright killing of the insects. These chemosterilants could be exposed along with the attractants in dispensers that allow access only to the insect, for example, through small holes. Insects responding to the attractant could thus be brought in contact with the chemosterilant, or possibly be induced to feed on it by admixing it with the lure, and then be free to fly off and mate with normal individuals of the opposite sex; such matings would, of course, result in no progeny.

With the ready commercial availability of a number of sex attractants of pest species, and others being identified rapidly, it is definitely within the realm of possibility that these sex attractants could be used to saturate the atmosphere to disrupt communication between the sexes (confusion) and prevent mating. Wright (*1303*, *1309*), Shorey (*1079*), and others have discussed this possibility, and small-scale field trials, as will be discussed later, have demonstrated the feasability of this approach in at least 3 species of insects (cabbage looper, red-banded leaf roller, gypsy moth). Inhibitors of such sex attractants may also find application by blocking the ability of sensory cells to pick up signals emitted by the opposite sex.

Wolf *et al.* (*1279*) have proposed a method for determining the density of traps required to reduce an insect population. It involves measuring the characteristics of a single trap and determining an empirical trap-density function which accounts for changes in trap performance.

*Achroia grisella*, lesser wax moth

After Kunike (*702*) had determined that the male lesser wax moth secretes a substance which attracts and excites the female, he suggested that the material might be used to control this beehive pest. He felt that if the substance could be synthesized it might be placed in the vicinity of the hives, surrounded by glue, and used to lure large numbers of virgin females to their death. Failing this, a number of live males surrounded by glue could be placed in the vicinity of the hives with the same effect.

*Anagasta kühniella*, Mediterranean flour moth

Laboratory studies (*1194*) on habituation of males to the female pheromone by continuous exposure to a benzene extract of virgin females indicated that disruption of normal male-to-female orientation in nature is possible, provided that the concentration of pheromone in the environment is high enough to ensure that almost all wild males are habituated to the concentration they might receive from wild females during emission in nature. Males in the laboratory failed to respond to an extract of the abdominal tips of virgin females following a 15-minute exposure to 10 $\mu$g of *cis*-9-tetradecen-1-ol acetate (*202*).

*Anthonomus grandis*, boll weevil

A total of 26,000 traps baited with males and placed in and around cotton fields (approximately 40,000 acres) infested with boll weevils in the Texas High Plains captured more than 4000 overwintered weevils per acre but failed to reduce populations because the confined weevils could not compete with the large number of native weevils in the cotton (*52*, *515*). Traps placed in and around a field of cotton in an area of low population density captured sufficient numbers of weevils to suppress the population until dispersal began. In order to control or eradicate the insect by this method, the population of overwintered weevils must be reduced throughout a large area (*515*, *516*).

In preliminary tests with 12 male-baited traps bordering a 35-acre cotton field, weevil captures were sufficiently high to prevent 90% of ordinary weevil damage to the developing cotton buds (squares) (*49*). Maximum field tests indicated that isolation, either natural or chemical, of 50 miles (preferably 100 miles) would be the minimum requirement for a pilot eradication experiment (*333*).

Approximately 80% suppression of boll weevil populations was attained in 17 of 34 cotton fields (1542 acres total) in west Texas in which yellow wing traps baited with male weevils were used. Wing traps situated around the fields were much more effective than traps placed in the fields (*518*).

A detailed discussion by Knipling and McGuire (*692*) of the theoretical aspects of the use of live, grass-reared male weevils in a control program proposed the use of specially designed cages, at the rate of 20 per acre, each containing 10 males, with destruction of the attracted females by appropriate means. With a natural population of 4 weevils per acre (2 males and 2 females), the caged males would outnumber natural males by 100:1. The reproductive potential should be reduced by about 96%. The level of 4 weevils per acre could be attained by the application of insecticides.

A reproductive-diapause pilot eradication program using traps baited with the synthetic pheromone (grandlure) over a 10,000-acre cotton-growing area in southern Mississippi is being planned as a result of preliminary investigations carried out in Mexico. It is designed to provide adequate technology to eradicate the pest (*905*).

*Argyrotaenia velutinana*, red-banded leaf roller

Field trapping tests conducted in New York apple orchards in 1968 by Glass *et al.* (*456*) using ice-cream carton traps and plastic freezer-container traps baited with 0.33 $\mu$g of the synthetic pheromone, riblure, showed that excellent male catches could be obtained in traps with $1\frac{1}{2}$-inch openings utilizing hollow polyethylene caps into which 10 $\mu$liter of synthetic pheromone were injected. The use of sand, dental roll, polyethylene tubing, and Saran-Wrap as wicks was less efficient. Even greater efficiency was obtained in 1969 field tests conducted in a 15-acre orchard by Roelofs *et al.* (*963*), using synthetic pheromone synergized with a closely related compound in polyethylene cap wicks placed in 1100 commercial foldup, adhesive-coated, paper traps (Sectar). Possible methods proposed to control the pest were disruption of male-female communication by atmospheric permeation with the synthetic substance, and combination of the synthetic pheromone with a virus, chemosterilant, or juvenile hormone in an integrated program.

Field tests with synthetic pheromone in combination with various homologs and analogs showed that male catches were greatly reduced by combining riblure with its trans isomer (*trans*-11-tetradecen-1-ol acetate) or other primary $C_{14}$ acetates or alcohols with unsaturation at carbon 11. *trans*-7-Tetradecen-1-ol acetate and *cis*-9-hexadecen-1-ol acetate were ineffective as inhibitors (*956*). Other strong inhibitors with the synthetic pheromone were *cis*-11-tetradecenyl alcohol, *cis*-9-tetradecenyl alcohol, 11-tetradecynyl alcohol, and 11-tetradecyn-1-ol acetate. Medium inhibitors were *trans*-9-tetradecen-1-ol acetate, *cis*-9-tetradecen-1-ol acetate, *trans*-7-tetradecen-1-ol acetate, *trans*-5-tetradecen-1-ol acetate, and *cis*-11-hexadecen-1-ol acetate. Questionable inhibitors were *cis*-9-hexadecen-1-ol acetate, *cis*-7-tetradecen-1-ol acetate, *cis*-7-hexadecen-1-ol acetate, 2-(11-tetradecynyloxy)tetrahydropyran, *cis*-7-dodecen-1-ol acetate, and 10-undecen-1-ol acetate. Methyl *cis*-11-tetradecenoate, *cis*-6-hexadecen-1-ol acetate, *cis*-5-dodecen-1-ol acetate, *cis*-7-decen-1-ol acetate, *cis*-5-tetradecen-1-ol acetate, and *n*-tetradecanol did not inhibit the synthetic lure (*954*, *958*). A concentration of 5% *trans*-11-tetradecen-1-ol acetate inhibited the response to 10 $\mu$g of synthetic pheromone. These inhibitors apparently interact with the antennal receptor sites of the male moth and modify the input to the central nervous system (*958*).

Using data from pheromone trapping experiments in both high and low populations in which 25 $\mu$liter of synthetic pheromone-dodecyl acetate (60:40) solution impregnated in polyethylene caps placed in Sectar traps were used, Roelofs *et al.* (*964*) discussed a program for calculating theoretical control of mating by sex pheromone trapping. The traps were at least twice as competitive as traps containing live females. Theoretical control for the summer moth flights was 48% for the high population and 99% for the low population. Records of fruit injury showed 32% damage for 1 orchard and 0% in another orchard.

*Attagenus megatoma*, black carpet beetle

Burkholder *et al.* (*225*) obtained a patent in 1970 covering the isolation, identification, and synthesis of the pheromone, megatomoic acid. According to the patent claims, this pheromone "can be employed as an attractant in traps to detect and control infestations of the insects." It may be used with an inert carrier or mixed with a toxicant to kill the beetles entering the traps and obtain control in selected areas known to be infested.

*Cadra cautella*, almond moth

Males in the laboratory failed to respond to an extract of the abdominal tips of a virgin female following a 15-minute exposure to 10 $\mu$g of *cis*-9-tetradecen-1-ol acetate (*202*).

*Dendroctonus brevicomis*, western pine beetle

As a result of field tests conducted with brevicomin, the use of this pheromone in sufficient concentration to confuse the insect, preventing congregation of the sexes, was mentioned as a distinct possibility (*45, 1220, 1285*). Whereas a mixture of brevicomin and 3-carene attracts numerous males, frontalin + 3-carene attracts numerous females. The use of a 3-component mixture of brevicomin, frontalin, and 3-carene "may be the ultimate answer to manipulation of *D. brevicomis* populations" (*866*).

*Dendroctonus frontalis*, southern pine beetle

In the early part of the season, few trees are being attacked by parent beetles and there are rather long intervals between attacks. "At this stage any upset of the delicate balance could be disastrous." Presumably spring and early summer control can break the emergence and pheromone-production patterns. The break in pheromone production during the winter and in the early development of the infestation has important control-operation implications (*423*). It is possible that the use of frontalin may succeed in obtaining population confusion, disrupting normal communication and development.

*Dendroctonus ponderosae*, mountain pine beetle

Schönherr (*1050*), in 1967, predicted that the synthetic pheromone would be prepared that could "lure the insects to traps where they can be liquidated in masses." In the same year, and the following year, it was clearly established that *trans*-verbenol (from the females) and α-pinene (from white pine resin) are the principal compounds stimulating aggregation of the beetles (*315*). In 1969, it was found that flying beetles could be induced to mass-attack live white pines baited with *trans*-verbenol and α-pinene. An attempt to protect baited trees by spraying the boles to a height of 20 feet was only partially successful. Trees simulated by 8 × 10 foot paperboard cylinders coated with adhesive and baited with the attractant lured and trapped large numbers of flying beetles. The effective attractant range of the synthetic pheromones was about 200–300 feet. A 50% reduction in the number of mass-attacked trees was achieved by using baited sticky traps to lure the beetles away from live white pines (*1285*).

*Diatraea saccharalis*, sugarcane borer moth

It was shown that virgin female moths can be used to achieve significant reductions in borer infestations and damage to sugarcane in small field plots when sticky traps, each baited with one female, were maintained continuously in plots at a rate of 400–800 traps/acre. Only 3.89% of 950 males, released 25–50 feet from 2 sticky traps, were recovered. The traps had been placed in an isolated sugarcane field and baited with 2 virgin females (*847*).

*Diparopsis castanea*, red bollworm

The use of bait stations, using filter papers soaked in various concentrations of aqueous tepa solutions to sterilize attracted males, was in-

vestigated in outdoor cages 6 feet square, with virgin females as bait.
Males released overnight were recaptured the following night and mated to
virgin females to see if a sterilizing dose had been picked up by the males.
It was decided that, if the sex attractant became available in large amount,
such sterilizing bait stations were a distinct possibility (*279, 280*).

*Diprion similis,* introduced pine sawfly

The insect is a pest of eastern white pine in the United States. Since it
was observed that large numbers of males swarmed toward females from
the surrounding area, investigations were conducted to determine whether
a male eradication program would be practical (*310*). The wooden traps
used consisted of a board (12 × 6 × 1 inches) with a 2½-inch screened
opening in the center. A virgin female was placed in the screened portion
and Tanglefoot was spread over the wooden portion. The traps were
suspended from trees in infested areas. Traps baited with 1 virgin female
attracted an average of 1000 males each (data from 8 traps), ranging from
542 to 1706; large numbers of males also fell to the ground. Some females
did not attract males for an unknown reason. The male response was rapid,
many approaching within 30 seconds after the traps were set up. Traps set
up at an angle of 90° to the wind direction at the edge of the woods were
consistently more attractive than those set in the dense portion of the
woods. Greatest activity took place from midmorning to midafternoon;
no male movement was observed in the early evening hours before sunset.
One trap with a virgin female exposed from 11:00 AM to 4:00 PM attracted
more than 7000 males during this period; she continued attracting males at
approximately 1000 per day until she died on the fifth day, after which
small numbers were caught for the next 3 days. Males were lured 200 feet
out of the forest over an open field.

Field control of the insect based on luring the males to an insecticide-
attractant mixture seems unlikely. Chemosterilant-attractant mixtures
should be more effective in reducing or eradicating a field population of
this insect (*286*).

*Grapholitha molesta,* Oriental fruit moth

Males are attracted from a distance of at least 6 feet by live females or
an extract of females (*446*). Since females have been found to mate an
average of 1.6 times, it appears that in practice it would be best if no
females were released along with irradiated males to direct the male mating
abilities from the target population, particularly when mating has already
been reduced by irradiation. However, the timely release of irradiated
females into sparse field populations along with traps baited with the
pheromone, might effectively occupy the field males as they emerge and
deny their services to the field females that subsequently emerge (*447*).

The use of the sex pheromone prepared synthetically for trapping and survey is being investigated (990).

*Ips confusus*, California five-spined ips

Long-term confinement in an atmosphere containing male frass odor caused a reduction in the response of both sexes to male frass extract, presumably because of adaptation of olfactory receptors. Attractive odor in the air still suppressed flight significantly after beetles had been conditioned up to 4 hours in contact with male frass (184). A confusion program could presumably result in population suppression of this insect.

*Laspeyresia pomonella*, codling moth

A detailed discussion by Knipling and McGuire (692) of the theoretical aspects of the use of live, virgin females or their extract in a control program proposed the use of large numbers of mass-reared females in a 1000-acre orchard following reduction of the natural population with insecticides to about 150 moths per acre. The maintenance of 250,000 caged females, or their equivalent in extract, in the area of treatment would be necessary each day. If the pheromone can be identified and prepared synthetically, the cost of the program would be much lower.

*Limonius canus*, Pacific Coast wireworm

Trapping tests with crude methylene chloride extract of virgin females around a wheat field showed that the extract could compete with natural females. The pheromone could be used over a 2- to 3-week period to attract males to their destruction, with almost no competition from virgin females during the first 7–10 days, since females in nature emerge considerably later than males (833).

*Lobesia botrana*, grapevine moth

*Paralobesia viteana*, grape berry moth

Götz (461–463), in 1939 and 1940, reported on field tests conducted in the vineyards of Geisenheim and Rüdesheim, Germany, with the purpose of determining whether the live females of these insects could capture sufficient numbers of males to prevent large-scale fertilization of females. Geisenheim traps, placed on posts approximately 150 cm high, consisted of 8 wooden boards each having 1 side covered with glued white paper and the other side covered with glued black paper; in the angle formed by those boards, there was a gauze-covered container with several females. In a nearby vineyard, trap-glasses containing grape marc wine, to which had been added 10 ml of wine vinegar and 5 grams of sugar per half-liter, were placed at 3-meter intervals for catch comparison. Although catches were dependent on weather and numbers and ages of females used, live females were approximately 38 times more attractive than the trap-glasses. Females 1 day old were more attractive than those 2 days of age, smaller numbers of males being attracted by females 9–13 days old. The largest numbers of males were attracted from the south and smallest numbers from the north. One trap in which 5 female *Paralobesia* moths had been placed, 4 of which had died mysteriously, caught 6 males on the white side and 20 males on the black side on July 7, 1937; on July 8, 9, and 10 the numbers caught were 37 and 79, 26 and 51, and 18 and 24, respectively. A heavy rainfall on July 10 knocked down the trap. Whereas the total number of males caught by this single trap was 271, only 212 *Paralobesia* moths of both sexes (predominantly females) were caught in 83 glass traps during the corresponding period. The trial also showed the moths to be positively phototropic. A single 6-boarded trap with alternate boards having two white papers and two black papers was baited with 5 2-day-old *Paralobesia* females and placed 25 m from the nearest vineyard; it attracted 250 males to the white sides and 90 males to the black sides.

Rüdesheim vineyards contained primarily *Lobesia*, since these prefer a dry climate and *Paralobesia* a damp climate. Traps with 4 wooden boards covered with white papers only were baited with 6 females, permitted to catch for 3 days, and then provided with an additional 11 fresh, 1-day-old females; by the following day, these traps had caught 10 times more males than they had the previous day. Boards facing west showed the least number of males and those facing north and east showed the most males. Other traps containing 14 females apiece were set up on a meadow surrounded on 3 sides by vineyards (the closest was 20 m away) and on the fourth side by a vegetable garden; a total of 1069 males were caught in 1 day, whereas 5 glass traps caught a total of 8 moths in this period. Boards facing west and north caught the least numbers of males in this test.

With both species of grape insects, flight occurred mainly during periods having temperatures of 11.7°–14.6°C, low wind velocity, and a relative humidity greater than 80%; very few males were caught during cool weather (*466, 467*). Götz concluded that the trapping method was practical for control of these insects only if the attractant could be identified and synthesized.

The findings with regard to *L. botrana* were confirmed in field tests by Chaboussou and Carles (*289*), using metal traps, and the same conclusions were reached regarding control of the insect by sex attraction. However, these authors found that equal numbers of males were attracted to both black and white traps baited with equal numbers of virgin females.

*Ostrinia nubilalis,* European corn borer

Mating of this insect was inhibited considerably by permeation of the atmosphere in an incubator with an estimated concentration of $3.8 \times 10^{-9}$

mole of *cis*-11-tetradecen-1-ol acetate per liter of air. This presumably involved sensory or central nervous system adaptation of the male moth to the compound. The use of 3.8 mg of synthetic pheromone reduced mating by 55%, and 6.5 mg pheromone reduced it by 79%; concentrations of 9.1 and 13 mg did not increase inhibition beyond the 79% level. A lower moth density increases the efficiency of inhibition by the synthetic pheromone. The inhibiting effect on males is reversible (*689*). Flight orientation of males to the location of virgin females or to *cis*-11-tetradecen-1-ol acetate in traps was inhibited by the presence of 14 μg of *trans*-11-tetradecen-1-ol acetate or 11-tetradecyn-1-ol acetate (*690a*).

*Pectinophora gossypiella*, pink bollworm moth

Field traps, about 6 per acre, baited with a methylene chloride extract of 50 virgin females failed to reduce a field infestation, even though 24,666 males were collected over a 6-month period (May–November) in a total of 49 traps. Traps were rebaited twice each week. Lack of field isolation and the use of an insufficient number of traps were probably responsible for the failure (*471*).

Tests were conducted in one-sixteenth acre field cages planted with cotton and artificially infested with pink bollworms. In the center of 1 cage 2 Frick traps were placed, 70 feet apart, baited with 25 female equivalents of extract plus 1000 female equivalents of propylure. Two other cages each contained 1 blacklight trap; 1 of the light traps was baited with extract plus propylure. A fourth cage was held as a check. Frick traps did not reduce the population during the 8 weeks of the test, whereas the baited light trap reduced the population by 46%; the unbaited traps reduced the population by 19% (*491*).

*Plodia interpunctella*, Indian meal moth

Males previously exposed in the laboratory for 15 minutes to 10 μg of *cis*-9-tetradecen-1-ol acetate or *cis*-11-tetradecen-1-ol acetate prior to exposure to a female equivalent of female extract failed to respond or responded only weakly to the latter. Fair inhibition of response was also obtained by prior exposure to decyl acetate, 10-undecen-1-ol acetate, dodecyl acetate, and *cis*-7-tetradecen-1-ol acetate, but not with hexyl acetate, octyl acetate, nonyl acetate, linalyl acetate, tetradecyl acetate, hexadecyl acetate, propylure, octadecyl acetate, tetradecanol, tetradecanoic acid, tetradecanal, or methyl palmitate. Some inhibition was obtained by prior exposure to 100 μg, but not to 10 μg, of methyl myristate. The inhibitory phenomenon may well prove to be of major significance in the development of a control program for this pest of stored grain (*202*).

*Porthetria dispar*, gypsy moth

In 1893, Kirkland (*672*) reported an unsuccessful attempt to use the attraction of females for males in gypsy moth control, although no details were given.

Beroza (*144*), in 1960, and Brown (*212*), in 1961, suggested that the control of gypsy moths by male confusion with gyplure might be successful, and, in 1963, Babson (*80*) proposed that gyplure be broadcast by air over infested areas to confuse or frustrate the males during the mating period. In a reply, Burgess (*220*) pointed out that such a test had actually been undertaken by the Plant Pest Control Division of the U.S. Department of Agriculture during the summer of 1961. Gyplure, in both liquid and granular formulations, had been distributed by aircraft over Rattlesnake Island, a 400-acre island with varying intensities of gypsy moth infestation located in Lake Winnepesaukee, New Hampshire. Subsequent field observations revealed that the commercial gyplure used in this test, although active in laboratory bioassay, had exhibited only weak attractiveness, and no adverse effect was obtained on male mating activity.

Plans were made to repeat the "confusion" test in the summer of 1964 with a large lot of laboratory-prepared gyplure (85% cis content) closely approximating that previously shown to be effective in both laboratory and field tests, but both granular (corn cob grits) and liquid (methylene chloride) formulations prepared especially for aerial application lured few males to field traps (*1248*).

Large numbers of males were trapped in the field with extremely small doses of disparlure. The traps attracted males for months without rebaiting. Disparlure thus has the potential for reducing populations of this pest through mass trapping or confusion (*61*).

Approximately 14,000 traps baited with 20 $\mu$g of disparlure were air-dropped over a 9-square-mile infested region in Pennsylvania. The traps were hollow cylinders, the insides of which had been coated with Stikem. In another approach, about 300,000 strips of hydrophobic paper impregnated with disparlure were dropped on each of several infested 40-acre regions in the Northeast. It was hoped that the odor from the attractant permeating the atmosphere would overload the male's guidance system so that he could not locate females. Both approaches appeared to be successful in areas of light infestation (*68*). Identification of a volatile substance in benzene extracts of virgin females that inhibits the response of males may lead to its use in control (*149*).

*Porthetria monacha*, nun moth

In 1930–1931, a Czechoslovakian entomologist named Dyk fastened small containers of female nun moth pupae to tree trunks in infested areas and surrounded the containers with sticky flypaper strips. Females

*Fig. XIII.1.* Male nun moths lured by live female. [From Farsky (*394*, pp. 52–56), Verlag Paul Parey, Hamburg.]

emerging in these containers lured large numbers of males (Fig. XIII.1), suggesting that this procedure could be used for control; this method became known as "Dyk's nun moth control" (*722*). It was tried in the same country in 1931 by Jacentkovski (*578*), who placed on a tree in a nun moth-infested forest a small matchbox containing a virgin female; 50 males were caught on the sticky strips during the first night of exposure. A total of 69 of these traps, baited with 85 virgin females during the period July 20–August 14, caught 9662 males, despite the presence of predators. Each female remained attractive for approximately 8 days. Although Dyk was able to lure large numbers of males to containers of females, he soon realized that control by this method was impractical. Since the female's attractiveness was strongest immediately after emergence and ceased with oviposition, and males were usually first to emerge, his idea was to collect female pupae, allow the adults to emerge in a warm laboratory, and expose these females in a male-infested area before normal female emergence in the field (*363*).

In 1937, Ambros (*18, 19*) reviewed Dyk's method and decided to try it, using cardboard matchboxes (10 × 15 cm) to hold his females and the easily available sticky paper strips on tree trunks. Female larvae and pupae were collected (by school children 7–14 years of age) and the larvae were reared to the pupal stage on pine twigs in open plots covered with gauze to prevent escape. Ambros reported that male catches were "spectacular," with traps tacked to the east or southeast side of trees catching best. Even traps that had previously contained a female for 24 hours remained attractive for 3 nights, but females held in place on the sticky strips fell prey to small animals and the strips did not catch well in rainy weather. It was found that large numbers of males could be caught before the main female emergence started, and Ambros felt that this method of control would be cheaper than using arsenic. This work was reviewed by Eckstein (369).

Elated by his previous results, Ambros (*20*) ran a large-scale trial (1916 square hectares) using 5 virgin females in each standardized cardboard trap. Traps were hung on trees 160 m apart at a height of 1.5 m from the ground and were surrounded by glue-covered paper sheets 12 cm wide and 60 cm–1 m long. A glue known as Liparin, manufactured by a Czechoslovakian firm, was used, since it was not affected by rain, remained tacky for the entire flight period (59 days), and did not harden in cool weather. Placing a wad of cotton with the females in each trap increased the effective period for a few days after the females died (8–14 days). In order to insure a supply of females sufficient to last for the entire flight period, female pupae were collected, stored, and transported at 2°C, and warmed as needed; although the cold pupae became stiff and hard, warming caused the emergence of normal females. Traps (approximately 2 for each hectare) were checked each morning for male catch. To Ambros' surprise and delight, a total of 150,104 males were destroyed in this test! Most of the males flew in from the east against a westerly wind. The effective radius for these traps was not greater than 300 m, as contrasted with 3.3–3.8 km reported previously by Collins and Potts (*304*) for gypsy moth traps. A total catch of 384,448 males in 49 days was reported by Ambros (*21*) in 1940 with 480 traps set in 756.57 square hectares. Results obtained with the natural attractant caused Ambros to doubt that an equally effective synthetic attractant for the nun moth could be found.

Modifications of Dyk's method were tried on a small scale by Hanno (*507*), utilizing gauze-covered cardboard containers of females. These containers were (1) tacked to a sticky oilpaper-covered board nailed to a tree trunk, (2) tacked to a large sheet of sticky paper nailed around a tree trunk, or (3) tacked to the tree trunk and surrounded by a wooden frame holding flypaper strips. During the period from July 14 to September 4, traps tested by method (3) in 3 sites caught only 18, 13, and 18 males, respectively, probably due to loss in tackiness of the strips from heat and wind; methods (1) and (2) proved to be more effective. A container of 2 live females set out July 19 by method (1) lured 204 males by July 22, when the females were removed; the empty trap then caught 59 males on July 23, 16 males on July 26, 1 male on July 28, and 3 males on August 8. Hanno could not substantiate Ambros' reports of best results with positions on the east or southeast side of trees, since this was dependent on wind direction. He was very optimistic about the possibility of using these trapping methods in both survey and control of nun moths, since 1 trapped male is prevented from fertilizing numerous females.

During the following year, Nolte (*828*) reported that Dyk's method of luring male nun moths with females surrounded by sticky papers was impractical for either survey or control of this insect, although it served to show the presence of the insects in an area and could cause male congregation. He tried Ambros' method using paper strips as well as square paste-covered boards, but the former were effective for only a short period and the latter were too expensive. Nolte found Hanno's method best, using paste-covered oilpaper sheets wrapped around tree trunks; the female was placed in the center in a small paraffined cardboard container covered with gauze held in place with a rubberband. On trees with highly cracked bark, Nolte recommended the insertion of wrapping paper under the oil paper to facilitate the application of paste. Each container should have at least 2 females, which should be renewed at least every 5 days. Trap sheets should be checked every day if possible, male counts made, and the paste renewed if necessary. Although Nolte found that captured males could be eaten by birds and mice, placing of the containers on the paste-covered sheets prevented ants from reaching the females. It was determined that placing a large number of traps in a given area did not affect the total number of males caught, since each trap then caught fewer numbers. Depending on the area used (Nolte surveyed 9 areas in 3 years), traps caught a total from 239 to 3560 males. In any 1 night, the largest number of males caught in a trap was 578.

To determine the effective distance of the lure, 24 traps were set at various distances from a wooded area surrounded by a flat, treeless terrain;

1 trap was set in the wooded area to indicate the start of male flight. Each trap was provided with fresh females daily. Flight began on July 21, 1938, and ended on August 22, during which time a total of 11 males were caught in traps 100 m away from the woods, 13 in traps at a distance of 200 m, and none at 300–1000 m. In 1939, Nolte ran a similar test using males marked with lacquer and released downwind in the afternoon from a distance up to 700 m in all directions. The results of this test were not definite, since a bad storm developed in the next day or 2, but a total of 20 males were attracted from 200 m, 2 males from 250 m, 1 from 500 m, 2 from 600 m, and 1 from 700 m. Flight was considerably affected by inclement weather and wind direction. Largest numbers of males were lured from the east and southeast, since the prevailing winds during flight were usually from the west and northwest. Trap position was important, since traps set in a depression seldom caught as well as those placed on flat ground. There was no correlation between the number of females per hectare and the number of males caught in the area, but flight numbers and the numbers of males attracted paralleled one another.

A field test conducted by Nolte (828) to determine whether the fertilization of wild females could be prevented by the use of a large number of traps was unsuccessful, as a normal number of fertile eggs was found the following January. Nolte concluded that such a test can be successful only if the male is caught before he can find a female, but that this is improbable, owing to the searching flight pattern of the male. He decided that Hanno's method is useful for survey only, and that Dyk's method cannot possibly be used for prognosis, since the size of the infestation is dependent on several factors and the total number of eggs laid is much greater than the number of males caught in a given area.

*Sanninoidea exitiosa*, peach tree borer

Sleeve cages containing varying numbers of virgin female moths were placed on the ground near a peach orchard infested with the borers and the number of males attracted over a 16-day period was observed. A total of 8 males were attracted to 3 females, 35 males to 15 females, 223 males to 25 females, 326 males to 35 females, and 128 males to 37 females. It was concluded that the female attractant is potent enough to achieve areawide population suppression through male annihilation if traps are distributed in infested areas at proper densities and are baited with sufficient females to compete with females in the natural population (581).

*Spodoptera frugiperda*, fall armyworm moth

The synthetically produced pheromone may be used in the future to suppress wild populations through mass trapping or confusion (43).

*Synanthedon pictipes*, lesser peach tree borer

Live virgin females, their extracts, or a synthetically prepared attractant could conceivably be used to suppress natural populations of the pest by overwhelming the attractiveness of the native female population (*39*).

*Trichoplusia ni*, cabbage looper

Numerous blacklight and barrel traps, each baited with 100 mg of synthetic attractant, were set along roads in and around infested lettuce plots at Red Rock and Picacho, Arizona, in 1967. Trapping results obtained indicated that significant control of this pest could be achieved by this method (*1277*). Catches obtained with pheromone-baited maze traps placed at the same locations and with pheromone-baited, double-cone traps placed around Riverside, California, also appeared to be very promising (*668*, *1070*).

The male looper population was reduced significantly in a 400-square mile area of cigar-wrapper tobacco in Florida through the use of blacklight traps baited with 100 mg of synthetic attractant. These tests were conducted during the period 1967–1969 (*444a*).

Laboratory experiments conducted to evaluate the male inhibition technique for behavioral control through permeation of the atmosphere with synthetic pheromone were likewise very promising. They established the lower threshold for male response to the pheromone at about $2 \times 10^{-14}$ gm per liter of air. Male responsiveness to a high concentration of $2 \times 10^{-9}$ gm per liter was completely inhibited when they had been conditioned for 25 minutes in air containing lower pheromone concentrations. A concentration of $5 \times 10^{-5}$ gm per liter was required to prevent males from orienting to and mating with females in a 2-liter container (*1094, 1194*). Following up with field tests, Shorey et al. (*443, 1090, 1094*) demonstrated that premating communication could be achieved through atmospheric saturation. About 17 mg of the synthetic pheromone was placed on each of 100 stakes cross-hatched at 3-m intervals in a 27 m³ plot and virgin females were placed in the center of the plot. The synthetic pheromone concentration was sufficiently high to prevent released males from orienting to the additional increment of pheromone released by the females. The pheromone concentration was about $1 \times 10^{-10}$ gm per liter. For large-area mating control, it was calculated that 0.2 gm per acre must be expended per night.

# BIBLIOGRAPHY

1. Abbott, C. E. The physiology of insect senses. *Entomol. Amer.* **16**, 225 (1936).
2. Abbott, C. E. On the olfactory powers of *Megarhyssa lunator* (Hymenoptera: Ichneumonidae). *Entomol. News* **47**, 263 (1936).
3. Acker, T. S. Courtship and mating behavior in *Agulla* species (Neuroptera: Raphidiidae). *Ann. Entomol. Soc. Amer.* **59**, 1 (1966).
4. Acree, F., Jr. The isolation of gyptol, the sex attractant of the female gypsy moth. *J. Econ. Entomol.* **46**, 313 (1953).
5. Acree, F., Jr. Studies on the chromatography of gyptyl azoate. *J. Econ. Entomol.* **46**, 900 (1953).
6. Acree, F., Jr. The chromatography of gyptol and gyptyl esters. *J. Econ. Entomol.* **47**, 321 (1954).
7. Acree, F., Jr., Beroza, M., Holbrook, R. F., and Haller, H. L. The stability of hydrogenated gypsy moth sex attractant. *J. Econ. Entomol.* **52**, 82 (1959).
8. Adeesan, C., Rahalkar, G. W., and Tamhankar, A. J. Effect of age and previous mating on the response of khapra beetle males to female sex pheromone. *Entomol. Exp. Appl.* **12**, 229 (1969).
9. Adeesan, C., Tamhankar, A. J., and Rahalkar, G. W. Sex pheromone gland in the potato tuberworm moth, *Phthorimaea operculella*. *Ann. Entomol. Soc. Amer.* **62**, 670 (1969).
10. Adler, V. E. Physical conditions important to the reproducibility of electroantennograms. *Ann. Entomol. Soc. Amer.* **64**, 300 (1971).
11. Adler, V. E., and Jacobson, M. Electroantennogram responses of adult male and female Japanese beetles to their extracts. *J. Econ. Entomol.* **64**, 1561 (1971).
12. Adlung, K. G. Field tests on the attraction of male nun moths (*Lymantria monacha* L.) and gypsy moths (*Lymantria dispar* L.) to gyplure, a synthetic sex attractant. *Z. Angew. Entomol.* **54**, 304 (1964).

13. Agee, H. R. Mating behavior of bollworm moths. *Ann. Entomol. Soc. Amer.* **62,** 1120 (1969).

14. Albert, P. J., Seabrook, W. D., and Paim, U. The antennae as the site of pheromone receptors in the eastern spruce budworm, *Choristoneura fumiferana* (Lepidoptera: Tortricidae). *Can. Entomol.* **102,** 1610 (1970).

14a. AliNiazee, M. T., and Stafford, E. M. Evidence of a sex pheromone in the omnivorous leaf roller, *Platynota stultana* (Lepidoptera: Tortricidae): Laboratory and field testing of male attraction to virgin females. *Ann. Entomol. Soc. Amer.* **64,** 1330 (1971).

15. Allen, N., and Hodge, C. R. Mating habits of the tobacco hornworm. *J. Econ. Entomol.* **48,** 526 (1955).

16. Allen, N., and Kinard, W. S. Production of tobacco hornworm moths from field-collected larvae. *J. Econ. Entomol.* **62,** 1068 (1969).

17. Allen, N., Kinard, W. S., and Jacobson, M. Procedure used to recover a sex attractant for the male tobacco hornworm. *J. Econ. Entomol.* **55,** 347 (1962).

18. Ambros, W. Nonnenfalterkontrolle auf biologischer Grundlage. *Zentralbl. Ges. Forstw.* **63,** 140 (1937).

19. Ambros, W. Einige spezielle Beobachtungen und Untersuchungen während der Nonnenkontrolle im Jahre 1937. *Zentralbl. Ges. Forstw.* **64,** 49 (1938).

20. Ambros, W. Die biologische Nonnenfalterkontrolle 1937. *Zentralbl Ges. Forstw.* **64,** 209 (1938).

21. Ambros, W. Einige Beobachtungen und Untersuchungen an der Nonne im Jahre 1938. *Zentralbl. Ges. Forstw.* **66,** 131 and 166 (1940).

22. Amin, E. S. The sex-attractant of the silkworm moth (*Bombyx mori*). *J. Chem. Soc., London* p. 3764 (1957).

23. Amoore, J. E. Elucidation of the stereochemical properties of the olfactory receptor sites. *Proc. Toilet Goods Ass., Sci. Sect.* **37,** Suppl., 13 (1962).

24. Amoore, J. E. Stereochemical theory of olfaction. *Nature (London)* **198,** 271 (1963).

25. Amoore, J. E. Stereochemical theory of olfaction. *Nature (London)* **199,** 912 (1963).

26. Amoore, J. E. Current status of the steric theory of odor. *Ann. N. Y. Acad. Sci.* **116,** 457 (1964).

27. Amoore J. E. Primary odor correlated with molecular shape by scanning computer. *J. Soc. Cosmet. Chem.* **21,** 99 (1970).

28. Amoore, J. E. "Molecular Basis of Odor." Thomas, Springfield, Illinois, 1970.

29. Amoore, J. E., Johnson, J. W., Jr., and Rubin, M. The stereochemical theory of odor. *Sci. Amer.* **210** (2), 42 (1964).

30. Anders, F., and Bayer, E. Versuche mit dem Sexualduftstoff aus den Sacculi laterales vom Seidenspinner (*Bombyx mori* L.). *Biol. Zentralbl.* 584 (1959).

31. Anderson, R. F. Host selection by the pine engraver. *J. Econ. Entomol.* **41,** 596 (1948).

32. Andersson, C. O., Bergström, G., Kullenberg, B., and Ställberg-Stenhagen, S. Studies on natural odouriferous compounds. I. Identification of macrocyclic lactones as odouriferous components of the scent of the solitary bees *Halictus calceatus* Scop. and *Halictus albipes* F. *Ark. Kemi* **26,** 191 (1966).

33. Anonymous. Love among the insects. *Time* **79,** 66 (1963).

34. Anonymous. Operation cockroach. *Agr. Res.* **12** (3), 8 (1964).

35. Anonymous. Natural attractant lures insects attacking stored products. *U.S., Dep. Agr. Press Release* **2724–65** (1965).

36. Anonymous. Sex attractant for the bollworm. *Int. Pest Contr.* **7** (6), 8 (1965).

37. Anonymous. Many sex attractants of insects may well prove to be longchain, unsaturated alcohols or their esters. *Chem. Eng. News* **43** (38), 42 (1965).
38. Anonymous. Do moths use radar? *Agr. Res.* **14** (8), 3 (1966).
39. Anonymous. Caged female insects used to trap males. *Agr. Res.* **15** (5), 5 (1966).
40. Anonymous. The sex attractant of a class of bark beetle has been isolated, identified, and synthesized. *Chem. Eng. News* **44** (33), 43 (1966).
41. Anonymous. *Rep. Anti-Locust Res. Cent. 1966* p. 9 (1967).
42. Anonymous. *Rep. Anti-Locust Res. Cent. 1966* p. 10 (1967).
43. Anonymous. Sex attractant for fall armyworm (produced artificially). *Agr. Res.* **16** (5), 15 (1967).
44. Anonymous. Beetles ravaging white pine trees. *N. Y. Times* **116**, 12 (1967).
45. Anonymous. SRI chemists identify another bark beetle sex attractant. *Chem. Eng. News* **45** (47), 21 (1967).
46. Anonymous. Stopping the codling moth. *Agr. Res.* **17** (2), 8 (1968).
47. Anonymous. Physiology, autocides and sterilants. *Insect Pathol. Res. Inst. Progr. Rev., 1965–1969* p. 24 (1969).
48. Anonymous. Male signals draw weevils to cotton fields. *Agr. Res.* **17** (7), 6 (1969).
49. Anonymous. Caged weevils sharpen detection tools. *Agr. Res.* **17** (10), 14 (1969).
50. Anonymous. Isolating and synthesizing natural sex attractants and finding man-made lures receive major attention. *Chem. Eng. News* **47** (18), 36 (1969).
51. Anonymous. Lure bags more bollworms. *Agr. Res.* **18** (3), 15 (1969).
52. Anonymous. Operation boll weevil. Trapping that 1 percent. *Agr. Res.* **18** (4), 8 (1969).
53. Anonymous. Teams ready weevil lure for field tests. *Chem. Eng. News* **48** (4), 40, 43 (1970).
54. Anonymous. European corn borer sex attractant. *Chem. Eng. News* **48** (6), 43 (1970).
55. Anonymous. Shape of smells. *Agr. Res.* **18** (7), 3 (1970).
56. Anonymous. Researchers identify the call of the corn borer. *Agr. Res.* **18** (8), 7 (1970).
57. Anonymous. Sex pheromones of three related moth species have been identified. *Chem. Eng. News* **48** (8), 41 (1970).
58. Anonymous. Isolated at last—the boll weevil's grandlure. *Agr. Res.* **18** (11), 3 (1970).
59. Anonymous. Team makes sex attractant of gypsy moth. *Chem. Eng. News* **48** (41), 35 (1970).
60. Anonymous. A pair of sex attractants has been isolated from female moths of the southern armyworm. *Chem. Eng. News* **48** (47), 43 (1970).
61. Anonymous. A powerful and persistent synthetic lure for trapping the gypsy moth. *Agr. Res.* **19** (5), 8 (1970).
62. Anonymous. Sex pheromone of the codling moth. *Chem. Eng. News* **48** (53), 37 (1970).
63. Anonymous. Several insect families use the same pheromone. *Chem. Eng. News* **49** (3), 19 (1971).
64. Anonymous. Pheromones: Chemical communicators. *Sciences* **11** (2), 14 (1971).
65. Anonymous. Luring the southern armyworm. *Agr. Res.* **19** (10), 8 (1971).
66. Anonymous. A human pheromone? *Lancet* p. 279 (1971).
67. Anonymous. Borer traps signal best time to spray. *Agr. Res.* **20** (1), 8 (1971).
68. Anonymous. Pheromones basis of moth control methods. *Chem. Eng. News* **49** (32), 79 (1971).

69. Anonymous. Housefly attractant isolated. *Chem. Eng. News* **49** (42), 9 (1971).
69a. Anonymous. Key to less pesticide use—citrus scale trap. *Agr. Res.* **20** (7), 5 (1972).
70. Anthon, E., Smith, L. O., and Garrett, S. D. Artificial diet and pheromone studies with peach twig borer. *J. Econ. Entomol.* **64,** 259 (1971).
71. Antram, C. B. Sexual attraction in Lepidoptera. *J. Bombay Natur. Hist. Soc.* **18,** 923 (1908).
72. Aplin, R. T., and Birch, M. C. Pheromones from the abdominal brushes of male noctuid Lepidoptera. *Nature (London)* **217,** 1167 (1968).
73. Aplin, R. T., and Birch, M. C. Identification of odorous compounds from male Lepidoptera. *Experientia* **26,** 1193 (1970).
74. Armstrong, T. The life history of the peach borer, *Synanthedon exitiosa* Say, in Ontario. *Sci. Agr.* **20,** 557 (1940).
75. Asakawa, M. Pesticides at the turning point, their aspects and some prospective researches. *Kagaku Kogyo* **18** (7), 689 (1967).
76. Ashpole, E. Sex scents to control insect pests. *Med. Surg. (Baroda, India)* **7** (9), 23 (1967).
77. Atema, J., and Engstrom, D. G. Sex pheromone in the lobster, *Homarus americanus. Nature (London)* **232,** 261 (1971).
78. Atkins, M. D. Scolytid pheromones—ready or not. *Can. Entomol.* **100,** 1115 (1968).
79. August, C. J. The role of male and female pheromones in the mating behavior of *Tenebrio molitor. J. Insect Physiol.* **17,** 739 (1971).
80. Babson, A. L. Eradicating the gypsy moth. *Science* **142,** 447 (1963).
81. Bakke, A. Pheromone in the bark beetle, *Ips acuminatus* Gyll. *Z. Angew. Entomol.* **59,** 49 (1967).
82. Bakke, A. Evidence of a population aggregating pheromone in *Ips typographus* (Coleoptera; Scolytidae). *Contrib. Boyce Thompson Inst.* **24,** 309 (1970).
83. Baldwin, W. F., Knight, A. G., and Lynn, K. R. A sex pheromone in the insect *Rhodnius prolixus* (Hemiptera: Reduviidae). *Can. Entomol.* **103,** 18 (1971).
84. Banerjee, A. C. Flight activity of the sexes of crambid moths as indicated by light-trap catches. *J. Econ. Entomol.* **60,** 383 (1967).
85. Banerjee, A. C. Sex attractants in sod webworms. *J. Econ. Entomol.* **62,** 705 (1969).
86. Banerjee, A. C., and Decker, G. C. Studies on sod webworms. I. Emergence rhythm, mating, and oviposition behavior under natural conditions. *J. Econ. Entomol.* **59,** 1237 (1966).
87. Barbier, M., Lederer, E., and Nomura, T. Synthèse de l'acide céto-9-décène-2-*trans*-oique (substance royale) et de l'acide céto-8-nonène-2-*trans*-oique. *C. R. Acad. Sci.* **251,** 1133 (1960).
88. Bar Ilan, A. Khapra beetle. *Riv. Parassitol.* **26,** 290 (1965).
89. Bar Ilan, A., Stanic, V., and Shulov, A. Attracting substance (pheromone) produced by virgin females of *Trogoderma granarium* Everts (Col., Dermestidae). *Riv. Parassitol.* **26,** 27 (1965).
90. Bariola, L. A., Cowan, C. B., Jr., Hendricks, D. E., and Keller, J. C. Efficacy of hexalure and light traps in attracting pink bollworm moths. *J. Econ. Entomol.* **64,** 323 (1971).
91. Barnes, M. M., Peterson, D. M., and O'Connor, J. J. Sex pheromone gland in the female codling moth, *Carpocapsa pomonella* (Lepidoptera: Olethreutidae). *Ann. Entomol. Soc. Amer.* **59,** 732 (1966).
92. Barnes, M. M., Robinson, D. W., and Forbes, A. G. Attractants for moths of the western grape leaf skeletonizer. *J. Econ. Entomol.* **47,** 58 (1954).

93. Barrett, C. G. Lepidoptera on stone walls and rocks. *Entomol. Mon. Mag.* **22,** 111 (1886).
94. Bartell, R. J., and Bellas, T. E. Light brown apple moth sex pheromone. *CSIRO Div. Entomol., Canberra, Aust., Annu. Rep.* p. 20 (1969–1970).
94a. Bartell, R. J., and Bellas, T. E. Synergism in sex pheromone of female light brown apple moth. *CSIRO Div. Entomol., Canberra, Aust., Annu. Rep.* p. 35 (1970–1971).
95. Bartell, R. J., and Lawrence, L. A. Sex pheromones of light brown apple moth. *CSIRO Div. Entomol., Canberra, Aust., Annu. Rep.* p. 33 (1969–1970).
96. Bartell, R. J., and Lawrence, L. A. Sex pheromone of cotton leaf worm. *CSIRO Div. Entomol., Canberra, Aust., Annu. Rep.* p. 33 (1969–1970).
96a. Bartell, R. J., and Lawrence, L. A. "Calling" behavior of female light brown apple moth. *CSIRO Div. Entomol., Canberra, Aust., Annu. Rep.* p. 35 (1970–1971).
97. Bartell, R. J., and Shorey, H. H. A quantitative bioassay for the sex pheromone of *Epiphyas postvittana* (Lepidoptera) and factors limiting male responsiveness. *J. Insect Physiol.* **15,** 33 (1969).
98. Bartell, R. J., and Shorey, H. H. Pheromone concentrations required to elicit successive steps in the mating sequence of males of the light brown apple moth, *Epiphyas postvittana. Ann. Entomol. Soc. Amer.* **62,** 1206 (1969).
99. Bartell, R. J., Shorey, H. H., and Barton-Browne, L. Pheromonal stimulation of the sexual activity of males of the sheep blowfly *Lucilia cuprina* (Calliphoridae) by the female. *Anim. Behav.* **17,** 576 (1969).
100. Barth, R. Bau und Funktion der Flügeldrüsen einiger Mikrolepidopteren. Untersuchungen an den Pyraliden *Aphomia gularis, Galleria mellonella, Plodia interpunctella, Ephestia elutella, E. kühniella. Z. Wiss. Zool.* **150,** 1 (1937).
101. Barth, R. Herkunft, Wirkung und Eigenschaften des weiblichen Sexualduftstoffes einiger Pyraliden. *Zool. Jahrb., Abt. Allg. Zool. Physiol. Tiere* **58,** 297 (1937–1938).
102. Barth, R. Die männlichen Duftorgane einiger Argynnis-Arten. (Vergleichende Untersuchungen an *Argynnis paphia, Adippe* und *Aglaia). Zool. Jahrb., Abt. Anat. Ontog. Tiere* **68,** 331 (1944).
103. Barth, R. Die Hautdrüsen des Männchens von *Opsiphanes invirae isagoras* Fruhst. (Lepidoptera, Brassolidae). *Zool. Jahrb., Abt. Anat. Ontog. Tiere* **72,** 216 (1952).
104. Barth, R. Das männliche Duftorgan von *Erebus odoratus* L. (Lepidoptera, Noctuidae). *Zool. Jahrb., Abt. Anat. Ontog. Tiere* **72,** 289 (1952).
105. Barth, R. Männliche Duftorgane Brasilianischer Lepidopteren. 22. *Hammaptera frondosata* Guer. (Geometridae, Larentiinae). *An. Acad. Brasil. Cienc.* **31,** 567 (1959).
106. Barth, R. H., Jr. Hormonal control of sex attractant production in the Cuban cockraoch. *Science* **133,** 1598 (1961).
107. Barth, R. H., Jr. Comparative and experimental studies on mating behavior in cockroaches. Doctoral Thesis, Harvard University, Cambridge, Massachusetts (1961).
108. Barth, R. H., Jr. The endocrine control of mating behavior in the cockroach *Byrsotria fumigata* (Guérin). *Gen. Comp. Endocrinol.* **2,** 53 (1962).
109. Barth, R. H., Jr. Endocrine-exocrine mediated behavior in insects. *Proc. Int. Congr. Zool., 16th, 1963* Vol. 3, p. 3 (1963).
110. Barth, R. H., Jr. The mating behavior of *Byrsotria fumigata* (Guérin) (Blattidae, Blaberinae). *Behavior* **23,** 1 (1964).
111. Barth, R. H., Jr. Insect mating behavior: Endocrine control of a chemical communication system. *Science* **149,** 882 (1965).

112. Barth, R. H., Jr. The comparative physiology of reproductive processes in cockroaches. I. Mating behavior and its endocrine control. *Advan. Reprod. Physiol.* **3,** 167 (1968).

113. Barth, R. H., Jr. The mating behavior of *Eurycotis floridana* (Walker) (Blattaria, Blattoidea, Blattidae, Polyzosteriinae). *Psyche* **75,** 274 (1968).

114. Barth, R. H., Jr. The mating behavior of *Periplaneta americana* (Linnaeus) and *Blatta orientalis* Linnaeus (Blattaria, Blattinae), with notes on 3 additional species of *Periplaneta* and interspecific action of female sex pheromones. *Z. Tierpsychol.* **27,** 722 (1970).

115. Barth, R. H., Jr., and Bell, W. J. Effects of juvenile hormone and JH analogues on the reproductive physiology of females of the cockroach, *Byrsotria fumigata* (Guérin). *Amer. Zool.* **9,** 1085 (1969).

116. Barth, R. H., Jr., and Bell, W. J. Reproductive physiology and behavior of *Byrsotria fumigata* gynandromorphs (Orthoptera (Dictyoptera): Blaberidae). *Ann. Entomol. Soc. Amer.* **64,** 874 (1971).

117. Bartlett, A. C., Hooker, P. A., and Hardee, D. D. Behavior of irradiated boll weevils. I. Feeding, attraction, mating, and mortality. *J. Econ. Entomol.* **61,** 1677 (1968).

118. Barton-Browne, L. B., Bartell, R. J., and Shorey, H. H. Pheromone-mediated behaviour leading to group oviposition in the blowfly *Lucilia cuprina*. *J. Insect Physiol.* **15,** 1003 (1969).

119. Barton-Browne, L. B., Soo Hoo, C. F., van Gerwen, A. C. M., and Sherwell, I. R. Mating flight behaviour in three species of *Oncopera* moths (Lepidoptera: Hepialidae). *J. Aust. Entomol. Soc.* **8,** 168 (1969).

120. Basavanna, G. P. C., and Thontadarya, T. S. Occurrence of thoracic glands in *Mylabris pustulata* Thunbg. (Coleoptera: Meloidae). *Curr. Sci.* **30,** 111 (1961).

121. Batiste, W. C. A timing sex-pheromone trap with special reference to codling moth collections. *J. Econ. Entomol.* **63,** 915 (1970).

122. Bayer, E. Quality and flavor by gas chromatography. *J. Gas Chromatogr.* **4,** 67 (1966).

123. Bayer, E., and Anders, F. Biologische Objekte als Detektoren zur Gaschromatographie. *Naturwissenschaften* **46,** 380 (1959).

124. Bedard, W. D., and Browne, L. E. A delivery-trapping system for evaluating insect chemical attractants in nature. *J. Econ. Entomol.* **62,** 1202 (1969).

125. Bedard, W. D., Silverstein, R. M., and Wood, D. L. Bark beetle pheromones. *Science* **167,** 1638 (1970).

126. Bedard, W. D., and Tilden, P. E. Sources of attractants evoking the mass attack of *Dendroctonus brevicomis* in the field. *Abstr. Nat. Meet., Entomol. Soc. Amer., Miami Beach, Fla.* p. 57 (1970).

127. Bedard, W. D., Tilden, P. E., Wood, D. L., Silverstein, R. M., Brownlee, R. G., and Rodin, J. O. Western pine beetle: Field response to its sex pheromone and a synergistic host terpene. *Science* **164,** 1284 (1969).

128. Bedoukian, P. Z. Purity, identity and quantification of pheromones. *In* "Advances in Chemoreception" (J. W. Johnson, Jr., D. G. Moulton, and A. Turk, eds.), Vol. 1, p. 19. Appleton, New York, 1970.

129. Beets, M. G. J. A molecular approach to olfaction. *Mol. Pharmacol.* **2,** 3 (1964).

130. Begg, J. A. The eastern field wireworm, *Limonius agonus* Say (Coleoptera: Elateridae), in southwestern Ontario. *Proc. Entomol. Soc. Ont.* **92, 1961,** 38 (1962).

131. Bell, W. J., and Barth, R. H., Jr. Quantitative effects of juvenile hormone on re-

production in the cockroach *Byrsotria fumigata. J. Insect Physiol.* **16,** 2303 (1970).
132. Bellas, T. E., Brownlee, R. G., and Silverstein, R. M. Synthesis of brevicomin, principal sex attractant in the frass of the female western pine beetle. *Tetrahedron* **25,** 5149 (1969).
132a. Bellas, T. E., and Fletcher, B. S. Sex attractant of the Queensland fruit fly. *CSIRO Div. Entomol., Canberra, Aust., Annu. Rept.* p. 22 (1970–1971).
133. Benjamin, D. M. The biology and ecology of the red-headed pine sawfly. *U.S., Dep. Agr., Tech. Bull.* **1118,** 1—57 (1955).
133a. Bennett, R. B., and Borden, J. H. Flight arrestment of tethered *Dendroctonus pseudotsugae* and *Trypodendron lineatum* (Coleoptera: Scolytidae) in response to olfactory stimuli. *Ann. Entomol. Soc. Amer.* **64,** 1273 (1971).
134. Benz, G., and Schmid, K. Stimulation der Eiablage unbegatteter Weibchen des Mondspinners *Actias selene* durch ein Männchen-Pheromon. *Experientia* **24,** 1279 (1968).
135. Berger, R. S. Isolation, identification, and synthesis of the sex attractant of the cabbage looper, *Trichoplusia ni. Ann. Entomol. Soc. Amer.* **59,** 767 (1966).
136. Berger, R. S. Sex pheromone of the cotton leafworm. *J. Econ. Entomol.* **61,** 326 (1968).
137. Berger, R. S., and Canerday, T. D. Specificity of the cabbage looper sex attractant. *J. Econ. Entomol.* **61,** 452 (1968).
138. Berger, R. S., Dukes, J. C., and Chow, Y. S. Demonstration of a sex pheromone in three species of hard ticks. *J. Med. Entomol.* **8,** 84 (1971).
139. Berger, R. S., McGough, J. M., and Martin, D. F. Sex attractants of *Heliothis zea* and *H. virescens. J. Econ. Entomol.* **58,** 1023 (1965).
140. Berger, R. S., McGough, J. M., Martin, D. F., and Ball, L. R. Some properties and the field evaluation of the pink bollworm sex attractant. *Ann. Entomol. Soc. Amer.* **57,** 606 (1964).
141. Bergström, G., Kullenberg, B., Ställberg-Stenhagen, S., and Stenhagen, E. Studies on natural odoriferous compounds. II. Identification of a 2,3-dihydro-farnesol as the main component of the marking perfume of male bumble bees of the species *Bombus terrestris* L. *Ark. Kemi* **28,** 454 (1968).
142. Bergström, G., and Lofqvist, J. Odour similarities between the slave-keeping ants *Formica sanguinea* and *Polyergus rufescens* and their slaves *Formica fusca* and *Formica rufibarbis. J. Insect Physiol.* **14,** 995 (1968).
143. Berisford, C. W. Field trapping the Nantucket pine tip moth with a pheromone extract. *Abstr. Nat. Meet., Entomol. Soc. Amer., Miami Beach, Fla.* p. 57 (1970).
144. Beroza, M. Insect attractants are taking hold. *Agr. Chem.* **15** (7), 37 (1960).
145. Beroza, M. Insect attractants. *Soap Chem. Spec.* **36** (2), 74 (1960).
146. Beroza, M. Insect sex attractants and their use. *Proc. Int. Cong. Endocrinol., 2nd, 1964* p. 203 (1965).
147. Beroza, M. Agents affecting insect fertility. *In* "A Symposium on Agents Affecting Fertility" (C. R. Austin and J. S. Perry, eds.), p. 136. Churchill, London, 1965.
148. Beroza, M. The future role of natural and synthetic attractants for pest control. *U.S. Dep. Agr., ARS* **33–110,** 34 (1966).
149. Beroza, M. Nonpersistent inhibitor of the gypsy moth sex attractant in extracts of the insect. *J. Econ. Entomol.* **60,** 875 (1967).
150. Beroza, M. Insect sex attractants. *Amer. Sci.* **59,** 320 (1971).
150a. Beroza, M., Bierl, B. A., Knipling, E. F., and Tardif, J. G. R. The activity of

the gypsy moth sex attractant disparlure vs. that of the live female moth. *J. Econ. Entomol.* **64,** 1527 (1971).

150b. Beroza, M., Bierl, B. A., Tardif, J. G. R., Cook, D. A., and Paszek, E. C. Activity and persistence of synthetic and natural sex attractants of the gypsy moth in laboratory and field trials. *J. Econ. Entomol.* **64,** 1499 (1971).

151. Beroza, M., and Green, N. Lures for insects. *Yearbook Agr.* (*U.S. Dep. Agr.*) p. 365 (1962).

152. Beroza, M., and Green, N. Synthetic chemicals as insect attractants. *Advan. Chem. Ser.* **41,** 11 (1963).

153. Beroza, M., and Jacobson, M. Chemical insect attractants. *World Rev. Pest Contr.* **2** (2), 36 (1963).

154. Beroza, M., and Jacobson, M. Trapping insects by their scents. *Chemistry* **38,** 6 (1965).

155. Beroza, M., Staten, R. T., and Bierl, B. A. Tetradecyl acetate and related compounds as inhibitors of attraction of the pink bollworm moth to the sex lure hexalure. *J. Econ. Entomol.* **64,** 580 (1971).

156. Bestmann, H. J., Kunstmann, R., and Schultz, H. Reaktionen mit Phosphinalkylenen. XV. Eine Synthese der "Königinnensubstanz" und der *trans*-10-Hydroxy-decen-(2)-säure-1(1) (royal jelly acid). *Justus Liebig's Ann. Chem.* **699,** 33 (1966).

157. Bestmann, H. J., Range, P., and Kunstmann, R. Pheromone. II. Synthese des Essigsäure-[*cis*-tetradecen-(9)-yl-esters]. *Chem. Ber.* **104,** 65 (1971).

158. Bierl, B. A., Beroza, M., and Collier, C. W. Potent sex attractant of the gypsy moth: Its isolation, identification, and synthesis. *Science* **170,** 87 (1970).

159. Bierl, B. A., Beroza, M., Cook, D. A., Tardif, J. G. R., and Paszek, E. C. Enhancement of the activity of extracts containing the gypsy moth sex attractant. *J. Econ. Entomol.* **64,** 297 (1971).

160. Birch, M. C. Aphrodisiac effects of scent produced by male noctuid lepidoptera. *Abstr. Int. Congr. Entomol., 13th 1968* p. 32 (1968).

161. Birch, M. C. Pre-courtship use of abdominal brushes by the nocturnal moth, *Phlogophora meticulosa* (L.) (Lepidoptera: Noctuidae). *Anim. Behav.* **18,** 310 (1970).

162. Birch, M. C. Structure and function of the pheromone-producing brush-organs in males of *Phlogophora meticulosa* (L.) (Lepidoptera: Noctuidae). *Trans. Roy. Entomol. Soc. London* **122,** 277 (1970).

163. Birch, M. C. Persuasive scents in moth sex life. *J. Amer. Mus. Natur. Hist.* **79** (9), 34 and 72 (1970).

164. Birch, M. C. Intrinsic limitations in the use of electroantennograms to bioassay male pheromones in Lepidoptera. *Nature* (*London*) **233,** 57 (1971).

165. Blight, M. M., Grove, J. F., and McCormick, A. Volatile neutral compounds emanating from laboratory-reared colonies of the desert locust, *Schistocerca gregaria. J. Insect. Physiol.* **15,** 11 (1969).

166. Block, B. C. Laboratory method for screening compounds as attractants to gypsy moth males. *J. Econ. Entomol.* **53,** 172 (1960).

167. Block, B. C. Behavioral studies of the responses of gypsy moth males (*Porthetria dispar* L.) to the female sex attractant and related compounds. *Abstr. East. Br. Meet., Entomol. Soc. Amer.* p. 11 (1961).

168. Block, B. C., Boeckh, J., and Schneider, D. Behavioral and electrophysiological

correlations to odorous sex attractants in the gypsy moth and silkworm moth. *Bull. Entomol. Soc. Amer.* **8**, 146 and 157 (1962).

169. Blum, M. S., Boch, R., Doolittle, R. E., Tribble, M. T., and Traynham, J. G. Honey bee sex attractant: Conformational analysis, structural specificity, and lack of masking activity of congeners. *J. Insect Physiol.* **17**, 349 (1971).

170. Bobb, M. L. Apparent loss of sex attractiveness by the female of the Virginia-pine sawfly, *Neodiprion pratti pratti. J. Econ. Entomol.* **57**, 829 (1964).

170a. Bobb, M. L. Influence of sex pheromones on mating behavior and populations of Virginia pine sawfly. *Environ. Entomol.* **1**, 78 (1972).

171. Boch, R., and Shearer, D. A. Identification of geraniol as the active component in the Nassanoff pheromone of the honey bee. *Nature (London)* **194**, 704 (1962).

172. Boch, R., and Shearer, D. A. Production of geraniol by honey bees of various ages. *J. Insect Physiol.* **9**, 431 (1963).

173. Boch, R., and Shearer, D. A. Identification of nerolic and geranic acids in the Nassanoff pheromone of the honey bee. *Nature (London)* **202**, 320 (1964).

174. Boch, R., and Shearer, D. A. Attracting honey bees to crops which require pollination. *Amer. Bee J.* **105**, 166 (1965).

175. Boch, R., and Shearer, D. A. Alarm in the beehive. *Amer. Bee J.* **105**, 206 (1965).

176. Bodenstein, W. G. Distribution of female sex pheromone in the gut of *Periplaneta americana* (Orthoptera: Blattidae). *Ann. Entomol. Soc. Amer.* **63**, 336 (1970).

177. Boeckh, J. Elektrophysiologische Untersuchungen an einzelnen Geruchsrezeptoren auf der Antenne des Totengräbers (Necrophorus, Coleoptera). *Verh. Deut. Zool. Ges. Wien* p. 297 (1962).

178. Boeckh, J. Elektrophysiologische Untersuchungen an einzelnen Geruchsrezeptoren auf den Antennen des Totengräbers (Necrophorus, Coleoptera). *Z. Vergl. Physiol.* **46**, 212 (1962).

179. Boeckh, J. Inhibition and excitation of single insect olfactory receptors, and their role as a primary sensory code. *Proc. Int. Symp. Olfaction Taste, 2nd, 1965* Vol. 2, p. 721 (1967).

180. Boeckh, J., Kaissling, K. E., and Schneider, D. Sensillen und Bau der Antennengeissel von *Telea polyphemus* (Vergleiche mit weiteren Saturniden: *Antheraea, Platysamia* und *Philosamia*). *Zool. Jahrb., Abt. Anat. Ontog. Tiere* **78**, 559 (1960).

181. Boeckh, J., Kaissling, K. E., and Schneider, D. Insect olfactory receptors. *Cold Spring Harbor Symp. Quant. Biol.* **30**, 263 (1965).

182. Boeckh, J., Priesner, E., Schneider, D., and Jacobson, M. Olfactory receptor response to the cockroach sexual attractant. *Science* **141**, 716 (1963).

183. Boeckh, J., Sass, H., and Wharton, D. R. A. Antennal receptors: Reactions to female sex attractant in *Periplaneta americana. Science* **168**, 589 (1970).

184. Borden, J. H. Factors influencing the response of *Ips confusus* (Coleoptera: Scolytidae) to male attractant. *Can. Entomol.* **99**, 1164 (1967).

185. Borden, J. H. Sex pheromone of *Dendroctonus pseudotsugae* (Coleoptera: Scolytidae): Production, bio-assay, and partial isolation. *Can. Entomol.* **100**, 597 (1968).

186. Borden, J. H., Nair, K. K., and Slater, C. E. Synthetic juvenile hormone: Induction of sex pheromone production in *Ips confusus. Science* **166**, 1626 (1969).

187. Borden, J. H., and Slater, C. E. Sex pheromone of *Trypodendron lineatum:* production in the female hindgut-malpighian tubule region. *Ann. Entomol. Soc. Amer.* **62**, 454 (1969).

188. Borden, J. H., and Wood, D. L. The antennal receptors and olfactory response

of *Ips confusus* (Coleoptera: Scolytidae) to male sex attractant in the laboratory. *Ann. Entomol. Soc. Amer.* **59,** 253 (1966).

189. Bornemissza, G. F. Sex attractant of male scorpion flies. *Nature (London)* **203,** 786 (1964).

190. Bornemissza, G. F. Specificity of male sex attractants in some Australian scorpion flies. *Nature (London)* **209,** 732 (1966).

191. Bornemissza, G. F. Observations on the hunting and mating behaviour of two species of scorpion flies (Bittacidae: Mecoptera). *Aust. J. Zool.* **14,** 371 (1966).

192. Bosman, T. The sex pheromones of insects. *S. Afr. J. Sci.* **66,** 228 (1970).

193. Bossert, W. H., and Wilson, E. O. The analysis of olfactory communication among animals. *J. Theor. Biol.* **5,** 443 (1963).

194. Bottrell, D. G., Reeves, R. E., Almand, L. K., Hardee, D. D., and Cross, W. H. Studies of boll weevil populations and their movement in the High and Rolling Plains of Texas using male-baited traps, 1968. *Tex., Agr. Exp. Sta., Misc. Publ.* **948,** 1–8 (1970).

195. Bounhiol. J. J. Physiologie de l'attraction sexuelle chez *Bombyx mori* (Lepidoptera). *Programme Int. Congr. Entomol., 12th, 1964* p. 68 (1965).

196. Boush, G. M., and Baerwald, R. J. Courtship behavior and evidence for a sex pheromone in the apple maggot parasite, *Opius alloeus* (Hymenoptera: Braconidae). *Ann. Entomol. Soc. Amer.* **60,** 865 (1967).

197. Bowers, W. S., and Bodenstein, W. G. Sex pheromone mimics of the American cockroach. *Nature (London)* **232,** 259 (1971).

198. Boyce, H. R. Peach tree borers (Lepidoptera: Aegeriidae) in Ontario. *Proc. Entomol. Soc. Ont.* **92, 1961,** 45 (1962).

199. Brader-Breukel, L. M. Modalités de l'attraction chez *Diparopsis watersi* (Roths.). Ph.D. Thesis, Eijksuniversiteit, Leiden, The Netherlands (1969).

200. Bradley, J. R., Jr., Clower, D. F., and Graves, J. B. Field studies of sex attraction in the boll weevil. *J. Econ. Entomol.* **61,** 1457 (1968).

201. Brady, U. E. Response of male *Sitotroga* to the sex pheromones of *Sitotroga* and *Pectinophora* females and to propylure and deet. Deet analysis in *Sitotroga* females. *J. Ga. Entomol. Soc.* **4,** 11 (1969).

202. Brady, U. E. Inhibition of the behavioral response of males of Indian-meal moths, *Plodia interpunctella,* and related species to female sex pheromones: Exposure to sex pheromones of unrelated species. *J. Ga. Entomol. Soc.* **4,** 41 (1969).

203. Brady, U. E., and Finley, L. H. Behavioral response of Bidrin-treated male Indian meal moths to the female sex pheromone. *J. Econ. Entomol.* **62,** 1117 (1969).

204. Brady, U. E., and Nordlund, D. A. *cis*-9, *trans*-12-Tetradecadien-1-yl acetate in the female tobacco moth *Ephestia elutella* (Hübner) and evidence for an additional component of the sex pheromone. *Life Sci.* **10,** Part II, 797 (1971).

205. Brady, U. E., and Smithwick, E. B. Production and release of sex attractant by the female Indian-meal moth, *Plodia interpunctella. Ann. Entomol. Soc. Amer.* **61,** 1260 (1968).

206. Brady, U. E., Tumlinson, J. H., III, Brownlee, R. G., and Silverstein, R. M. Sex stimulant and attractant in the Indian meal moth and in the almond moth. *Science* **171,** 802 (1971).

207. Brieger, G., and Butterworth, F. M. *Drosophila melanogaster:* Identity of male lipid in reproductive system. *Science* **167,** 1262 (1970).

208. Bronskill, J. F. Permanent whole-mount preparation of lepidopterous genitalia

for complete visibility of the sex pheromone gland. *Ann. Entomol. Soc. Amer.* **63,** 898 (1970).

209. Bronskill, J. F., Roelofs, W. L., Chapman, P. J., and Lienk, S. E. The sex pheromone as a taxonomic principle. *Proc. Int. Congr. Entomol., 13th, 1968* Vol. 1, p. 115 (1971).

210. Brower, L. P., van Zandt Brower, J., and Cranston, F. P. Courtship behavior of the queen butterfly, *Danaus gilippus berenice* (Cramer). *Zoologica (New York)* **50,** 1 (1965).

211. Brower, L. P., and Jones, M. A. Precourtship interaction of wing and abdominal sex glands in male *Danaus* butterflies. *Proc. Roy. Entomol. Soc. London, Ser. A* **40,** 147 (1965).

212. Brown, W. L., Jr. Mass insect control programs: Four case histories. *Psyche* **68,** 75 (1961).

213. Brown, W. L., Jr., Eisner, T., and Whittaker, R. H. Allomones and kairomones: transspecific chemical messengers. *Bio Science* **20,** 21 (1970).

214. Bruce, H. M. Pheromones and behavior in mice. *Acta Neurol. Psychiat. Belg.* **69,** 529 (1969).

215. Bruce, H. M. Pheromones. *Brit. Med. Bull.* **26,** 10 (1970).

216. Buchta, E., and Fuchs, F. 10.13-Dioxo- und 7.10.13-Trioxo-hexadecansäure sowie 10.13-Dihydroxy-hexadecansäuremethylester und Hexadecantriol-(1.10.13). *Naturwissenschaften* **48,** 454 (1961).

217. Buchta, E., and Fuchs, F. 10.13-Dihydroxy-hexadecansäure-methylester und Hexadecantriol-(1.10.13). *Justus Liebig's Ann. Chem.* **704,** 115 (1967).

218. Bull, D. L., Stokes, R. A., Hardee, D. D., and Gueldner, R. C. Gas chromatographic determination of the components of the synthetic boll weevil sex pheromone (grandlure). *J. Agr. Food Chem.* **19,** 202 (1971).

219. Burgess, E. D. Development of gypsy moth sex-attractant traps. *J. Econ. Entomol.* **43,** 325 (1950).

220. Burgess, E. D. Gypsy moth control. *Science* **143,** 526 (1964).

221. Burke, H. E. Notes on the carpenter worm (*Prionoxystus robiniae* Peck) and a new method of control. *J. Econ. Entomol.* **14,** 369 (1921).

222. Burkhardt, D. Ultramikroelektroden aus Glas, ihre Herstellung und Verwendung bei elektrophysiologischen Messungen. *Glas-Instrum.-Tech.* **3,** 115 (1959).

223. Burkholder, W. E. Pheromone research with stored-product Coleoptera. *In* "Control of Insect Behavior by Natural Products" (D. L. Wood, R. M. Silverstein, and M. Nakajima, eds.), p. 1. Academic Press, New York, 1970.

224. Burkholder, W. E., and Dicke, R. J. Evidence of sex pheromones in females of several species of Dermestidae. *J. Econ. Entomol.* **59,** 540 (1966).

225. Burkholder, W. E., Silverstein, R. M., Rodin, J. O., and Gorman, J. E. Sex attractant of the black carpet beetle. U.S. Pat. 3,501,566 (1970).

226. Buschinger, A. "Locksterzeln" begattungsbereiter ergatoider Weibchen von *Harpagoxenus sublaevis* Nyl. (Hymenoptera, Formicidae). *Experientia* **24,** 297 (1968).

227. Butenandt, A. Zur Kenntnis der Sexual-Lockstoffe bei Insekten. *Jahrb. Preuss. Akad. Wiss.* p. 97 (1939).

228. Butenandt, A. Untersuchungen über Wirkstoffe aus dem Insektenreich. *Angew. Chem.* **54,** 89 (1941).

229. Butenandt, A. Über Wirkstoffe des Insektenreiches. II. Zur Kenntnis der Sexual-Lockstoffe. *Naturwiss. Rundsch.* p. 457 (1955).

230. Butenandt, A. Fettalkohole als Sexual-Lockstoffe der Schmetterlinge. *Fette, Seifen, Anstrichm.* **64,** 187 (1962).

231. Butenandt, A. Bombycol, the sex-attractive substance of the silkworm, *Bombyx mori. J. Endocrinol.* **27,** 9 (1963).

232. Butenandt, A., Beckmann, R., and Hecker, E. Über den Sexuallockstoff des Seidenspinners. I. Der biologische Test und die Isolierung des reinen Sexuallockstoffes Bombykol. *Hoppe-Seyler's Z. Physiol. Chem.,* **324,** 71 (1961).

233. Butenandt, A., Beckmann, R., and Stamm, D. Über den Sexuallockstoff des Seidenspinners. II. Konstitution und Konfiguration des Bombykols. *Hoppe-Seyler's Z. Physiol. Chem.* **324,** 84 (1961).

234. Butenandt, A., Beckmann, R., Stamm, D., and Hecker, E. Über den Sexuallockstoff des Seidenspinners *Bombyx mori.* Reindarstellung und Konstitution. *Z. Naturforsch. B.* **14,** 283 (1959).

235. Butenandt, A., Guex, W., Hecker, E., Ruegg, R., and Schwieter, U. Insektenlockmittel. Ger Pat. 1,108,976 (1961).

236. Butenandt, A., and Hecker, E. Dimethylamine, the supposed sex-attractant of the silkworm moth *(Bombyx mori). Proc. Chem. Soc., London* p. 53 (1958).

237. Butenandt, A., and Hecker, E. Synthese des Bombykols, des Sexuallockstoffes des Seidenspinners, und seiner geometrischen Isomeren. *Angew. Chem.* **73,** 349 (1961).

238. Butenandt, A., and Hecker, E. La synthèse de bombycol. *Nucleus (Paris)* **5,** 325 (1964).

239. Butenandt, A., Hecker, E., Hopp, M., and Koch, W. Über den Sexuallockstoff des Seidenspinners. IV. Die Synthese des Bombykols und der *cis-trans*-Isomeren Hexadecadien-(10,12)-ole-(1). *Justus Liebig's Ann. Chem.* **658,** 39 (1962).

240. Butenandt, A., Hecker, E., and Zachau, H. G. Über die vier geometrischen Isomeren des 2,4-Hexadienols-(1). *Chem. Ber.* **88,** 1185 (1955).

241. Butenandt, A., Hecker, E., and Zayed, S. M. A. D. Über den Sexuallockstoff des Seidenspinners. III. Ungesättigte Fettsäuren aus den Hinterleibsdrüsen (Sacculi laterales) des Seidenspinner-weibchens *(Bombyx mori* L.). *Hoppe-Seyler's Z. Physiol. Chem.* **333,** 114 (1963).

242. Butenandt, A., and Tam, N. D. Über einen geschlechtsspezifischen Duftstoff der Wasserwanze *Belostoma indica* Vitalis *(Lethocerus indicus* Lep.). *Hoppe-Seyler's Z. Physiol. Chem.* **308,** 277 (1957).

243. Butenandt, A., Truscheit, E., Eiter, K., and Hecker, E. Verfahren zur Herstellung von Hexadecadien-(10,12)-ol-(1). Ger. Pat. 1,096,345 (1961).

244. Butler, C. G. Pheromones in sexual processes in insects. *In* "Insect Reproduction," Symp. No. 2, p. 66. Roy. Entomol. Soc., London, 1964.

245. Butler, C. G. Sex attraction in *Andrena flavipes* Panser (Hymenoptera: Apidae), with some observations on nest-site restriction. *Proc. Roy. Entomol. Soc. Ser. A* **40,** 77 (1965).

246. Butler, C. G. Mandibular gland pheromone of worker honeybees. *Nature (London)* **212,** 530 (1966).

247. Butler, C. G. Insect pheromones. *Biol. Rev. Cambridge Phil. Soc.* **42,** (1967).

248. Butler, C. G. A sex attractant acting as an aphrodisiac in the honeybee *(Apis mellifera* L.). *Proc. Roy. Entomol. Soc. London Ser. A* **42,** 71 (1967).

249. Butler, C. G. Some pheromones controlling honeybee behaviour. *Proc. Congr. Int. Union Study Social Insects, 6th, 1969* p. 19 (1969).

250. Butler, C. G. Pheromones of queen honeybees. *Rep. Rothamsted Exp. Sta.,* **1969** Part 1, p. 256 (1970).

251. Butler, C. G. Chemical communication in insects: Behavioral and ecologic aspects. *In* "Communication by Chemical Signals" (J. W. Johnston, Jr., D. G. Moulton, and A. Turk, eds.), p. 35. Appleton, New York, 1970.

251a. Butler, C. G. The mating behaviour of the honeybee (*Apis mellifera* L.). *J. Entomol., A* **46**, 1 (1971).

252. Butler, C. G., and Calam, D. H. Pheromones of the honey bee—the secretion of the Nassanoff gland of the worker. *J. Insect Physiol.* **15**, 237 (1969).

253. Butler, C. G., Calam, D. H., and Callow, R. K. Attraction of *Apis mellifera* drones by the odours of the queens of two other species of honeybees. *Nature (London)* **213**, 423 (1967).

254. Butler, C. G., and Callow, R. K. Honeybee pheromones. *Chem. Ind. (London)* p. 883 (1965).

255. Butler, C. G., and Callow, R. K. Pheromones of the honeybee (*Apis mellifera* L.): The inhibitory scent of the queen. *Proc. Roy. Entomol. Soc. London Ser. A*, **43**, 62 (1968).

256. Butler, C. G., Callow, R. K., and Chapman, J. R. 9-Hydroxydec-*trans*-2-enoic acid, a pheromone stabilizing honeybee swarms. *Nature (London)* **201**, 733 (1964).

257. Butler, C. G., Callow, R. K., and Johnston, N. C. The isolation and synthesis of queen substance, 9-oxodec-*trans*-2-enoic acid, a honeybee pheromone. *Proc. Roy. Soc., Ser B* **155**, 417 (1961).

258. Butler, C. G., and Fairey, E. M. Pheromones of the honeybee: Biological studies of the mandibular gland secretion of the queen. *J. Apicult. Res.* **3**, 65 (1964).

259. Butler, C. G., and Simpson, J. Pheromones of the honeybee (*Apis mellifera*). An olfactory pheromone from the Koschewnikow gland of the queen. *Ved. Pr. Vyzk. Ustavu Vcelarskeho* p. 33 (1965).

260. Butler, C. G., and Simpson, J. Pheromones of the queen honeybee (*Apis mellifera* L.) which enable her workers to follow her when swarming. *Proc. Roy. Entomol. Soc. London, Ser A* **42**, 149 (1967).

260a. Butt, B. A. Personal communication (1962).

261. Butt, B. A., Beroza, M., McGovern, T. P., and Freeman, S. K. Synthetic chemical sex stimulants for the codling moth. *J. Econ. Entomol.* **61**, 570 (1968).

262. Butt, B. A., and Hathaway, D. O. Female sex pheromone as attractant for male codling moths. *J. Econ. Entomol.* **59**, 476 (1966).

263. Butz, A., and Aronoff, A. Adult male response to the sex attractant of the virgin female American cockroach. *Abstr. Nat. Meet., Entomol. Soc. Amer., Miami Beach, Fla.* p. 51 (1970).

264. Cade, S. C., Hrutfiord, B. F., and Gara, R. I. Identification of a primary attractant for *Gnathotrichus sulcatus* isolated from western hemlock logs. *J. Econ. Entomol.* **63**, 1014 (1970).

265. Caillot, Y., and Boisson, C. Developpement larvaire du Belostome (*Lethocerus indicus* Lep.); Insecte Hemiptere, Hydrocoryse, Cryptocerate. *Ann. Sci. Natur. Zool. Biol. Anim.* [11] **16**, 51 (1954).

266. Calam, D. H. Species and sex-specific compounds from the heads of male bumblebees (*Bombus* spp.). *Nature (London)* **221**, 856 (1969).

267. Callahan, P. S. A photographic analysis of moth flight behavior, with special reference to the theory for electromagnetic radiation as an attractance force between the sexes. *Programme Int. Congr. Entomol., 12th, 1964* p. 48 (1965).

268. Callahan, P. S. Far infra-red emission and detection by night-flying moths. *Nature (London)* **206**, 1172 (1965).

269. Callahan, P. S. Intermediate and far infrared sensing of nocturnal insects. I.

Evidence for a far infrared (FIR) electromagnetic theory of communication and sensing in moths and its relationship to the limiting biosphere of the corn earworm. *Ann. Entomol. Soc. Amer.* **58**, 727 (1965).

270. Callahan, P. S. Intermediate and far infrared sensing of nocturnal insects. II. The compound eye of the corn earworm, *Heliothis zea*, and other moths as a mosaic-electromagnetic thermal radiometer. *Ann. Entomol. Soc. Amer.* **58**, 746 (1965).

271. Callahan, P. S. A photographic analysis of moth flight behavior with special reference to the theory for electromagnetic radiation as an attractive force between the sexes and to host plants. *Proc. Int. Congr. Entomol., 12th, 1964* p. 302 (1965).

272. Callahan, P. S. Are arthropods infrared and microwave detectors? *Proc. N. Cent. Br. Entomol. Soc. Amer.* [*N.S.*] **20**, 20 (1965).

273. Callahan, P. S. Are arthropods infrared and microwave detectors? *U.S., Dep. Agr., Entomol. Res. Div., Spec. Rep.* **V–302**, 1–36 (1965).

274. Callahan, P. S. Do insects communicate by radio? *Animals* **8**, 198 (1966).

275. Callahan, P. S. Electromagnetic communication in insects—elements of the terrestrial infrared environment, including generation, transmission, and detection by moths. *U.S. Dep. Agr. ARS* **ARS 33–110**, 156 (1966).

276. Callahan, P. S. Insect molecular bioelectronics: A theoretical and experimental study of insect sensillae as tubular waveguides, with particular emphasis on their dielectric and thermoelectret properties. *Misc. Publ. Entomol. Soc. Amer.* **5**, 315 (1967).

277. Callow, R. K., Chapman, J. R., and Paton, P. N. Pheromones of the honeybee: Chemical studies of the mandibular gland secretion of the queen. *J. Apicult. Res.* **3**, 77 (1964).

277a. Camors, F. B., Jr., and Payne, T. L. Response of *Heydenia unica* (Hymenoptera: Pteromalidae) to *Dendroctonus frontalis* (Coleoptera: Scolytidae) pheromones and a host-tree terpene. *Ann. Entomol. Soc. Amer.* **65**, 31 (1972).

278. Campbell, W. H. Sexual attraction in Lepidoptera. *J. Bombay Natur. Hist. Soc.* **18**, 511 (1908).

279. Campion, D. G. Chemosterilization of the red bollworm (*Diparopsis castanea* Hmps.) by mutagenic agents. *Nature (London)* **214**, 1031 (1967).

280. Campion, D. G., and Outram, I. Factors affecting male probing activity of the red bollworm *Diparopsis castanea* (Hmps.) in relation to sterility induced by tris-(1-aziridinyl) phosphine oxide (tepa). *Ann. Appl. Biol.* **60**, 191 (1967).

281. Cantelo, W. W., and Skot, O. A cage for holding female tobacco hornworms at light traps. *J. Econ. Entomol.* **64**, 1322 (1971).

282. Cantelo, W. W., and Smith, J. S., Jr. Collections of tobacco hornworm moths in traps equipped with one- or four-blacklight lamps baited with adult virgin females. *J. Econ. Entomol.* **64**, 555 (1971).

282a. Cantelo, W. W., and Smith, J. S., Jr. Attraction of tobacco hornworm moths to blacklight traps baited with virgin females. *J. Econ. Entomol.* **64**, 1511 (1971).

283. Carlson, D. A., Mayer, M. S., Silhacek, D. L., James, J. D., Beroza, M., and Bierl, B. A. Sex attractant pheromone of the house fly: Isolation, identification and synthesis. *Science* **174**, 76 (1971).

284. Carthy, J. D. Insect communication. *In* "Insect Behavior" (P. T. Haskell, ed.), Symp. No. 3, pp. 68–80. Roy Entomol. Soc., London, 1966.

285. Cartwright, W. B. Sexual attraction of the female Hessian fly (*Phytophaga destructor* Say.). *Can. Entomol.* **54**, 154 (1922).

286. Casida, J. E., Coppel, H. C., and Watanabe, T. Purification and potency of the sex attractant from the introduced pine sawfly, *Diprion similis*. *J. Econ. Entomol.* **56,** 18 (1963).
287. Castek, K. L., Barbour, J. F., and Rudinsky, J. A. Isolation and purification of the attractant of the striped ambrosia beetle. *J. Econ. Entomol.* **60,** 658 (1967).
288. Chaboussou, F. Piégeage sexuel des males d'eudémis (*Lobesia botrana* Schiff.) par attirance des femelles vierges. Relation avec le niveau de population dans le vignoble. *Ann. Epiphyt.* **13,** 331 (1962).
289. Chaboussou, F., and Carles, J. P. Observations sur le piégeage sexuel des mâles d'eudémis (*Lobesia botrana* Schiff.). *Rev. Zool. Agr. Appl.* **61,** 81 (1962).
290. Champlain, A. B. The curious mating habit of *Megarhyssa atrata* (Fab.) (Hymen.: Ichneumonoidea). *Entomol. News* **32,** 241 (1921).
291. Chapman, J. A. Evidence for a sex attractant in the elaterid beetle, *Hemicrepidius morio* (LeConte). *Can. Entomol.* **96,** 909 (1964).
292. Chapman, J. A. The effect of attack by the ambrosia beetle *Trypodendron lineatum* (Olivier) on log attractiveness. *Can. Entomol.* **98,** 50 (1966).
293. Chapman, J. A. Response behaviour of scolytid beetles and odour meteorology. *Can. Entomol.* **99,** 1132 (1967).
294. Chapman, J. A., and Dyer, E. D. A. Cross attraction between the Douglas-fir beetle [*Dendroctonus pseudotsugae* Hopk.] and the spruce beetle [*D. obesus* (Mann.)]. *Can. Dep. Fish. Forestry, Ottawa, Bi-Mon. Res. Notes* **25** (4), 31 (1969).
295. Chapman, J. R. Photolysis of the tetrahydropyranyl ether of 3-hydroxy-2,2,4,4-tetramethylcyclobutanone tosylhydrazone. *Tetrahedron Lett.* p. 113 (1966).
296. Chararas, C. Recherches sur l'attractivité chez les Scolytidae. Étude sur l'attractivité sexuelle chez *Carphoborus minimus* Fabr. Coléoptère Scolytidae typiquement polygame. *C. R. Acad. Sci., Ser. D* **262,** 2492 (1966).
297. Chararas, C. Recherches sur le comportement sexuel de *Pityokteines spinidens* Reit. (*Coléoptère* Scolytidae *polygame*) et étude des facteurs qui agissent sur le pouvoir attractif du mâle a l'égard de la femelle. *C. R. Acad. Sci., Ser. D* **266,** 1852 (1969).
298. Chararas, C., and Berton, A. Recherches sur les constituants odorants des exhalaisons terpéniques de diverses essences et sur leur action a l'égard d'*Ips sexdentatus* Boerner (Coléoptère, Scolytidae). *C. R. Acad. Sci., Ser. D* **264,** 1471 (1967).
299. Chararas, C., and M'Sadda, K. Attraction chimique et attraction sexuelle chez *Orthotomicus erosus* Woll. (Coléoptère Scolytidae). *C. R. Acad. Sci., Ser. D* **271,** 1904 (1970).
300. Chauvin, R. Etude du comportement des insectes. *Annee Biol.* [4] **1,** 207 (1962).
301. Chen, S. H., and Young, B. Observations on the mating behavior of *Bombyx mori*. *Sinensia* **14,** 45 (1943).
302. Cleveland, M. L., and Murdock, L. L. Natural sex attractant of the lesser peach tree borer. *J. Econ. Entomol.* **57,** 761 (1964).
303. Cole, L. R. Observations on the finding of mates by male *Phaeogenes invisor* and *Apanteles medicaginis* (Hymenoptera: Ichneumonoidae). *Anim. Behav.* **18,** 184 (1970).
304. Collins, C. W., and Potts, S. F. Attractants for the flying gypsy moths as an aid in locating new infestations. *U.S., Dep. Agr., Tech. Bull.* **336,** 1–44 (1932).
305. Collins, M. M., and Weast, R. D. "Wild Silk Moths of the United States," pp. 81 and 89–91. Collins Radio Co., Cedar Rapids, Iowa, 1961.

306. Comfort, A. Likelihood of human pheromones. *Nature (London)* **230,** 432 and 479 (1971).

307. Cone, W. W., McDonough, L. M., Maitlen, J. C., and Burdajewicz, S. Pheromone studies of the two-spotted spider mite. I. Evidence of a sex pheromone. *J. Econ. Entomol.* **64,** 355 (1971).

308. Cone, W. W., Predki, S., and Klostermeyer, E. C. Pheromone studies of the two-spotted spider mite. II. Behavioral response of males to quiescent deutonymphs. *J. Econ. Entomol.* **64,** 379 (1971).

308a. Cone, W. W., and Pruszynski, S. Pheromone studies of the two-spotted spider mite. III. Response of males to different plant tissues, age, searching area, sex ratios, and solvents in bioassay trials. *J. Econ. Entomol.* **65,** 74 (1972).

309. Connin, R. V., and Hoopingarner, R. A. Sexual behavior and diapause of the cereal leaf beetle, *Oulema melanopus. Ann. Entomol. Soc. Amer.* **64,** 655 (1971).

310. Coppel, H. C., Casida, J. E., and Dauterman, W. C. Evidence for a potent sex attractant in the introduced pine sawfly, *Diprion similis* (Hymenoptera: Diprionidae). *Ann. Entomol. Soc. Amer.* **53,** 510 (1960).

311. Cory, E. N. The peach-tree borer, *Sanninoidea exitiosa* Say. *Md., Agr. Exp. Sta., Bull.* **176,** 181 (1913).

312. Coster, J. E. Production of aggregating pheromones in re-emerged parent females of the southern pine beetle. *Ann. Entomol. Soc. Amer.* **63,** 1186 (1970).

313. Coster, J. E., and Gara, R. I. Studies on the attack behavior of the southern pine beetle. II. Response to attractive host material. *Contrib. Boyce Thompson Inst.* **24,** 69 (1968).

313a. Coster, J. E., and Vité, J. P. Effects of feeding and mating on pheromone release in the southern pine beetle. *Ann. Entomol. Soc. Amer.* **65,** 263 (1972).

314. Cowan, B. D., and Rogoff, W. M. Variation and heritability of responsiveness of individual male house flies, *Musca domestica,* to the female sex pheromone. *Ann. Entomol. Soc. Amer.* **61,** 1215 (1968).

315. Cox, R. G., Anderson, J. W., and McDowell, H. G. Cooperative research on control of the mountain pine beetle in western white pine. *West. Forest Pest Comm. Progr. Rep.* No. 7, pp. 1–22 (1971).

316. Crescitelli, F., and Geissman, T. A. Invertebrate pharmacology: Selected topics. *Annu. Rev. Pharmacol.* **2,** 143 (1962).

317. Cross, W. H., Hardee, D. D., Nichols, F., Mitchell, H. C., Mitchell, E. B., Huddleston, P. M., and Tumlinson, J. H. Attraction of female boll weevils to traps baited with males or extracts of males. *J. Econ. Entomol.* **62,** 154 (1969).

318. Cross, W. H., and Mitchell, H. C. Mating behavior of the female boll weevil. *J. Econ. Entomol.* **59,** 1505 (1966).

319. Crossley, A. C., and Waterhouse, D. F. The ultrastructure of a phermone-secreting gland in the male scorpion-fly *Harpobittacus australis* (Bittacidae: Mecoptera). *Tissue & Cell* **1,** 273 (1969).

320. Cummings, B. F. Scent organs in Trichoptera. *Proc. Zool. Soc. London* p. 459 (1914).

321. Curtis, R. F., Ballantine, J. A., Keverne, E. B., Bonsall, R. W., and Michael, R. P. Identification of primate sexual pheromones and the properties of synthetic attractants. *Nature (London)* **232,** 396 (1971).

322. Cuthbert, F. P., Jr., and Reid, W. J., Jr. Studies of sex attractant of banded cucumber beetle. *J. Econ. Entomol.* **57**, 247 (1964).

323. Dahl, E., Emanuelsson, H., and von Mecklenburg, C. Pheromone transport and reception in an amphipod. *Science* **170**, 739 (1970).

324. Dahm, K. H., Finn, W. E., and Röller, H. The sex attractant of *Achroia grisella* Fabr. *Abstr. Int. Symp. Chem. Natur. Prod., 7th, 1970* p. 558 (1970).

325. Dahm, K. H., Meyer, D., Finn, W. E., Reinhold, V., and Röller, H. The olfactory and auditory mediated sex attraction in *Achroia grisella* (Fabr.). *Naturwissenschaften* **58**, 265 (1971).

326. Dahm, K. H., Richter, I., Meyer, D., and Röller, H. The sex attractant of the Indian-meal moth, *Plodia interpunctella* (Hübner). *Life Sci.* **10**, Part II, 531 (1971).

327. Dalla-Torre, K. W. von. Die Duftapparate der Schmetterlinge. *Kosmos* **10**, 354 and 410 (1885).

328. Daniel, D. M. *Macrocentrus ancylivorus* Rohwer, a polyembryonic braconid parasite of the Oriental fruit moth. *N. Y., State Agr. Exp. Sta., Geneva, Tech. Bull.* **187**, 1–101 (1932).

329. Daterman, G. E. Laboratory mating of the European pine shoot moth, *Rhyacionia buoliana. Ann. Entomol. Soc. Amer.* **61**, 920 (1968).

330. Daterman, G. E. Reproductive biology of the European pine shoot moth *Rhyacionia buoliana* (Schiffermuller) (Lepidoptera: Olethreutidae), with special reference to mating behavior, sex attraction, and fecundity. Ph.D. Thesis, Oregon State University, Corvallis (1969).

331. Daterman, G. E. An improved technique for mating European pine shoot moth, *Rhyacionia buoliana* (Lepidoptera: Olethreutidae) in the laboratory. *Can. Entomol.* **102**, 541 (1970).

331a. Daterman, G. E. Laboratory bioassay for sex pheromone of the European pine shoot moth, *Rhyacionia buoliana. Ann. Entomol. Soc. Amer.* **65**, 119 (1972).

332. Daterman, G. E., and McComb, D. Female sex attractant for survey trapping European pine shoot moth. *J. Econ. Entomol.* **63**, 1406 (1970).

333. Davich, T. B., Hardee, D. D., and Alcala, M. J. Long-range dispersal of boll weevils determined with wing traps baited with males. *J. Econ. Entomol.* **63**, 1706 (1970).

334. Davis, F. M., and Henderson, C. A. Attractiveness of virgin female moths of the southwestern corn borer. *J. Econ. Entomol.* **60**, 279 (1967).

335. Day, A. C., and Whiting, M. C. The structure of the sex attractant of the American cockroach. *Proc. Chem. Soc., London*, p. 368 (1964).

336. Day, A. C., and Whiting, M. C. On the structure of the sex attractant of the American cockroach *J. Chem. Soc., C* p. 464 (1966).

337. De, R. K. Sex pheromones in stored-grain insects. *Abstr. Nat. Meet., Entomol. Soc. Amer., Miami Beach, Fla.* p. 55 (1970).

338. Dean, G. J. W., Clements, S. A., and Paget, J. Observations on sex attraction and mating behaviour of the tsetse fly *Glossina morsitans orientalis* Vanderplank. *Bull. Entomol. Res.* **59**, 355 (1968).

339. Dean, R. W., and Roelofs, W. L. Synthetic sex pheromone of the red-banded leaf roller moth as a survey tool. *J. Econ. Entomol.* **63**, 684 (1970).

340. Debolt, J. W. Dispensers for baiting traps equipped with blacklight lamps with

synthetic sex pheromone of the female cabbage looper. *J. Econ. Entomol.* **63,** 141 (1970).

341. Deegener, P. Das Duftorgan von *Hepialus hectus*. *Z. Wiss. Zool.* **71,** 276 (1902).

342. Deegener, P. Das Duftorgan von *Phassus schamyl* Chr. *Z. Wiss. Zool.* **78,** 245 (1904).

343. Delmas, R. Notes sur la biologie de *Pristiphora conjugata*. *Bull. Biol. Fr. Belg.* **60,** 447 (1926).

344. Dethier, V. G., "Chemical Insect Attractants and Repellents," pp. 21–25. Mc-Graw-Hill (Blakiston) New York, 1947.

345. Dethier, V. G. The physiology of olfaction in insects. *Ann. N.Y. Acad. Sci.* **58,** 139 (1954).

346. Dethier, V. G. "The Physiology of Insect Senses." Wiley, New York, 1963.

347. Dethier, V. G., and Chadwick, L. E. Chemoreception in insects. *Physiol. Rev.* **28,** 220 (1948).

348. Devakul, V., and Maarse, H. A second compound in the odorous gland liquid of the giant water bug *Lethocerus indicus* (Lep. and Serv.). *Anal. Biochem.* **7,** 269 (1964).

349. Dickins, G. R. The scent glands of certain Phycitidae (Lepidoptera). *Trans. Roy. Entomol. Soc. London* **85,** 331 (1936).

350. Doane, C. C. Evidence for a sex attractant in females of the red pine scale. *J. Econ. Entomol.* **59,** 1539 (1966).

351. Doane, C. C. Aspects of mating behavior of the gypsy moth. *Ann. Entomol. Soc. Amer.* **61,** 768 (1968).

352. Doane, J. Movement on the soil surface, of adult *Ctenicera aeripennis destructor* (Brown) and *Hypolithus bicolor* Esch. (Coleoptera: Elateridae), as indicated by funnel pitfall traps, with notes on captures of other arthropods. *Can. Entomol.* **93,** 636 (1961).

353. Dodson, C. H., Dressler, R. L., Hills, H. G., Adams, R. M., and Williams, N. H. Biologically active compounds in orchid fragrances. *Science* **164,** 1243 (1969).

354. Donovan, C. A. Some clinical observations on sexual attraction and deterrence in dogs and cattle. *Vet. Med. & Small Anim. Clin.* **62,** 1047 (1967).

355. Doolittle, R. E., Beroza, M., Keiser, I., and Schneider, E. L. Deuteration of the melon fly attractant, cue-lure, and its effect on olfactory response and infra-red absorption. *J. Insect Physiol.* **14,** 1697 (1968).

356. Doolittle, R. E., Blum, M. S., and Boch, R. *cis*-9-Oxo-2-decenoic acid: Synthesis and evaluation as a honey bee pheromone and masking agent. *Ann. Entomol. Soc. Amer.* **63,** 1180 (1970).

357. Downes, J. A. Observations on the mating behaviour of the crab hole mosquito *Deinocerites cancer* (Diptera: Culicidae). *Can. Entomol.* **98,** 1169 (1966).

358. Duane, J. P., and Tyler, J. E. Operation Saturnid. *Interchem. Rev.* **9,** 25 (1950).

359. Dufay, C. Sur l'attraction sexuelle chez *Lasiocampa quercus* L. *Bull. Soc. Entomol. Fr.* **62,** 61 (1957).

360. Duges. As cited by Rau and Rau (*911*).

361. Dustan, G. G. Mating behaviour of the Oriental fruit moth, *Grapholitha molesta* (Busck) (Lepidoptera: Olethreutidae). *Can. Entomol.* **96,** 1087 (1964).

362. Dyer, E. D. A., and Taylor, D. W. Attractiveness of logs containing female spruce beetles, *Dendroctonus obesus* (Coleoptera: Scolytidae). *Can. Entomol.* **100,** 769 (1968).

363. Dyk, A. (1933). As cited by Farsky (*394*).

364. Dyson, G. M. Some aspects of the vibration theory of odour. *Perfum. Essent. Oil Rec.* **19,** 456 (1928).

365. Dyson, G. M. Raman effect and the concept of odour. *Perfum. Essent. Oil Rec.* **28,** 13 (1937).

366. Dyson, G. M. The scientific basis of odour. *Chem. Ind. (London)* **16,** 647 (1938).

367. Dyson, G. M. Odour and chemical constitution. *Nature (London)* **173,** 831 (1954).

368. Earle, N. W. Demonstration of a sex attractant in the buck moth, *Hemileuca maia* (Lepidoptera: Saturniidae). *Ann. Entomol. Soc. Amer.* **60,** 277 (1967).

369. Eckstein, K. Auf neuen Wegen der Schädlingsbekämpfung. *Forstl. Wochenschr. Silva* **25,** 245 (1937).

369a. Economopoulos, A. P., Giannakakis, A., Tzanakakis, M. E., and Voyadjoglou, A. V. Reproductive behavior and physiology of the olive fruit fly. I. Anatomy of the adult rectum and odors emitted by adults. *Ann. Entomol. Soc. Amer.* **64,** 1112 (1971).

370. Edelsten, H. M., and Fryer, J. C. F. Non-specific assembling scents in macrolepidoptera. *Entomol. Rec.* **56,** 7 (1944).

371. Edmunds, M. Evidence for sexual selection in the mimetic butterfly *Hypolimnas misippus* L. *Nature (London)* **221,** 488 (1969).

372. Eidmann, H. Morphologische und physiologische Untersuchungen am weiblichen Genitalapparat der Lepidopteren. I. Morphologischer Teil. *Z. Angew. Entomol.* **15,** 1 (1929).

373. Eidmann, H. Morphologische und physiologische Untersuchungen am weiblichen Genitalapparat der Lepidopteren. II. Physiologischer Teil. *Z. Angew. Entomol.* **18,** 57 (1931).

374. Eisner, T., and Kafatos, F. C. Defense mechanisms of arthropods. X. A pheromone promoting aggregation in an aposematic distasteful insect. *Psyche* **69,** 53 (1962).

375. Eiter, K. Insect sex attractants. *Chem. Pflanzenschutz Schaedlingsbek.* p. 497 (1970).

375a. Eiter, K. Insektensexuallockstoffe. *Prog. Chem. Organic Natur. Prod.* **28,** 204 (1970).

376. Eiter, K., Truscheit, E., and Boness, M. Synthesen von D,L-10-Acetoxy-hexa-decen-(7-*cis*)-ol-(1), 12-Acetoxy-octadecen-(9-*cis*)-ol-(1) ("Gyplure") und 1-Acetoxy-10-propyl-tridecadien-(5-*trans*, 9). *Justus Liebig's Ann. Chem.* **709,** 29 (1967).

377. Ellertson, F. E. *Pleocoma oregonensis* Leach as a pest in sweet cherry orchards. *J. Econ. Entomol.* **49,** 431 (1956).

378. Eltringham, H. On the abdominal brushes in certain male noctuid moths. *Trans. Roy. Entomol. Soc. London* p. 1 (1925).

379. Embody, D. R. Possible methods for measuring the effective range of the sex-lure trap for pink bollworm. *U.S. Dep. Agr., ARS* **ARS 81–43,** 1—5 (1971).

380. Emerson, A. E. The mechanism of tandem behavior following the colonizing flight in termites. *Anat. Rec.* **57,** Suppl., 61 (1933).

381. Emmerich, H., and Barth, R. H., Jr. Effect of farnesyl methyl ether on reproductive physiology in the cockroach, *Byrsotria fumigata* (Guérin). *Z. Naturforsch. B* **23,** 1019 (1968).

382. Engelmann, F. Hormonal control of mating behavior in an insect. *Experientia* **16,** 69 (1960).

383. Engelmann, F., and Barth, R. H., Jr. Endocrine control of female receptivity in *Leucophaea maderae* (Blattaria). *Ann. Entomol. Soc. Amer.* **61,** 503 (1968).

384. Evans, L. J. Female *Arctia caia* L. attracting other Arctiinae. *Entomol. Rec.* 64, 86 (1952).

385. Evans, R. M. Hormones and related substances. The sex attractants of the gypsy moth. *Mfg. Chem.* 32, 175 (1961).

386. Evans, R. M. Hormones and related substances. Insect attractants. *Mfg. Chem.* 33, 472 (1962).

387. Evers, A. Über die Funktion der Excitatoren beim Liebesspiel der Malachiidae. *Entomol. Bl.* (*Krefeld*) 52, 165 (1956).

388. Ewing, A. W., and Manning, A. The effect of exogenous scent on the mating of *Drosophila melanogaster. Anim. Behav.* 11, 596 (1966).

389. Fabre, J. H. "Souvenirs Entomologiques," 8th ed., Paris, 1904.

390. Fabre, J. H. Hochzeitsflüge der Nachtpfauenaugen. *Kosmos* 3, 45 (1906).

391. Fabre, J. H. "The Life of the Caterpillar." Dodd, Mead, New York, 1916.

392. Falls, O. Sex attraction in *Samia cecropia. Trans. Kans. Acad. Sci.* 36, 215 (1933).

393. Farbenfabriken Bayer A. G. 10, 12-Hexadecadiene derivatives. Brit. Pat. 945,298 (1963).

394. Farsky, O. Nonnenkontroll- und Vorbeugungsmethode nach Professor Forst.-Ing. Ant. Dyk. *Anz. Schädlingskunde* 14, 52 and 65 (1938).

395. Fattig, P. W. The Mutillidae or velvet ants of Georgia. *Emory Univ. Mus. Bull.* 1, 1 (1943).

396. Fatzinger, C. W., and Asher, W. C. Mating behavior and evidence for sex phero-mones of *Dioryctria abietella. Abstr. Nat. Meet., Entomol. Soc. Amer., Miami Beach, Fla.* p. 57 (1970).

397. Fatzinger, C. W., and Asher, W. C. Mating behavior and evidence for a sex pheromone of *Dioryctria abietella* (Lepidoptera: Pyralidae (Phycitinae)). *Ann. Entomol. Soc. Amer.* 64, 612 (1971).

398. Federley, H. Die Bedeutung der Kreuzung für die Evolution. *Jena. Z. Naturwiss.* 67, 364 (1932).

399. Fellin, D. G. Further observations on trapping male *Pleocoma* with female-baited traps (Coleoptera: Scarabeidae). *Pan-Pac. Entomol.* 44, 69 (1968).

400. Féron, M. Attraction chimique du mâle de *Ceratitis capitata* Wied. (Dipt. Try-petidae) pour la femelle. *C. R. Acad. Sci.* 248, 2403 (1959).

401. Féron, M. L'Instinct de reproduction chez la mouche Mediterraneenne des fruits *Ceratitis capitata* Wied. (Dipt. Trypetidae); comportement sexuelcomportement de ponte. *Rev. Pathol. Veg. Entomol. Agr. Fr.* 41, 1 (1962).

401a. Findlay, J. A. Personal communication (1970).

402. Findlay, J. A., and Macdonald, D. R. Investigation of the sex-attractant of the spruce budworm moth. *Chem. Can.* p. 3 (1966).

403. Findlay, J. A., and Macdonald, D. R. A chemical sex attractant for the spruce budworm. *Can. Dep. Fish. Forestry, Ottawa Bi-Mon. Res. Notes* 23, 12 (1967).

404. Findlay, J. A., Macdonald, D. R., and Tang, C. S. A synthetic attractant for the male spruce budworm moth *Choristoneura fumiferana* (Clem.). *Experientia* 23, 377 (1967).

405. Finger, A., Heller, D., and Shulov, A. Olfactory response of the khapra beetle (*Trogoderma granarium* Everts.) larva to factors from larvae of the same species. *Ecology* 46, 542 (1965).

406. Fink, D. E. The biology of *Macrocentrus ancylivora* Rohwer, an important para-site of the strawberry leaf roller (*Ancylis comptana* Froehl). *J. Agr. Res.* 32, 1121 (1926).

407. Flaschenträger, B. Über Anlockungsstoffe von Baumwollschädlingen. *Angew. Chem.* **61**, 252 (1949).

408. Flaschenträger, B., and Amin, E. S. Chemical attractants for insects: Sex- and food-odours of the cotton leaf worm and the cut worm. *Nature (London)* **165**, 394 (1950).

409. Flaschenträger, B., Amin, E. S., and Jarczyk, H. J. Ein Lockstoffanalysator (Odouranalyser) für Insekten. *Mikrochim. Acta*, p. 385 (1957).

410. Fleming, W. E. Attractants for the Japanese beetle. *U.S., Dep. Agr., Tech. Bull.* **1399**, 46 (1969).

411. Fletcher, B. S. Storage and release of a sex pheromone by the Queensland fruit fly, *Dacus tryoni* (Diptera: Trypetidae). *Nature (London)* **219**, 631 (1968).

412. Fletcher, B. S. The structure and function of the sex pheromone glands of the male Queensland fruit fly, *Dacus tryoni*. *J. Insect Physiol.* **15**, 1309 (1969).

412a. Fletcher, B. S. Bioassay of sex pheromone of the male Queensland fruit fly. *CSIRO Div. Entomol., Canberra, Aust., Annu. Rep.* p. 33 (1970–1971).

412b. Fletcher, B. S. Factors affecting sexual responsiveness of Queensland fruit fly. *CSIRO Div. Entomol., Canberra, Aust., Annu. Rep.* p. 33 (1970–1971).

412c. Fletcher, B. S. Pheromone production by irradiated males of the Queensland fruit fly. *CSIRO Div. Entomol., Canberra, Aust., Annu. Rep.* p. 34 (1970–1971).

413. Fletcher, L. W., Claborn, H. V., Turner, J. P., and Lopez, E. Difference in response of two strains of screw-worm flies to the male pheromone. *J. Econ. Entomol.* **61**, 1386 (1968).

414. Fletcher, L. W., O'Grady, J. J., Jr., Claborn, H. V., and Graham, O. H. A pheromone from male screw-worm flies. *J. Econ. Entomol.* **59**, 142 (1966).

415. Fletcher, L. W., and Turner, J. P. Selection for mating aggressiveness in female screw-worms. *J. Econ. Entomol.* **63**, 1611 (1970).

416. Forbes, R. S., and Daviault, L. The biology of the mountain-ash sawfly, *Pristiphora geniculata* (Htg.) (Hymenoptera: Tenthredinidae), in eastern Canada. *Can. Entomol.* **96**, 1117 (1965).

417. Forbush, E. H., and Fernald, C. H. "The Gypsy Moth." Mass. State Bd. Agr., Boston, Massachusetts, 1896.

418. Ford, E. B. Zygaenidae attracted by the female of *Lasiocampa quercus*, L. *Proc. Entomol. Soc. London* **1**, 20 (1926).

419. Ford, E. B. "Moths." Collins, London, 1955.

420. Forell, A. "The Senses of Insects." Methuen, London, 1908.

421. Francia, F. C., and Graham, K. Aspects of orientation behavior in the ambrosia beetle *Trypodendron lineatum* (Olivier). *Can. J. Zool.* **45**, 985 (1967).

422. Francke-Grosmann, H. Some new aspects in forest entomology. *Annu. Rev. Entomol.* **8**, 415 (1963).

423. Franklin, R. T. Southern pine beetle population behavior. *J. Ga. Entomol. Soc.* **5**, 175 (1970).

424. Franz, J. Die Tannentriebwickler *Cacoecia murinana* Hb. Beiträge zur Bionomie und Oekologie. *Z. Angew. Entomol.* **27**, 345 (1940).

425. Free, J. B., and Butler, C. G. "Bumblebees," pp. 39–41. Collins, London, 1959.

426. Free, J. B., and Simpson, J. The alerting pheromones of the honeybee. *Z. Vergl. Physiol.* **61**, 361 (1968).

427. Freiling, H. H. Duftorgane der weiblichen Schmetterlinge nebst Beiträgen zur Kenntnis der Sinnesorgane auf dem Schmetterlingsflügel und der Duftpinsel der Männchen von *Danais* und *Euploea*. *Z. Wiss. Zool.* **92**, 210 (1909).

428. Friedman, L., and Miller, J. G. Odor incongruity and chirality. *Science* **172**, 1044 (1971).
429. Gad, A. M., Zayed, S., Hassan, M. M., and Hussein, T. M. Investigations on the chemical composition of the fat obtained from the Egyptian cotton leaf worm *Prodenia litura* F. *J. Chem. UAR* **7**, 141 (1964).
430. Gade, G. Observations on the habits of *Pimpla (Rhyssa) lunator*. *Bull. Brooklyn Entomol. Soc.* **7**, 103 (1884).
431. Ganyard, M. C., Jr. Cross attraction and inhibition of attraction between Indian meal moths and almond moths. *Abstr. Nat. Meet., Entomol. Soc. Amer., Miami Beach, Fla.* p. 55 (1970).
432. Gara, R. I. Studies on the flight behavior of *Ips confusus* (Lec.) (Coleoptera: Scolytidae) in response to attractive material. *Contrib. Boyce Thompson Inst.* **22**, 51 (1963).
433. Gara, R. I. A field olfactometer for studying the response of the southern pine beetle to attractants. *J. Econ. Entomol.* **60**, 1180 (1967).
434. Gara, R. I. Studies on the attack behavior of the southern pine beetle. I. The spreading and collapse of outbreaks. *Contrib. Boyce Thompson Inst.* **23**, 349 (1967).
435. Gara, R. I., and Coster, J. E. Studies on the attack behavior of the southern pine beetle. III. Sequence of tree infestation within stands. *Contrib. Boyce Thompson Inst.* **24**, 77 (1968).
436. Gara, R. I., Vité, J. P., and Cramer, H. H. Manipulation of *Dendroctonus frontalis* by use of a population aggregating pheromone. *Contrib. Boyce Thompson Inst.* **23**, 55 (1965).
437. Gary, N. E., Queen honey bee attractiveness as related to mandibular gland secretion. *Science* **133**, 1479 (1961).
438. Gary, N. E. Chemical mating attractants in the queen honey bee. *Science* **136**, 773 (1962).
439. Gary, N. E. Observations of mating behaviour in the honeybee. *J. Apicult. Res.* **2**, 3 (1963).
440. Gary, N. E. Pheromones of the honey bee, *Apis mellifera* L. *In* "Control of Insect Behavior by Natural Products" (D. L. Wood, R. M. Silverstein, and M. Nakajima, eds.), p. 29. Academic Press, New York, 1970.
441. Gaston, L. K., Fukuto, T. R., and Shorey, H. H. Sex pheromones of noctuid moths. IX. Isolation techniques and quantitative analysis for the pheromones, with special reference to that of *Trichoplusia ni* (Lepidoptera: Noctuidae). *Ann. Entomol. Soc. Amer.* **59**, 1062 (1966).
442. Gaston, L. K., and Shorey, H. H. Sex pheromones of noctuid moths. IV. An apparatus for bioassaying the pheromones of six species. *Ann. Entomol. Soc. Amer.* **57**, 779 (1964).
443. Gaston, L. K., Shorey, H. H., and Saario, C. A. Insect population control by the use of sex pheromones to inhibit orientation between the sexes. *Nature (London)* **213**, 1155 (1967).
444. Gaston, L. K., Shorey, H. H., and Saario, C. A. Sex pheromones of noctuid moths. XXIII. Rate of evaporation of a model compound of *Trichoplusia ni* sex pheromone from different substrates at various temperatures and its application to insect orientation. *Ann. Entomol. Soc. Amer.* **64**, 381 (1971).
444a. Gentry, C. R., Dickerson, W. A., Jr., Henneberry, T. J., and Baumhover, A. H. Evaluation of pheromone-baited blacklight traps for controlling cabbage loopers on shade-grown tobacco in Florida. *U.S. Dep. Agr. Prod. Res. Rep.* **133**, 1–12 (1971).

445. Gentry, C. R., Lawson, F. R., and Hoffman, J. D. A sex attractant in the tobacco budworm. *J. Econ. Entomol.* **57**, 819 (1964).

446. George, J. A. Sex pheromone of the Oriental fruit moth, *Grapholitha molesta* (Busck) (Lepidoptera: Tortricidae). *Can. Entomol.* **97**, 1002 (1965).

447. George, J. A. Effects of gamma radiation on fertility, mating, and longevity of males of the Oriental fruit moth, *Grapholitha molesta* (Lepidoptera: Tortricidae). *Can. Entomol.* **99**, 850 (1967).

448. George, J. A., and Howard, M. G. Insemination without spermatophores in the Oriental fruit moth, *Grapholitha molesta* (Lepidoptera: Tortricidae). *Can. Entomol.* **100**, 190 (1968).

449. Gerig, L. Nouveaux résultats de recherches scientifiques sur le comportement des reines et des faux bourdons. *J. Suisse Apicult.* **62**, 76 (1965).

450. Gerig, L. Einige weitere Beobachtungen über das Verhalten der Drohnen auf ihren Sammelplatzen. *Schweiz. Bienen-Ztg.* **92**, 528 and 578 (1969).

451. Germer, F. Untersuchungen über den Bau und die Lebensweise der Lymexyloniden, speziell des *Hylecoetus dermestoides* L. *Z. Wiss. Zool.* **101**, 683 (1912).

452. Gilbert, L. I. Physiology of growth and development: Endocrine aspects. *In* "The Physiology of Insecta" (M. Rockstein, ed.), Vol. 1, p. 149. Academic Press, New York, 1964.

453. Gillett, A. Airborne factor affecting the grouping behaviour of locusts. *Nature (London)* **218**, 782 (1968).

454. Girault, A. A. The lesser peach tree borer. *U.S., Dep. Agr., Bur. Entomol. Bull.* **68**, Part IV, 31 (1907).

455. Gladney, W. J. Mate-seeking by female *Amblyomma maculatum* (Acarina: Ixodidae) on a bovine. *Nature (London)* **232**, 401 (1971).

456. Glass, E. H., Roelofs, W. L., Arn, H., and Comeau, A. Sex pheromone trapping red-banded leaf roller moths and development of a long-lasting polyethylene wick. *J. Econ. Entomol.* **63**, 370 (1970).

457. Goodrich, B. S. Cuticular lipids of adults and puparia of the Australian sheep blowfly *Lucilia cuprina* (Wied.). *J. Lipid Res.* **11**, 1 (1970).

458. Goonewardene, H. F., Zepp, D. B., and Grosvenor, A. E. Virgin female Japanese beetles as lures in field traps. *J. Econ. Entomol.* **63**, 1001 (1970).

459. Görnitz, K. Anlockversuche mit dem weiblichen Sexualduftstoff des Schwammspinners (*Lymantria dispar*) und der Nonne (*Lymantria monacha*). *Anz. Schädlingskunde* **22**, 145 (1949).

460. Gossard, H. A., and King, J. L. The peach tree borer. *Bull. Ohio Agr. Exp. Sta.* **329**, 55 (1918).

461. Götz, B. Untersuchungen über die Wirkung des Sexualduftstoffes bei den Traubenwicklern *Clysia ambiguella* und *Polychrosis botrana*. *Z. Angew. Entomol.* **26**, 143 (1939).

462. Götz, B. Über weitere Versuche zur Bekämpfung der Traubenwickler mit Hilfe des Sexualduftstoffes. *Anz. Schadlingskunde* **15**, 109 (1939).

463. Götz, B. Sexualduftstoffe als Lockmittel in der Schädlingsbekämpfung. *Umschau* **44**, 794 (1940).

464. Götz, B. Lockflüssigkeiten zur Beobachtung des Traubenwicklermottenfluges. *Wein Rebe* **22**, 15 (1940).

465. Götz, B. Neue Apparate zum Studium der Insektenphysiologie. *Umschau* **45**, 779 (1941).

466. Götz, B. Der Sexualduftstoff als Bekämpfungsmittel gegen die Traubenwickler im Freiland. *Wein Rebe* **23**, 75 (1941).

467. Götz, B. Beiträge zur Analyse des Mottenfluges bei den Traubenwicklern *Clysia ambiguella* und *Polychrosis botrana*. *Wein Rebe* **23**, 207 (1941).
468. Götz, B. Die Sexualduftstoffe an Lepidopteren. *Experientia* **7**, 406 (1951).
469. Graham, H. M., Glick, P. A., and Martin, D. F. Nocturnal activity of adults of six lepidopterous pests of cotton as indicated by light-trap collections. *Ann. Entomol. Soc. Amer.* **57**, 328 (1964).
470. Graham, H. M., and Martin, D. F. Use of cyanide in pink bollworm sex-lure traps. *J. Econ. Entomol.* **56**, 901 (1963).
471. Graham, H. M., Martin, D. F., Ouye, M. T., and Hardman, R. M. Control of pink bollworms by male annihilation. *J. Econ. Entomol.* **59**, 950 (1966).
472. Graham, H. M., McGough, J. M., and Jacobson, M. Influence of solvents on effectiveness of sex lure for pink bollworm. *J. Econ. Entomol.* **59**, 761 (1966).
473. Graham, K. Anaerobic induction of primary chemical attractancy for ambrosia beetles. *Can. J. Zool.* **46**, 905 (1968).
474. Grant, G. G. Evidence for a male sex pheromone in the noctuid, *Trichoplusia ni*. *Nature (London)* **227**, 1345 (1970).
475. Grant, G. G. Feeding activity of adult cabbage loopers on flowers with strong olfactory stimuli. *J. Econ. Entomol.* **64**, 315 (1971).
476. Grant, G. G. Scent apparatus of the male cabbage looper. *Ann. Entomol. Soc. Amer.* **64**, 347 (1971).
476a. Grant, G. G. Electroantennogram responses to the scent brush secretions of several male moths. *Ann. Entomol. Soc. Amer.* **64**, 1428 (1971).
477. Gravitz, N., and Willson, C. A sex pheromone from the citrus mealybug. *J. Econ. Entomol.* **61**, 1458 (1968).
478. Green, E. E. Notes on assembling of males of certain moths in Ceylon. *Entomol. Mon. Mag.* **35**, 258 (1899).
479. Green, N., Beroza, M., and Hall, S. A. Recent developments in chemical attractants for insects. *Advan. Pest Contr. Res.* **3**, 129 (1960).
480. Green, N., Jacobson, M., Henneberry, T. J., and Kishaba, A. N. Insect sex attractants. VI. 7-Dodecen-1-ol acetates and congeners. *J. Med. Chem.* **10**, 533 (1967).
481. Green, N., Jacobson, M., and Keller, J. C. Hexalure, an insect sex attractant discovered by empirical screening. *Experientia* **25**, 682 (1969).
482. Green, N., and Keller, J. C. Attractant for pink bollworm moths. U.S. Pat. 3,586,712 (1971).
482a. Green, N., Warthen, J. D., Jr., and Mangum, C. L. Analysis of hexalure as related to its attractancy to pink bollworm moths. *J. Econ. Entomol.* **64**, 1381 (1971).
483. Greene, G. L., Janes, M. J., and Mead, F. W. Fall armyworm, *Spodoptera frugiperda*, males captured at three Florida locations in traps baited with virgin females. *Fla. Entomol.* **54**, 165 (1971).
484. Greet, D. N. Observations on sexual attraction and copulation in the nematode *Panagrolaimus rigidus* (Schneider). *Nature (London)* **204**, 96 (1964).
485. Greet, D. N., Green, C. D., and Poulton, M. E. Extraction, standardization and assessment of the volatility of the sex attractants of *Heterodera rostochiensis* Woll. and *H. schachtii* Schm. *Ann. Appl. Biol.* **61**, 511 (1968).
486. Griffith, P. H., and Susskind, C. Electromagnetic communication versus olfaction in the corn earworm moth, *Heliothis zea*. *Ann. Entomol. Soc. Amer.* **63**, 903 (1970).
487. Grosch, D. S. The importance of antennae in the mating reaction of male *Habro-*

*bracon. J. Comp. Physiol. Psychol.* **40,** 23 (1947).

488. Grosch, D. S. Experimental studies on the mating reaction of male *Habrobracon. J. Comp. Physiol. Psychol.* **41,** 188 (1948).

489. Guerra, A. A. New techniques to bioassay the sex attractant of pink bollworms with olfactometers. *J. Econ. Entomol.* **61,** 1252 (1968).

490. Guerra, A. A. Catches of male pink bollworms in several modifications of the Frick trap baited with sex attractant. *J. Econ. Entomol.* **62,** 1514 (1969).

491. Guerra, A. A., Garcia, R. D., and Leal, M. P. Suppression of populations of pink bollworms in field cages with traps baited with sex attractant. *J. Econ. Entomol.* **62,** 741 (1969).

492. Guerra, A. A., and Ouye, M. T. Catch of male pink bollworms in traps baited with sex attractant. *J. Econ. Entomol.* **60,** 1046 (1967).

493. Guex, W., Rüegg, R., and Schwieter, U. Verfahren zur Herstellung von ungesättigten, gegebenenfalls veresterten Alkoholen. Ger. Pat. 1,111,615 (1961).

494. Gupta, A. P. A critical review of the studies on the so-called stink or repugnatorial glands of Heteroptera with further comments. *Can. Entomol.* **93,** 482 (1961).

495. Gupta, A. P. Musculature and mechanism of the nymphal scent-apparatus of *Riptortus linearis* H. S. (Heteroptera: Alydidae) with comments on the number, variation and homology of the abdominal scent glands in other Heteroptera. *Proc. Entomol. Soc. Wash.* **66,** 12 (1964).

496. Haas, A. Neue Beobachtungen zum Problem der Flugbahnen bei Hummelmännchen. *Z. Naturforsch* **1,** 596 (1946).

497. Hass A. Arttypische Flugbahnen von Hummelmännchen. *Z. Vergl. Physiol.* **31,** 281 (1949).

498. Haas, A. Die Mandibeldrüse als Duftorgan bei einigen Hymenopteren. Naturwissenschaften **39,** 484 (1952).

499. Haase, E. Dufteinrichtungen indischer Schmetterlinge. Zool. Anz. **11,** 475 (1888).

500. Haeger, J. S., and Phinizee, J. The biology of the crabhole mosquito, *Deinocerites cancer* Theobald. *Rep. 30th Annu. Meet., Fla. Anti-Mosquito Ass.* p. 34 (1959).

501. Hall, S. A. Wiles of sex attractants used in war on pests. *Yearbook Agr., (U.S. Dep. Agr.)* p. 77 (1968).

502. Haller, H. L., Acree, F., Jr., and Potts, S. F. The nature of the sex attractant of the female gypsy moth. *J. Amer. Chem. Soc.* **66,** 1659 (1944).

503. Hamm, A. H. Persistent odour of *Bombyx quercus. Entomol. Mon. Mag.* **6,** 74 (1895).

504. Hammad, S. M., and Jarczyk, H. J. Contributions to the biology and biochemistry of the cotton leaf-worm, *Prodenia litura* F. III. The morphology and histology of the sexual scent glands of the female moth of *Prodenia litura* F. *Bull. Soc. Roy. Entomol. Egypte* **42,** 253 (1958).

505. Hammond, A. M. A sex pheromone in the rice stalk borer. *Abstr. Nat. Meet. Entomol. Soc. Amer., Miami Beach, Fla.* p. 51 (1970).

506. Hammond, A. M., and Hensley, S. D. A bioassay for the sex attractant in the sugarcane borer. *Ann. Entomol. Soc. Amer.* **63,** 64 (1970).

506a. Hammond, A. M., and Oliver, B. F. A sex pheromone in the rice stalk borer. *Ann. Entomol. Soc. Amer.* **64,** 1469 (1971).

507. Hanno, K. Anlockversuche bei *Lymantria monacha* L. *Z. Angew. Entomol.* **25,** 628 (1939).

508. Happ, G. M. Multiple sex pheromones of the mealworm beetle, *Tenebrio molitor*

L. *Nature (London)* **222**, 180 (1969).

509. Happ, G. M. Maturation of the response of male *Tenebrio molitor* to the female sex pheromone. *Ann. Entomol. Soc. Amer.* **63**, 1782 (1970).

510. Happ, G. M., and Happ, C. M. Fine structure and histochemistry of the spermathecal gland in the mealworm beetle, *Tenebrio molitor*. *Tissue & Cell* **2**, 443 (1970).

511. Happ, G. M., Schroeder, M. E., and Wang, J. C. H. Effects of male and female scent on reproductive maturation in young female *Tenebrio molitor*. *J. Insect Physiol.* **16**, 1543 (1970).

512. Happ, G. M., and Wheeler, J. Bioassay, preliminary purification, and effect of age, crowding, and mating on the release of sex pheromone by female *Tenebrio molitor*. *Ann Entomol. Soc. Amer.* **62**, 846 (1969).

513. Hardee, D. D. Pheromone production by male boll weevils as affected by food and host factors. *Contrib. Boyce Thompson Inst.* **24**, 315 (1970).

514. Hardee, D. D., Cleveland, T. C., Davis, J. W., and Cross, W. H. Attraction of boll weevils to cotton plants and to males fed on three diets. *J. Econ. Entomol.* **63**, 990 (1970).

515. Hardee, D. D., Cross, W. H., Huddleston, P. M., and Davich, T. B. Survey and control of the boll weevil in west Texas with traps baited with males. *J. Econ. Entomol.* **63**, 1041 (1970).

516. Hardee, D. D., Cross, W. H., and Mitchell, E. B. Male boll weevils are more attractive than cotton plants to boll weevils. *J. Econ. Entomol.* **62**, 165 (1969).

517. Hardee, D. D., Cross, W. H., Mitchell, E. B., Huddleston, P. M., Mitchell, H. C., Merkl, M. E., and Davich, T. B. Biological factors influencing responses of the female boll weevil to the male sex pheromone in field and large-cage tests. *J. Econ. Entomol.* **62**, 161 (1969).

518. Hardee, D. D., Lindig, O. H., and Davich, T. B. Suppression of populations of boll weevils over a large area in West Texas with pheromone traps in 1969. *J. Econ. Entomol.* **64**, 928 (1971).

518a. Hardee, D. D., McKibben, G. H., Gueldner, R. C., Mitchell, E. B., Tumlinson, J. H., and Cross, W. H. Boll weevils in nature respond to grandlure, a synthetic pheromone. *J. Econ. Entomol.* **65**, 97 (1972).

519. Hardee, D. D., Mitchell, E. B., and Huddleston, P. M. Procedure for bioassaying the sex attractant of the boll weevil. *J. Econ. Entomol.* **60**, 169 (1967).

520. Hardee, D. D., Mitchell, E. B., and Huddleston, P. M. Laboratory studies of sex attraction in the boll weevil. *J. Econ. Entomol.* **60**, 1221 (1967).

520a. Hardee, D. D., Wilson, N. M., Mitchell, E. E., and Huddleston, P. M. Factors affecting activity of grandlure, the pheromone of the boll weevil, in laboratory bioassays. *J. Econ. Entomol.* **64**, 1454 (1971).

521. Hardy, E. Scents from insects. Butterfly and moth perfumes and their extraction. *Perfum. Essent. Oil Rec.* **38**, 403 (1947).

522. Hardy, E. Do insects locate each other by scent or wireless? BEAMA (*Brit. Elec. Allied Mfr. Ass.*) J. **56**, 257 (1949).

523. Hartman, H. B., and Roth, L. M. Stridulation by a cockroach during courtship behavior. *Nature (London)* **213**, 1243 (1967).

524. Hartman, H. B., and Roth, L. M. Stridulation by the cockroach *Nauphoeta cinerea* during courtship behavior. *J. Insect Physiol.* **13**, 579 (1967).

525. Hatanaka, A. Chemie über den Sexuallockstoff der Insekten. *Bochu-Kagaku* **28**, 110 (1963).

526. Hauser, G. Physiologische und histologische Untersuchungen über das Geruchs-organ der Insekten. *Z. Wiss. Zool.* **34,** 367 (1880).

527. Hayes, J. T., and Wheeler, A. G., Jr. Evidence for a sex attractant in *Hemicrepidius decoloratus* (Coleoptera: Elateridae). *Can. Entomol.* **100,** 207 (1968).

528. Hecker, E. Isolation and characterization of the sex attractant of the silk worm moth (*Bombyx mori* L.). *Proc Int. Congr. Entomol., 10th, 1956* Vol. 2, p. 293 (1958).

529. Hecker, E. Sexuallockstoffe—hochwirksame Parfüms der Schmetterlinge. I. *Umschau* **59,** 465 (1959).

530. Hecker, E. Sexuallockstoffe—hochwirksame Parfüms der Schmetterlinge. II. *Umschau* **59,** 499 (1959).

531. Hecker, E. Chemie und Biochemie des Sexuallockstoffes des Seidenspinners (*Bombyx mori*). *Proc. Int. Congr. Entomol., 11th, 1960* Vol. 3B, p. 69 (1961).

532. Hegdekar, B. M., and Dondale, C. D. A contact sex pheromone and some response parameters in lycosid spiders. *Can. J. Zool.* **47,** 1 (1969).

533. Henderson, H. E., and Warren, F. L. The sex pheromone of the false codling moth *Cryptophlebia leucotreta* Meyr., (formerly *Argyroploce leucotreta* Meyr.). Synthesis and bioassay of *trans*-dodec-7-en-1-yl acetate and related compounds. *J. S. Afr. Chem. Inst.* **23,** 9 (1970).

534. Hendricks, D. E. Use of virgin female tobacco budworms to increase catch of males in blacklight traps and evidence that trap location and wind influence catch. *J. Econ. Entomol.* **61,** 1581 (1968).

534a. Hendricks, D. E., Hollingsworth, J. P., and Hartstack, A. W., Jr. Catch of tobacco budworm moths influenced by color of sex-lure traps. *Environ. Entomol.* **1,** 48 (1972).

535. Henneberry, T. J., and Howland, A. F. Response of male cabbage loopers to blacklight with or without the presence of the female sex pheromone. *J. Econ. Entomol.* **59,** 623 (1966).

536. Henneberry, T. J., Howland, A. L., and Wolf, W. W. Combinations of blacklight and virgin females as attractants to cabbage looper moths. *J. Econ. Entomol.* **60,** 152 (1967).

537. Henneberry, T. J., Howland, A. F., and Wolf, W. W. Recovery of released male cabbage looper moths in traps equipped with blacklight lamps and baited with virgin females. *J. Econ. Entomol.* **60,** 532 (1967).

538. Henneberry, T. J., and McGovern, W. L. Effect of gamma radiation on mating competitiveness and behavior of *Drosophila melanogaster* males. *J. Econ. Entomol.* **56,** 739 (1963).

539. Henneberry, T. J., Shorey, H. H., and Kishaba, A. N. Mating frequency and copulatory aberrations; response to female sex pheromone, and longevity of male cabbage loopers treated with tepa. *J. Econ. Entomol.* **59,** 573 (1966).

540. Henzell, R. F. Phenol, an attractant for the grass grub beetle (*Costelytra zealandica*) (White) (Scarabaeidae: Coleoptera). *N. Z. J. Agr. Res.* **13,** 294 (1970).

541. Henzell, R. F., and Lowe, M. D. Sex attractant of the grass grub beetle. *Science* **168,** 1005 (1970).

542. Henzell, R. F., Lowe, M. D., Taylor, H. J., and Boston, E. Laboratory demonstration of a chemical sex attractant in *Costelytra zealandica* (White) (Scarabaeidae: Coleoptera). *N. Z. J. Sci.* **12,** 252 (1969).

543. Henzell, R. F., Taylor, H. J., and Lowe, M. D. Studies on laboratory handling and bioassay of the sex attractant of the grass grub, *Costelytra zealandica* (White) (Scarabaeidae: Coleoptera). *N. Z. J. Sci.* **13,** 460 (1970).

544. Herbert, R. B. Arthropod hormones, pheromones, and defensive secretions. *J. Leeds Univ. Union Chem. Soc.* **10**, 35 (1968).

545. Hering, M. "Biologie der Schmetterlinge," pp. 145–159. Springer-Verlag, Berlin and New York, 1926.

546. Hertel, G. D., Hain, F. P., and Anderson, R. F. Response of *Ips grandicollis* (Coleoptera: Scolytidae) to the attractant produced by attacking male beetles. *Can. Entomol.* **101**, 1084 (1969).

547. Hesse, R., and Doflein, F. "Tierbau und Tierleben," Vol. 1, pp. 644–645. Teubner, Leipzig, 1910.

548. Hesse, R., and Doflein, F. "Tierbau und Tierleben," Vol. 2, pp. 437–438. Teubner, Leipzig, 1910.

549. Hillaby, J. Smell chemistry of moths studied. *N.Y. Times* **112**, 35 (1963).

550. Hocking, B. Smell in insects: A bibliography with abstracts (to December 1958). *Def. Res. Bd. Can., Tech. Rep.* **8**, 1–266 (1960).

551. Hoffman, J. D. The influence of virgin females on catches of black light traps. *Bull. Entomol. Soc. Amer.* **10**, 170 (1964).

552. Hoffman, J. D., Lawson, F. R., and Peace, B. Attraction of blacklight traps baited with virgin female tobacco hornworm moths. *J. Econ. Entomol.* **59**, 809 (1966).

553. Hoffman-LaRoche & Co. Unsaturated alcohols and compositions containing same. Brit. Pat. 911,107 (1962).

554. Holbrook, R. F., Beroza, M., and Burgess, E. D. Gypsy moth (*Porthetria dispar*) detection with the natural female sex lure. *J. Econ. Entomol.* **53**, 751 (1960).

555. Holland, W. J. "The Moth Book. A Guide to the Moths of North America." Dover, New York, 1968.

556. Hölldobler, B. Untersuchungen zum Verhalten der Ameisenmännchen während der imaginalen Lebenszeit. *Experientia* **20**, 329 (1964).

557. Hölldobler, B. Sex pheromone in the ant *Xenomyrmex floridanus*. *J. Insect Physiol.* **17**, 1497 (1971).

558. Hölldobler, B., and Maschwitz, U. Der Hochzeitsschwarm der Rossameise *Camponotus herculeanus* L. (Hym. Formicidae). *Z. Vergl. Physiol.* **50**, 551 (1965).

559. Hope, J. A., Horler, D. F., and Rowlands, D. G. A possible pheromone of the bruchid, *Acanthoscelides obtectus* (Say). *J. Stored Prod. Res.* **3**, 387 (1967).

560. Horler, D. F. ( − )-Methyl *n*-tetradeca-*trans*-2,4,5-trienoate, an allenic ester produced by the male dried bean beetle *Acanthoscelides obtectus* (Say). *J. Chem. Soc., C* p. 859 (1970).

560a. Howell, J. F., and Davis, H. G. Protecting codling moths captured in sex-attractant traps from predaceous yellow jackets. *Environ. Entomol.* **1**, 122 (1972).

560b. Howell, J. F., and Thorp, K. D. Influence of mating on attractiveness of female codling moths. *Environ. Entomol.* **1**, 125 (1972).

561. Howland, A. F., Debolt, J. W., Wolf, W. W., Toba, H. H., Gibson, T., and Kishaba, A. N. Field and laboratory studies of attraction of the synthetic sex pheromone to male cabbage looper moths. *J. Econ. Entomol.* **62**, 117 (1969).

562. Howland, A. F., Henneberry, T. J., and Wolf, W. W. Comparison of cabbage looper and other moth species caught in blacklight traps baited or unbaited with unmated females. *J. Econ. Entomol.* **64**, 977 (1971).

563. Hoyt, C. P., Osborne, G. O., and Nulcock, A. P. Production of an insect sex attractant by symbiotic bacteria. *Nature (London)* **230**, 472 (1971).

564. Hughes, I. W. Reproductive behavior of the potato tuberworm, *Phthorimaea*

*operculella* (Zeller), (Lepidoptera: Gelechiidae) and the effect of the chemosterilant metepa. Ph.D. Dissertation, University of Florida, Gainesville (1967).

565. Hughes, P. R., and Pitman, G. B. A method for observing and recording the flight behavior of tethered bark beetles in response to chemical messengers. *Contrib. Boyce Thompson Inst.* **24**, 329 (1970).

566. Hurd, P. D. The California velvet ants of the genus *Dasymutilla* Ashmead (Hymenoptera: Mutillidae). *Bull. Calif. Insect Surv.* **1**, 89 (1951).

567. Ignoffo, C. M., Berger, R. S., Graham, H. M., and Martin, D. F. Sex attractant of *Trichoplusia ni* (Hübner) (Lepidoptera: Noctuidae). *Science* **141**, 902 (1963).

568. Ikan, R., Bergmann, E. D., Yinon, U., and Shulov, A. Identification, synthesis and biological activity of an "assembling scent" from the beetle *Trogoderma granarium. Nature (London)* **223**, 317 (1969).

569. Ikan, R., Gottlieb, R., and Bergmann, E. D. The pheromone of the queen of the Oriental hornet, *Vespa orientalis. J. Insect Physiol.* **15**, 1709 (1969).

570. Ikan, R., Yinon, U., and Shulov, A. The biological activity of natural and synthetic compounds on stored grain beetles. *Isr. J. Chem.* **7**, 14P (1969).

571. Illies, J. "Wir Beobachten und Züchten Insekten," p. 90. Franckh'sche Verlagshandlung, Stuttgart, 1964.

572. Illig, K. G. Duftorgane der männlichen Schmetterlinge. *Zoologica (Stuttgart)* **15** (38), 1–34 (1902).

573. Inhoffen, H. H. Versuch zur Isolierung eines Sexual-Lockstoffes. *Arch. Pharm. (Weinheim)* **284**, 337 (1951).

574. Inoue, Y., and Ohno, M. Insect attractants. *Kagaku No Ryoiki* **15**, 823 (1961).

575. Ishii, S. Aggregation of the German cockroach, *Blattella germanica* (L.). *In* "Control of Insect Behavior by Natural Products" (D. L. Wood, R. M. Silverstein, and M. Nakajima, eds.), p. 93. Academic Press, New York, 1970.

576. Ishii, S., and Kuwahara, Y. An aggregation pheromone of the German cockroach *Blattella germanica* L. (Orthoptera: Blattellidae). *Appl. Entomol. Zool.* **2**, 203 (1967).

577. Ishii, S., and Kuwahara, Y. Aggregation of German cockroach (*Blatella germanica*) nymphs. *Experientia* **24**, 88 (1968).

578. Jacentkovski, J. A. (1932). As cited by Farsky (*394*).

579. Jacentkovski, J. A. (1934). As cited by Farsky (*394*).

580. Jacklin, S. W., and Yonce, C. E. Induced shift of the diurnal emergence and calling of the peach tree borer. *J. Econ. Entomol.* **62**, 21 (1969).

581. Jacklin, S. W., Yonce, C. E., and Hollon, J. P. The attractiveness of female to male peach tree borers. *J. Econ. Entomol.* **60**, 1291 (1967).

582. Jacobson, M. Attractants for the gypsy moth. U.S. Pat. 2,900,756 (1959).

583. Jacobson, M. Synthesis of a highly potent gypsy moth sex attractant. *J. Org. Chem.* **25**, 2074 (1960).

584. Jacobson, M. Method of attracting male gypsy moth with 12-acetoxy-1-hydroxy-9-octadecene. U.S. Pat. 3,018,219 (1962).

585. Jacobson, M. 12-Acetoxy-1-hydroxy-9-octadecene and method for producing the same. U.S. Pat. 3,050,551 (1962).

586. Jacobson, M. Insect sex attractants. III. The optical resolution of *dl*-10-acetoxy-*cis*-7-hexadecen-1-ol. *J. Org. Chem.* **27**, 2670 (1962).

587. Jacobson, M. Recent progress in the chemistry of insect sex attractants. *Advan. Chem. Ser.* **41**, 1 (1963).

588. Jacobson, M. The sensitivity of insects to sexual olfactory stimuli. *Amer. Heart J.* **68,** 577 (1964).
589. Jacobson, M. The chemistry and biological activity of the sex attractants of lepidopterous insects. *Abstr. Pap., Int. Union Pure Appl. Chem. Symp. Chem. Natur. Prod., 1964* pp. 134–135 (1964).
590. Jacobson, M. The chemistry and biological activity of the insect sex attractants. *Proc. Int. Congr. Entomol. 12th, 1964* p. 505 (1965).
591. Jacobson, M. The use of attractants and repellents as alternate methods of pest control. *In* "Research in Pesticides" (C. O. Chichester, ed.), p. 251. Academic Press, New York, 1965.
592. Jacobson, M. Isolation and characterization of insect attractant lipids. *J. Amer. Oil Chem. Soc.* **42,** 681 (1965).
593. Jacobson, M. Natural insect attractants and repellents, new tools in pest control. *Advan. Chem. Ser.* **53,** 17 (1966).
594. Jacobson, M. Insect attractants. *In* "McGraw-Hill Yearbook of Science and Technology," p. 212. McGraw-Hill, New York, 1967.
595. Jacobson, M. Masking the effects of insect sex attractants. *Science* **154,** 422 (1966).
596. Jacobson, M. Chemical insect attractants and repellents. *Annu. Rev. Entomol.* **11,** 403 (1966).
597. Jacobson, M. Sex pheromone of the pink bollworm moth: Biological masking by its geometrical isomer. *Science* **163,** 190 (1969).
598. Jacobson, M. Methodology for isolation, identification and synthesis of sex pheromones in the Lepidoptera. *In* "Control of Insect Behavior by Natural Products" (D. L. Wood, R. M. Silverstein, and M. Nakajima, eds.), p. 111. Academic Press, New York, 1970.
599. Jacobson, M., and Beroza, M. Chemical insect attractants. *Science* **140,** 1367 (1963).
600. Jacobson, M., and Beroza, M. Sex attractant of the American cockroach. *Science* **142,** 1258 (1963).
601. Jacobson, M., and Beroza, M. Insect attractants. *Sci. Amer.* **211** (2), 20 (1964).
602. Jacobson, M., and Beroza, M. American cockroach sex attractant. *Science* **147,** 748 (1965).
603. Jacobson, M., Beroza, M., and Jones, W. A. Isolation, identification, and synthesis of the sex attractant of gypsy moth. *Science* **132,** 1011 (1960).
604. Jacobson, M., Beroza, M., and Jones, W. A. Insect sex attractants. I. The isolation, identification and synthesis of the sex attractant of the gypsy moth. *J. Amer. Chem. Soc.* **83,** 4819 (1961).
605. Jacobson, M., Beroza, M., and Yamamoto, R. T. Isolation and identification of the sex attractant of the American cockroach. *Science* **139,** 48 (1963).
606. Jacobson, M., Green, N., Warthen, D., Harding, C., and Toba, H. H. Sex pheromones of the Lepidoptera. Recent progress and structure-activity relationships. *In* "Chemicals Controlling Insect Behavior" (M. Beroza, ed.), p. 3. Academic Press, New York, 1970.
607. Jacobson, M., and Harding, C. Insect sex attractants. IX. Chemical conversion of the sex attractant of the cabbage looper to the sex pheromone of the fall armyworm. *J. Econ. Entomol.* **61,** 394 (1968).
608. Jacobson, M., and Jones, W. A. Insect sex attractants. II. The synthesis of a highly potent gypsy moth sex attractant and some related compounds. *J. Org. Chem.* **27,** 2523 (1962).

609. Jacobson, M., Lilly, C. E., and Harding, C. Sex attractant of sugar beet wireworm: Identification and biological activity. *Science* **159**, 208 (1968).
610. Jacobson, M., Ohinata, K., and Chambers, D. L. Sex pheromones of the Mediterranean fruit fly. *Abstr. Nat. Meet., Entomol. Soc. Amer., Miami Beach, Fla.*, p. 51 (1970).
611. Jacobson, M., Redfern, R. E., Jones, W. A., and Aldridge, M. H. Sex pheromones of the southern armyworm moth: Isolation, identification and synthesis. *Science* **170**, 542 (1970).
612. Jacobson, M., Schwarz, M., and Waters, R. M. Gypsy moth sex attractants: a reinvestigation. *J. Econ. Entomol.* **63**, 943 (1970).
613. Jacobson, M., and Smalls, L. A. Masking of the American cockroach sex attractant. *J. Econ. Entomol.* **59**, 414 (1966).
614. Jacobson, M., and Smalls, L. A. Sex attraction masking in the cynthia moth. *J. Econ. Entomol.* **60**, 296 (1967).
615. Jacobson, M., Sonnet, P. E., Adler, V. E., and Cook, D. Inactivity of a preparation reported to be highly active as a gypsy moth sex attractant. *Ann. Entomol. Soc. Amer.* **63**, 614 (1970).
616. Jacobson, M., Toba, H. H., DeBolt, J., and Kishaba, A. N. Insect sex attractants. VIII. Structure-activity relationships in sex attractant for male cabbage loopers. *J. Econ. Entomol.* **61**, 84 (1968).
617. Jantz, O. K., and Beroza, M. Caproic acid as an attractant for *Olcella parva*. *J. Econ. Entomol.* **60**, 290 (1967).
618. Jantz, O. K., and Rudinsky, J. A. Laboratory and field methods for assaying olfactory responses of the Douglas-fir beetle, *Dendroctonus pseudotsugae* Hopkins. *Can. Entomol.* **97**, 935 (1965).
619. Jantz, O. K., and Rudinsky, J. A. Studies of the olfactory behavior of the Douglas-fir beetle, *Dendroctonus pseudotsuqae* Hopkins. *Oreg., Agr. Exp. Sta., Tech. Bull.* **94**, 1–38 (1966).
620. Jeannel, R. "Introduction to Entomology." Hutchinson, London, 1960.
621. Jefferson, R. N., and Rubin, R. E. Sex pheromones of noctuid moths. XVII. A clarification of the description of the female sex pheromone gland of *Prodenia litura*. *Ann. Entomol. Soc. Amer.* **63**, 431 (1970).
622. Jefferson, R. N., Rubin, R. E., McFarland, S. U., and Shorey, H. H. Sex pheromones of noctuid moths. XXII. The external morphology of the antennae of *Trichoplusia ni, Heliothis zea, Prodenia ornithogalli*, and *Spodoptera exigua*. *Ann. Entomol. Soc. Amer.* **63**, 1227 (1970).
623. Jefferson, R. N., Shorey, H. H., and Gaston, L. K. Sex pheromones of noctuid moths. X. The morphology and histology of the female sex pheromone gland of *Trichoplusia ni* (Lepidoptera: Noctuidae). *Ann. Entomol. Soc. Amer.* **59**, 1166 (1966).
624. Jefferson, R. N., Shorey, H. H., and Rubin, R. E. Sex pheromones of noctuid moths. XVI. The morphology of the female sex pheromone glands of eight species. *Ann. Entomol. Soc. Amer.* **61**, 861 (1968).
625. Jefferson, R. N., Sower, L. L., and Rubin, R. E. The female sex pheromone gland of the pink bollworm, *Pectinophora gossypiella* (Lepidoptera: Gelechiidae). *Ann. Entomol. Soc. Amer.* **64**, 311 (1971).
626. Jones, F. M. The mating of the Psychidae (Lepidoptera). *Trans. Amer. Entomol. Soc.* **53**, 293 (1927).
627. Jones, P. A., Coppel, H. C., and Casida, J. E. Collection of additional sex at-

tractant from the virgin female introduced pine sawfly. *J. Econ. Entomol.* **58,** 465 (1965).

628. Jones, W. A. The isolation, identification and synthesis of the sex pheromone of the southern armyworm moth. Ph.D. Thesis, Dept. of Chem., American University, Washington, D.C. (1970).

629. Jones, W. A., and Jacobson, M. Insect sex attractants. IV. The determination of gyplure in its mixtures by adsorption and gas chromatography. *J. Chromatogr.* **14,** 22 (1964).

630. Jones, W. A., and Jacobson, M. Insect sex attractants. V. The synthesis of some additional compounds related to gyplure. *J. Med. Chem.* **7,** 373 (1964).

631. Jones, W. A., and Jacobson, M. Isolation of *N,N*-diethyl-*m*-toluamide (Deet) from female pink bollworm moths. *Science* **159,** 99 (1968).

632. Jones, W. A., Jacobson, M., and Martin, D. F. Sex attractant of the pink bollworm moth: Isolation, identification, and synthesis. *Science* **152,** 1516 (1966).

633. Kafka, W. A. Molekulare Wechselwirkungen bei der Erregung einzelner Riechzellen. *Z. Vergl. Physiol.* **70,** 105 (1970).

634. Kafka, W. A. Molekül und Riechzelle. *Umschau* **13,** 464 (1971).

635. Kafka, W. A. Specificity of odor molecule interaction in single cells. *In* "Gustation and Olfaction" (G. Ohloff and A. F. Thomas, eds.), p. 61. Academic Press, New York, 1971.

636. Kaissling, K. E. Kinetics of olfactory receptor potentials. *In* "Olfaction and Taste" (C. Pfaffmann, ed.), p. 52. Rockefeller Univ. Press, New York, 1969.

636a. Kaissling, K. E. Insect olfaction. *In* "Handbook of Sensory Physiology. Chemical Senses" (L. M. Beidler, ed.), p. 351. Springer-Verlag, Berlin and New York, 1971.

637. Kaissling, K. E., and Priesner, E. Die Riechschwelle des Seidenspinners. *Naturwissenschaften* **57,** 23 (1970).

638. Kaissling, K. E., and Renner, M. Antennale Rezeptoren für Queen Substance und Sterzelduft bei der Honigbiene. *Z. Vergl. Physiol.* **59,** 357 (1968).

639. Kalmus, H. The chemical senses. *Sci. Amer.* **198** (4), 97 and 104 (1958).

640. Kammerer, P. "Geschlecht, Fortpflanzung und Fruchtbarkeit: Eine Biologie der Zeugung," p. 94. Dreimasken Verlag, Munich, 1927.

641. Kangas, E., Oksanen, H., and Perttunen, V. Responses of *Blastophagus piniperda* L. (Col., Scolytidae) to *trans*-verbenol, *cis*-verbenol, and verbenone, known to be population pheromones of some American bark beetles. *Ann. Entomol. Fenn.* **36,** 75 (1970).

642. Kangas, E., Perttunen, V., and Oksanen, H. Studies on the olfactory stimuli guiding the bark beetle *Blastophagus piniperda* L. (Coleoptera, Scolytidae) to the host tree. *Ann. Entomol. Fenn.* **33,** 181 (1967).

643. Kannowski, P. B., and Johnson, R. L. Male patrolling behaviour and sex attraction in ants of the genus *Formica*. *Anim. Behav.* **17,** 425 (1969).

644. Karlson, P. Pheromones. *Ergeb. Biol.* **22,** 212 (1960).

645. Karlson, P. The chemistry of insect hormones and insect pheromones. *Pure Appl. Chem.* **14,** 75 (1967).

646. Karlson, P. Insect hormones and insect pheromones as possible agents for insect control: A critical review. *J. S. Afr. Chem. Inst.* **22,** Spec. Issue, S41 (1969).

647. Karlson, P., and Butenandt, A. Pheromones (ectohormones) in insects. *Annu. Rev. Entomol.* **4,** 39 (1959).

648. Karlson, P., and Lüscher, M. "Pheromone," ein Nomenklaturvorschlag für eine Wirkstoffklasse. *Naturwissenschaften* **46,** 63 (1959).

649. Karlson, P., and Lüscher, M. "Pheromones": A new term for a class of biologically active substances. (*London*) *Nature* **183**, 55 (1959).
650. Kasang, G. Synthese und biologische Aktivität bombykol- und gyptolartiger Duftstoffe. Ph.D. Dissertation, Freie University, Berlin (1965).
651. Kasang, G. Tritium-Markierung des Sexuallockstoffes Bombykol. *Z. Naturforsch. B.* **23**, 1331 (1968).
652. Kasang, G. Bombykol reception and metabolism on the antennae of the silkmoth *Bombyx mori*. In "Gustation and Olfaction" (G. Ohloff and A. F. Thomas, eds.), p. 245. Academic Press, New York, 1971.
653. Kaufmann, T. Observations on the biology and behavior of the evergreen bagworm moth, *Thyridopteryx ephemeraeformis* (Lepidoptera: Psychidae). *Ann. Entomol. Soc. Amer.* **61**, 38 (1968).
654. Kay, R. E. Acoustic signalling and its possible relationship to assembling and navigation in the moth, *Heliothis zea*. *J. Insect Physiol.* **15**, 989 (1969).
655. Keller, J. C., Mitchell, E. B., McKibben, G., and Davich, T. B. A sex attractant for female boll weevils from males. *J. Econ. Entomol.* **57**, 609 (1964).
656. Keller, J. C., Sheets, L. W., Green, N., and Jacobson, M. *cis*-7-Hexadecen-1-ol acetate (hexalure), a synthetic sex attractant for pink bollworm males. *J. Econ. Entomol.* **62**, 1520 (1969).
657. Kellogg, V. L. Some silkworm moth reflexes. *Biol. Bull.* **12**, 152 (1907).
658. Kelner-Pillault, S. Attirence sexuelle chez *Mantis religiosa* (Orth.). *Bull. Soc. Entomol. Fr.* **62**, 9 (1957).
659. Kelsey, J. M. *Costelytra zealandica* (White) sex attraction. *N. Z. Entomol.* **3** (5), 57 (1967).
660. Kershaw, S. H. *Phragmatobia fuliginosa* L. females attracted by scent of male *Panaxia dominula* L. *Entomol. Rec.* **65**, 219 (1953).
661. Kettlewell, H. B. D. The assembling scent of *Arctia villica* and *Parasemia plantaginis*. *Entomol. Rec.* **54**, 62 (1942).
662. Kettlewell, H. B. D. Further observations on non-specific assembling scents in macro-Lepidoptera. *Entomol. Rec.* **55**, 107 (1943).
663. Kettlewell, H. B. D. Female assembling scents with reference to an important paper on the subject. *Entomologist* **79**, 8 (1946).
664. Kettlewell, H. B. D. A further case of similar assembling scents. *Entomologist* **88**, 19 (1955).
665. Kettlewell, H. B. D. *Xanthorhoe montanata* and *X. spadicearia* assembling to females of *Biston betularia*. *Entomologist* **89**, 130 (1956).
666. Kettlewell, H. B. D. The radiation theory of female assembling in the Lepidoptera. *Entomologist* **94**, 59 (1961).
667. Keys, R. E., and Mills, R. B. Demonstration and extraction of a sex attractant from female Angoumois grain moths. *J. Econ. Entomol.* **61**, 46 (1968).
668. Killinen, R. G., and Ost, R. W. Pheromone-maze trap for cabbage looper moths. *J. Econ. Entomol.* **64**, 310 (1971).
669. King, H. Sex attractant principles of moths. *Proc. S. London Entomol. Natur. Hist. Soc. 1946-1947* p. 106 (1947).
670. Kinzer, G. W., Fentiman, A. F., Jr., Foltz, R. I., and Rudinsky, J. A. Bark beetle attractants: 3-methyl-2-cyclohexen-1-one isolated from *Dendroctonus pseudotsugae*. *J. Econ. Entomol.* **64**, 970 (1971).
671. Kinzer, G. W., Fentiman, A. F., Jr., Page, T. F., Jr., Foltz, R. L., Vité, J. P., and Pitman, G. B. Bark beetle attractants: Identification, synthesis and field bioassay of a new compound isolated from *Dendroctonus*. *Nature* (*London*) **221**, 477 (1969).

672. Kirkland, A. H. (1893). As cited by Farsky (*394*).
673. Kirkland, A. H. (1896). As cited by Forbush and Fernald (*417*).
674. Kirschenblatt, J. D. Klassifikation einiger biologisch aktiven Stoffe, die von den tierischen Organismen ausgearbeitet werden. *Trav. Soc. Natur. Leningrad* **73** (4), 225 (1957).
675. Kirschenblatt, J. D. Telergones and their biological significance. *Usp. Sovrem. Biol.* **46,** 322 (1958).
676. Kirschenblatt, J. D. Terminology of some biologically active substances and validity of the term "pheromones." *Nature (London)* **195,** 916 (1962).
677. Kishaba, A. N., Toba, H. H., Wolf, W. W., and Vail, P. V. Response of laboratory-reared male cabbage looper to synthetic sex pheromone in the field. *J. Econ. Entomol.* **63,** 178 (1970).
678. Kishaba, A. N., Wolf, W. W., Toba, H. H., Howland, A. F., and Gibson, T. Light and synthetic pheromone as attractants for male cabbage loopers *J. Econ. Entomol.* **63,** 1417 (1970).
679. Kittredge, J. A., Terry, M., and Takahashi, F. T. Sex pheromone activity of the molting hormone, crustecdysone, on male crabs. *Fish. Bull.* **69,** 337 (1971).
680. Klassen, W., and Earle, N. W. Permanent sterility induced in boll weevils with busulfan without reducing production of pheromone. *J. Econ. Entomol.* **63,** 1195 (1970).
681. Kliefoth, R. A., Vité, J. P., and Pitman, G. B. A laboratory technique for testing bark beetle attractants. *Contrib. Boyce Thompson Inst.* **22,** 283 (1964).
682. Kliewer, J. W., and Miura, T. Sex attractants of mosquitoes. *Abstr. Pac. Br. Meet., Entomol. Soc. Amer.* p. 11 (1964).
683. Kliewer, J. W., Miura, T., Husbands, R. C., and Hurst, C. H. Sex pheromones and mating behavior of *Culiseta inornata* (Diptera: Culicidae). *Ann. Entomol. Soc. Amer.* **59,** 530 (1966).
684. Kloek, J. The smell of some steroid sex-hormones and their metabolites. Reflections and experiments concerning the significance of smell for the mutual relation of the sexes. *Psychiatr., Neurol., Neurochir.* **64,** 309 (1961).
685. Klopping, H. L. Olfactory theories and the odors of small molecules. *J. Agr. Food Chem.* **19,** 999 (1971).
686. Klopping, H. L., and Meade, A. B. Molecular shape and size vs. attractancy to insects in some chemically unrelated compounds. *J. Agr. Food Chem.* **19,** 147 (1971).
687. Klum, J. A. Isolation of a sex pheromone of the European corn borer. *J. Econ. Entomol.* **61,** 484 (1968).
688. Klun, J. A., and Brindley, T. A. *cis*-11-Tetradecenyl acetate, a sex stimulant of the European corn borer. *J. Econ. Entomol.* **63,** 779 (1970).
689. Klun, J. A., and Robinson, J. F. Inhibition of European corn borer mating by *cis*-11-tetradecenyl acetate, a borer sex stimulant. *J. Econ. Entomol.* **63,** 1281 (1970).
690. Klun, J. A., and Robinson, J. F. Sex attractancy in the European corn borer. *Abstr. Nat. Meet., Entomol. Soc. Amer., Miami Beach, Fla.* p. 55 (1970).
690a. Klun, J. A., and Robinson, J. F. European corn borer moth: Sex attractant and sex attraction inhibitors. *Ann. Entomol. Soc. Amer.* **64,** 1083 (1971).
691. Knipling, E. F. Alternate methods of controlling insect pests. *FDA (Food Drug Admin.) Pap.* **3** (1), 16 (1969).
692. Knipling, E. F., and McGuire, J. U., Jr. Population models to test theoretical

effects of sex attractants used for insect control. *U.S., Dep. Agr., Inform. Bull.* **308**, 1–20 (1966).

693. Koch, W. Synthesis of bombykol, the sex attractant of the silkworm moth, and its geometric isomers. Doctoral Dissertation, Fac. Sci., University of Munich, Germany (1962).

694. Komarek, I., and Pfeffer, A. Eine neue biologische Kontrolle der Forstschädlinge. *Proc. Int. Congr. Entomol., 1938 7th*, Vol 3, p. 200 (1939).

695. Korotkova, O. A. Attracting substances (insect attractants). *Zh. Vses. Khim. Obshchest.* **9**, 518 (1964).

696. Kullenberg, B. Some observations on scents among bees and wasps (Hymenoptera). *Entomol. Tidskr.* **74**, 1 (1953).

697. Kullenberg, B. Field experiments with chemical sexual attractants on aculeate Hymenoptera males. *Zool. Bidr. Uppsala* **31**, 253 (1956).

698. Kullenberg, B. Studies in *Ophrys* pollination. *Zool. Bidr. Uppsala* **34**, 1 (1961).

699. Kullenberg, B. Pheromoner—kemiska retningsmedel och budbärare mellan individer, en mybenämnd grupp av biologiskt aktiva ämnen. *Sartryck Zool. Revy* p. 60 (1964).

700. Kullenberg, B. Forskning över kemisk stimulation, en viktig sida av ekologin. *Sartryck Zool. Revy* No. 1, p. 15 (1965).

701. Kullenberg, B., Bergstrom, G., and Ställberg-Stenhagen, S. Volatile components of the cephalic marking secretion of male bumble bees. *Acta Chem. Scand.* **24**, 1481 (1970).

702. Kunike, G. Zur Biologie der kleinen Wachsmotte, *Achroea grisella* Fabr. *Z. Angew. Entomol.* **16**, 304 (1930).

703. Kurihara, N. Chemical approaches to pest control. *Kagaku No Ryoiki* **22**, No. 7, 640 (1968).

704. Kuwahara, Y., Hara, H., Ishii, S., and Fukami, H. The sex pheromone of the Mediterranean flour moth. *Agr. Biol. Chem.* **35**, 447 (1971).

705. Kuwahara, Y., Kitamura, C., Takahashi, F., and Fukami, H. Studies on sex pheromones of Pyralididae. I. Changes in the quantity of the sex pheromone in the female almond moth, *Cadra cautella* Walker (Phycitinae). *Bochu-Kagaku* **33**, 158 (1968).

706. Kuwahara, Y., Kitamura, C., Takahashi, S., Hara, H., Ishii, S., and Fukami, H. Sex pheromone of the almond moth and the Indian meal moth: cis-9,trans-12-Tetradecadienyl acetate. *Science* **171**, 801 (1971).

707. Kwiatkowska, J. Pheromones and the problem of communication in the animal kingdom. *Postepy Hig. Med. Dosw.* **23**, 489 (1969).

708. Ladd, T. L., Jr. Preliminary field studies of sex attraction in the Japanese beetle. *Abstr. East. Br. Meet., Entomol. Soc. Amer. Abstr.* (1968).

709. Ladd, T. L., Jr. Sex attraction in the Japanese beetle. *J. Econ. Entomol.* **63**, 905 (1970).

710. Laithwaite, E. R. A radiation theory of the assembling of moths. *Entomologist* **93**, 113 and 133 (1960).

711. Lane, M. C. Recent progress in the control of wireowrms. *World's Grain Exhibition Conf., Regina, Can.* **2**, 529 (1933).

711a. Lange, R., Hoffmann, D., and Weissinger, M. Untersuchungen über das Flugverhalten männlicher Falter von *Rhyacionia buoliana* (Lep., Tortricidae). *Ocologia (Berlin)* **6**, 156 (1971).

712. Lanier, G. N. Sex pheromones: Abolition of specificity in hybrid bark beetles. *Science* **169**, 71 (1970).

713. Law, J. H., and Regnier, F. E. Pheromones. *Annu. Rev. Biochem.* **40**, 533 (1971).

714. Law, J. H., Wilson, E. O., and McCloskey, J. A. Biochemical polymorphism in ants. *Science* **149**, 544 (1965).

715. Lawson, F. R. Theory of control of insect populations by sexually sterile males. *Ann. Entomol. Soc. Amer.* **60**, 713 (1967).

716. Lebedeva, K. V. Attractants. *Zh. Vses. Khim. Obshchest.* **13**, 272 (1968).

717. Lederer, E. Odeurs et parfums des animaux. *Fortschr. Chem. Org. Naturst.* **6**, 87 (1950).

718. Lefebvre, A. Note sur le sentiment olfactif des antennes. *Ann. Soc. Entomol. Fr.* [1] **7**, 395 (1838).

719. Lehman, R. S. Experiments to determine the attractiveness of various aromatic compounds to adults of the wireworm, *Limonius* (*Pheletes*) *canus* Lec. and *Limonius* (*Pheletes*) *californicus* Mann. *J. Econ. Entomol.* **25**, 949 (1932).

720. Lehmensick, R., and Liebers, R. Beiträge zur Biologie der Microlepidopteren. (Untersuchungen an *Plodia interpunctella* Hb.). *Z. Angew. Entomol.* **24**, 582 (1938).

721. Leitereg, T. G., Guadagni, D. G., Harris, J., Mon, T. R., and Teranishi, R. Chemical and sensory data supporting the difference between the odors of the enantiomeric carvones. *J. Agr. Food Chem.* **19**, 785 (1971).

722. Lemarie, J. Neue Kontrollmethode des Nonnenvorkommens. *Anz. Schädlingskunde* **9**, 43 (1933).

723. Lener, W. Pheromone produced by *Oncopeltus fasciatus*, the large milkweed bug. *Amer. Zool.* **7**, 805 (1967).

724. Lener, W. Pheromone secretion in the large milkweed bug, *Oncopeltus fasciatus*. *Amer. Zool.* **9**, 1143 (1969).

724a. Leppla, N. C. Calling behavior during pheromone release in the female pink bollworm moth. *Ann. Entomol. Soc. Amer.* **65**, 281 (1972).

725. Levinson, H. Z., and Bar Ilan, A. R. Function and properties of an assembling scent in the khapra beetle *Trogoderma granarium*. *Riv. Parassitol.* **28**, 27 (1967).

726. Levinson, H. Z., and Bar Ilan, A. R. Control of the khapra beetle, *Trogoderma granarium*, by chemical and physical lures. *Abstr. Int. Congr. Entomol., 13th, 1968* p. 147 (1971).

727. Levinson, H. Z., and Bar Ilan, A. R. Lack of an intraspecific attractant in male *Trogoderma granarium*. *Riv. Parassitol.* **31**, 70 (1970).

728. Levinson, H. Z., and Bar Ilan, A. R. Olfactory and tactile behaviour of the khapra beetle, *Trogoderma granarium*, with special reference to its assembling scent. *J. Insect Physiol.* **16**, 561 (1970).

729. Levinson, H. Z., and Bar Ilan, A. R. Behaviour of the khapra beetle *Trogoderma granarium* towards the assembling scent released by the female. *Experientia* **26**, 846 (1970).

730. Levinson, H. Z., and Bar Ilan, A. R. Assembling and alerting scents produced by the bedbug *Cimex lectularius* L. *Experientia* **27**, 102 (1971).

730a. Lewis, W. J., Snow, J. W., and Jones, R. L. A pheromone trap for studying populations of *Cardiochiles nigriceps*, a parasite of *Heliothis virescens*. *J. Econ. Entomol.* **64**, 1417 (1971).

731. Lhoste, J., and Roche, A. Organes odoriférants des males de *Ceratitis capitata*. *Bull. Soc. Entomol. Fr.* **65**, 206 (1960).

732. Lilly, C. E. Response of males of *Limonius californicus* (Mann.) (Coleoptera:

Elateridae) to a sex attractant separable by paper chromatography. *Can. Entomol.* **91**, 145 (1959).

733. Lilly, C. E., and McGinnis, A. J. Reactions of male click beetles in the laboratory to olfactory pheromones. *Can. Entomol.* **97**, 317 (1965).

734. Lilly, C. E., and McGinnis, A. J. Quantitative responses of males of *Limonius californicus* (Coleoptera: Elateridae) to female sex pheromone. *Can. Entomol.* **100**, 1071 (1968).

734a. Lilly, C. E., and Shorthouse, J. D. Responses of males of the 10-lined June beetle, *Polyphylla decemlineata* (Coleoptera: Scarabaeidae), to female sex pheromone. *Can. Entomol.* **103**, 1757 (1971).

735. Lindquist, O. H., and Bowser, R. L. A biological study of the leaf miner, *Chrysopeleia ostryaella* Chambers (Lepidoptera: Cosmopterygidae), on ironwood in Ontario. *Can. Entomol.* **98**, 252 (1966).

736. Lodha, K. R., Treece, R. E., and Koutz, F. R. Studies on the mating behavior of the face fly. *J. Econ. Entomol.* **63**, 207 (1970).

737. Lyons, L. A. The European pine sawfly, *Neodiprion sertifer* (Geoff.) (Hymenoptera: Diprionidae). A review with emphasis on studies in Ontario. *Proc. Entomol. Soc. O.nt.* **94, 1963,** 5 (1964).

738. MacConnell, J. G., and Silverstein, R. M. Chemical ecology. *Chemistry* **44** (7), 6 (1971).

739. Macdonald, C. G., and Whittle, C. P. Mass spectrometry of unsaturated alcohols. *CSIRO Div. Entomol., Canberra, Aust., Annu. Rep.* p. 20 (1960–1970).

740. Macfarlane, J. H., and Earle, N. W. Morphology and histology of the female sex pheromone gland of the salt-marsh caterpillar, *Estigmene acrea. Ann. Entomol. Soc. Amer.* **63**, 1327 (1970).

741. Madden, J. L., and Simpson, R. F. Attraction of Sirex wasp to trees. *CSIRO Div. Entomol., Canberra, Aust., Annu. Rep.* p. 34 (1969–1970).

742. Madsen, H. F. Codling moth attractants. *Pest Art. News Summ.* **13**, 333 (1967).

743. Magnus, D. Beobachtungen zur Balz und Eiablage des Kaisermantels *Argynnis paphia* L. (Lep., Nymphalidae). *Z. Tierpsychol.* **7**, 435 (1950).

744. Makino, K., Satoh, K., and Inagami, K. Bombixin, a sex attractant discharged by female moth, *Bombyx mori. Biochim. Biophys. Acta* **19**, 394 (1956).

745. Maksimović, M. Testing of trap method for capturing the gypsy moth males. *Zast. Bilja* **52–53,** 177 (1959).

746. Maksimović, M. Sex attractant traps with female odour of the gypsy moth used for forecasting the increase of population of gypsy moth. *Proc. Int. Congr. Entomol., 12th 1964* p. 69 (1965).

747. Maksimović, M., and Marović, R. Influence of the form of traps upon the attractiveness of sexual odour of gypsy moth females. *Zast. Bilja* **18**, 115 (1967).

748. Manley, C., and Farrier, M. H. Attraction of the Nantucket pine tip moth, *Rhyacionia frustrana*, to incandescent light and blacklight, and to virgin females. *Ann. Entomol. Soc. Amer.* **62**, 443 (1969).

749. Manning, A. Sexual behaviour. *In* "Insect Behaviour" (P. T. Haskell, ed.), Symp. No. 3, p. 59. Roy. Entomol. Soc., London, 1966.

750. Manning, A. Antennae and sexual receptivity in *Drosophila melanogaster* females. *Science* **158**, 136 (1967).

751. Marikovskii, P. J. Sense organs of insects. *Priroda (Moscow)* No. 9, p. 30 (1969).

752. Marshall, J. The location of olfactory receptors in insects: A review of experimental evidence. *Trans. Roy. Entomol. Soc. London* **83**, 49 (1935).

753. Mathieu, J. M. Mating behavior of five species of Lucanidae (Coleoptera: Insecta). *Can. Entomol.* **101**, 1054 (1969).
754. Matsui, M., and Liau, C. E. Synthesis of 2,2-dimethyl-3-isopropylcyclopropyl propionate. *Agr. Biol. Chem.* **29**, 655 (1965).
755. Matthes, D. Excitatoren und Paarungsverhalten mittleuropäischer Malachiiden (Coleopt., Malacodermata). *Z. Morphol. Oekol. Tiere* **51**, 375 (1962).
756. Matthes, D. Intraspezifische Sekretwirkung bei Insekten. *Forsch. Fortschr.* **37**, 36 (1963).
757. Matthews, L. H., and Knight, M. "The Senses of Animals," p. 201. Philosophical Library, Inc., New York, 1963.
758. Mayer, A. G. On the mating instinct in moths. *Psyche* **9**, 15 (1900).
759. Mayer, A. G., and Soule, C. G. Some reactions of caterpillars and moths. *J. Exp. Zool.* **3**, 415 (1906).
760. Mayer, M. S. An olfactory attractant for the house fly. *Abstr. Nat. Meet., Entomol. Soc. Amer., Miami Beach, Fla.* p. 62 (1970).
761. Mayer, M. S., and James, J. D. Response of male *Musca domestica* to a specific olfactory attractant and its initial chemical purification. *J. Insect Physiol.* **17**, 833 (1971).
762. Mayer, M. S., and Thaggard, C. W. Investigations of an olfactory attractant specific for males of the housefly, *Musca domestica*. I. *J. Insect Physiol.* **12**, 891 (1966).
763. Mayr, E. The role of the antennae in the mating behavior of female *Drosophila*. *Evolution* **4**, 149 (1950).
764. McCluskey, R., Wright, C. G., and Yamamoto, R. T. Effect of starvation on the responses of male American cockroaches to sex and food stimuli. *J. Econ. Entomol.* **62**, 1465 (1969).
765. McDonough, L. M., and George, D. A. Gas chromatographic determination of the cis-trans isomer content of olefins. *J. Chromatogr. Sci.* **8**, 158 (1970).
766. McDonough, L. M., George, D. A., and Butt, B. A. Artificial sex pheromones for the codling moth. *J. Econ. Entomol.* **62**, 243 (1969).
766a. McDonough, L. M., George, D. A., Butt, B. A., Gamey, L. N., and Stegmeier, M. C. Field tests of artificial and natural sex pheromones for the codling mcth. *J. Econ. Entomol.* **65**, 108 (1972).
767. McDonough, L. M., George, D. A., Butt, B. A., Jacobson, M., and Johnson, G. R. Isolation of a sex pheromone of the codling moth. *J. Econ. Entomol.* **62**, 62 (1969).
768. McDonough, L. M., George, D. A., and Landis, B. J. Partial structure of two sex pheromones of the corn earworm, *Heliothis zea*. *J. Econ. Entomol.* **63**, 408 (1970).
769. McFadden, M. W., and Lam, J. J. Jr. Influence of population level and trap spacing on capture of tobacco hornworm moths in blacklight traps with virgin females. *J. Econ. Entomol.* **61**, 1150 (1968).
770. McGovern, T. P., Beroza, M., Ohinata, K., Miyashita, D., and Steiner, L. F. Volatility and attractiveness to the Mediterranean fruit fly of trimedlure and its isomers, and a comparison of its volatility with that of seven other insect attractants. *J. Econ. Entomol.* **59**, 1450 (1966).
771. McIndoo, N. E. The olfactory sense of Coleoptera. *Biol. Bull.* **28**, 407 (1915).
772. McKibben, G. H., Hardee, D. D., Davich, T. B., Gueldner, R. C., and Hedin, P. A. Slow-release formulations of grandlure, the synthetic pheromone of the boll weevil. *J. Econ. Entomol.* **64**, 317 (1971).
773. McMullen, L. H., and Atkins, M. D. On the flight and host selection of the

Douglas-fir beetle, *Dendroctonus pseudotsugae* Hopk. (Coleoptera: Scolytidae). *Can. Entomol.* **94**, 1309 (1962).

774. McNew, G. L. The Boyce Thompson Institute program in forest entomology that led to the discovery of pheromones in bark beetles. *Contrib. Boyce Thompson Inst.* **24**, 251 (1970).

775. Mehrotra, K. N., Singh, V. S., and Mookherjee, P. B. Evidence for the presence of a sex pheromone in the females of almond moth, *Cadra cautella*. *Indian J. Entomol.* **29**, 304 (1967).

776. Meinwald, J., Chalmers, A. M., Pliske, T. E., and Eisner, T. Pheromones. III. Identification of *trans,trans*-10-hydroxy-3,7-dimethyl-2,6-decadienoic acid as a major component in "hairpencil" secretion of the male monarch butterfly. *Tetrahedron Lett.* p. 4893 (1968).

777. Meinwald, J., Chalmers, A. M., Pliske, T. E., and Eisner, T. Identification and synthesis of *trans,trans*-3,7-dimethyl-2,6-decadien-1,10-dioic acid, a component of the pheromonal secretion of the male monarch butterfly. *Chem. Commun.* p. 86 (1969).

778. Meinwald, J., and Meinwald, Y. C. Structure and synthesis of the major components in the hairpencil secretion of a male butterfly, *Lycorea ceres ceres* (Cramer). *J. Amer. Chem. Soc.* **88**, 1305 (1966).

779. Meinwald, J., Meinwald, Y. C., and Mazzocchi, P. H. Sex pheromone of the queen butterfly: Chemistry. *Science* **164**, 1174 (1969).

780. Meinwald, J., Meinwald, Y. C., Wheeler, J. W., Eisner, T., and Brower, L. P. Major components in the exocrine secretion of a male butterfly (*Lycorea*). *Science* **151**, 583 (1966).

781. Meinwald, J., Wheeler, J. W., Nimetz, A. A., and Liu, J. S. Synthesis of some 1-substituted 2,2-dimethyl-3-isopropylidenecyclopropanes. *J. Org. Chem.* **30**, 1038 (1965).

782. Mell, R. "Biologie und Systematik der Südchinesischen Sphingiden." Friedländer, Berlin, 1922.

783. Menon, M. Hormone-pheromone relationships in the beetle, *Tenebrio molitor*. *J. Insect Physiol.* **16**, 1123 (1970).

783a. Mertins, J. W., and Coppel, H. C. Sexual behavior in gynandromorphs of *Diprion similis* (Hymenoptera: Diprionidae). *Ann. Entomol. Soc. Amer.* **64**, 1191 (1971).

783b. Mertins, J. W., and Coppel, H. C. Previously undescribed abdominal glands in the female introduced pine sawfly, *Diprion similis* (Hymenoptera: Diprionidae). *Ann. Entomol. Soc. Amer.* **65**, 33 (1972).

784. Michael, R. P., and Keverne, E. B. Pheromones in the communication of sexual status in primates. *Nature (London)* **218**, 746 (1968).

785. Michael, R. P., and Keverne, E. B. Primate sex pheromones of vaginal origin. *Nature (London)* **225**, 84 (1970).

786. Michael, R. P., Keverne, E. B., and Bonsall, R. W. Pheromones: Isolation of male sex attractants from a female primate. *Science* **172**, 964 (1971).

787. Micklem, H. S. The proposed biological term "pheromone." *Nature (London)* **83**, 1835 (1959).

788. Miller, T., Jefferson, R. N., and Thomson, W. W. Sex pheromones of noctuid moths. XI. The ultrastructure of the apical region of the female sex pheromone gland of *Trichoplusia ni*. *Ann. Entomol. Soc. Amer.* **60**, 707 (1967).

789. Minks, A. K. Sex pheromonen en hun betekenis voor de geintegreerde bestrijding van insektenplagen. *Vakbl. Biol.* **49**, 41 (1969).

789a. Minks, A. K. Decreased sex pheromone production in an in-bred stock of the summerfruit tortrix moth, *Adoxophyes orana. Entomol. Exp. Appl.* **14,** 361 (1971).

790. Minks, A. K., and Noordink, J. P. W. Sex attraction of the summerfruit tortrix moth, *Adoxophyes orana:* Evaluation in the field. *Entomol. Exp. Appl.* **14,** 57 (1971).

790a. Minks, A. K., Noordink, J. P. W., and Van den Anker, C. A. Recapture by sex traps of *Adoxophyes orana,* released from one point in an apple orchard. *Meded. Fak. Landbouw. Wetensch. Gent* **36,** 274 (1971).

791. Minnich, D. E. The chemical senses of insects. *Quart. Rev. Biol.* **4,** 100 (1929).

792. Mitchell, W. C., and Mau, R. F. L. Response of the female southern green stink bug and its parasite, *Trichopoda pennipes,* to male stink bug pheromones. *J. Econ. Entomol.* **64,** 856 (1971).

793. Moeck, H. A. An olfactometer for the bioassay of attractants for scolytids. *Can. Entomol.* **102,** 792 (1970).

794. Moeck, H. A. Ethanol as the primary attractant for the ambrosia beetle *Trypodendron lineatum* (Coleoptera: Scolytidae). *Can. Entomol.* **102,** 985 (1970).

795. Moncrieff, R. W. "The Chemical Senses." Leonard Hill, London, 1944.

796. Moncrieff, R. W. The gypsy moth may provide the perfumer with a new fixative. *Mfg. Chem.* **16,** 167 (1945).

797. Monnerot-Dumaine, J. Les attractifs des insectes. *Presse Med.* **73,** 754 (1965).

798. Moore, B. P. Pheromones and insect control. *Aust. J. Sci.* **28,** 243 (1965).

799. Moore, B. P. Chemical communication in insects. *Sci. J.* **3** (9), 44 (1967).

800. Moore, B. P., and Brown, W. V. Micro-ozonolysis technique for structure determination. *CSIRO Div. Entomol., Canberra, Aust., Annu. Rep.* p. 27 (1969–1970).

801. Moorhouse, J. E., Yeadon, R., Beevor, P. S., and Nesbitt, B. F. Method for use in studies of insect chemical communication. *Nature (London)* **223,** 1174 (1969).

802. Morita, H., and Yamashita, S. Receptor potentials recorded from sensilla basiconica on the antenna of the silkworm larvae, *Bombyx mori. J. Exp. Biol.* **38,** 851 (1961).

803. Morse, R. A. Swarm orientation in honeybees. *Science* **141,** 357 (1963).

803a. Morse, R. A., and Boch, R. Pheromone concert in swarming honey bees (Hymenoptera: Apidae). *Ann. Entomol. Soc. Amer.* **64,** 1414 (1971).

804. Morse, R. A., Gary, N. E., and Johansson, T. S. K. Mating of virgin queen honey bees (*Apis mellifera* L.) following mandibular gland extirpation. *Nature (London)* **194,** 605 (1962).

805. Müller, D. G., Jaenicke, L., Donike, M., and Akintobi, T. Sex attractant in a brown alga: Chemical structure. *Science* **171,** 815 (1971).

806. Müller, F. Wo hat der Moschusduft der Schwärmer seinen Sitz? *Kosmos* **3,** 84 (1878).

807. Müller, F. Notes on Brazilian entomology. *Trans. Roy. Entomol. Soc. London* p. 211 (1878).

808. Müller, F. As cited by Rau and Rau (*911*).

809. Müller, K. Histologische Untersuchungen über den Entwicklungsbeginn bei einem Kleinschmetterling (*Plodia interpunctella*). *Z. Wiss. Zool.* **151,** 192 (1938).

810. Müller-Schwarze, D. Complexity and relative specificity in a mammalian pheromone. *Nature (London)* **223,** 525 (1969).

811. Munakata, K. Chemical research on moth attractants. *Tampakushitsu, Kakusan, Koso* **8** (1), 20 (1963).

812. Munakata, K. Sex attractants for insects. *Kagaku To Seibutsu* (4), 177 (1965).

813. Murr, L. Über den Geruschsinn der Mehlmottenschlupfwespe *Habrobracon juglandis* Ashmead. *Z. Vergl. Physiol.* **11**, 210 (1930).

814. Murphy, M. R., and Schneider, G. E. Olfactory bulb removal eliminates mating behavior in the male golden hamster. *Science* **167**, 302 (1970).

815. Murvosh, C. M., Fye, R. L., and LaBrecque, G. C. Studies on the mating behavior of the house fly, *Musca domestica* L. *Ohio J. Sci.* **64**, 264 (1964).

816. Murvosh, C. M., LaBrecque, G. C., and Smith, C. N. Sex attraction in the house fly, *Musca domestica* L. *Ohio J. Sci.* **65**, 68 (1965).

817. Muto, F. Insect attractants. *Noyaku Seisan Gijutsu* **19**, 1 (1968).

818. Myers, J. The structure of the antennae of the Florida queen butterfly, *Danaus gilippus berenice* (Cramer). *J. Morphol.* **125**, 315 (1968).

819. Myers, J., and Brower, L. P. A behavioural analysis of the courtship pheromone receptors of the queen butterfly, *Danaus gilippus berenice*. *J. Insect Physiol.* **15**, 2117 (1969).

820. Myers, K. Oviposition and mating behaviour of Queensland fruit fly [*Dacus* (*Strumeta*) *tryoni* (Frogg.)] and the solanum fruit fly [*Dacus* (*Strumeta*) *cacuminatus* (Hering)]. *Aust. J. Sci. Res.*, Ser B **5**, 264 (1952).

820a. Nagata, K., Tamaki, Y., Noguchi, H., and Yushima, T. Changes in sex pheromone activity in adult females of the smaller tea tortrix moth *Adoxophyes fasciata*. *J. Insect Physiol.* **18**, 339 (1972).

821. Nagy, B. Vergleich verschiedener Fangmethoden in der Signalization der Hanfmotte (*Grapholitha sinana* Feld.). *Abstr. Int. Congr. Entomol., 13th, 1968* p. 178 (1971).

822. Nakagawa, S., Farias, G. J., and Steiner, L. F. Response of female Mediterranean fruit flies to male lures in the relative absence of males. *J. Econ. Entomol.* **63**, 227 (1970).

823. Nakajima, M. Studies on sex pheromones of the stored grain moths. *In* "Control of Insect Behavior by Natural Products" (D. L. Wood, R. M. Silverstein, and M. Nakajima, eds.), p. 209. Academic Press, New York, 1970.

824. Nakazema, S. Notes on the response of the silkworm (*Bombyx mori* L.) to odors. III. On the olfactory sense of the adult moth. *Bull. Miyazaki Coll. Agr. Forestry* No. 3, p. 129 (1931).

825. Nesbitt, B. F. Sex attractant of the red bollworm. *Rep. Trop. Prod. Inst., London, 1968/69* p. 31 (1970).

826. Nichols, J. O. The gypsy moth in Pennsylvania—its history and eradication. *Pa., Agr. Exp. Sta., Misc. Bull.* **4404**, 1–82 (1961).

827. Nolte, D. J., May, I. R., and Thomas, B. M. The gregarisation pheromone of locusts. *Chromosoma* **29**, 462 (1970).

828. Nolte, H. W. Neue Erfahrungen zur Dykschen Nonnenanlockmethode. *Zentralbl. Ges. Forstw.* **66**, 197 and 252 (1940).

829. Norris, M. J. Contributions towards the study of insect fertility, II: Experiments on the factors influencing fertility in *Ephestia kühniella* Z. (Lepidoptera, Phycitidae). *Proc. Zool. Soc. London* p. 903 (1933).

830. Norris, M. J. Aggregation response in ovipositing females of the desert locust, with special reference to the chemical factor. *J. Insect. Physiol.* **16**, 1493 (1970).

831. Nutting, W. L. Colonizing flights and associated activities of termites. I. The desert damp-wood termite *Paraneotermes simplicicornis* (Kalotermitidae). *Psyche* **73**, 131 (1966).

831a. Ohno, M. Chemistry of sex attractants of insects. *Bull. Inst. Chem. Res., Kyoto Univ.* **45**, 207 (1967).

832. Oksanen, H., Perttunen, V., and Kangas, E. Studies on the chemical factors involved in the olfactory orientation of *Blastophagus piniperda* (Coleoptera: Scolytidae). *Contrib. Boyce Thompson Inst.* **24**, 299 (1970).

833. Onsager, J. A., McDonough, L. M., and George, D. A. A sex pheromone in the Pacific Coast wireworm. *J. Econ. Entomol.* **61**, 691 (1968).

834. Osborne, G. O., and Hoyt, C. P. Preliminary note on a chemical attractant for the grass grub beetle *Costelytra zealandica* (White) from the flowers of elder (*Sambucus nigra* L.). *N. Z. J. Sci.* **11**, 137 (1968).

835. Osborne, G. O., and Hoyt, C. P. A chemical attractant for males of the grass grub beetle *Costelytra zealandica* (White) (Col. Scarabaeidae). *Bull. Entomol. Res.* **59**, 81 (1969).

836. Osborne, G. O., and Hoyt, C. P. Phenolic resins as chemical attractants for males of the grass grub beetle, *Costelytra zealandica* (Coleoptera: Scarabaeidae). *Ann. Entomol. Soc. Amer.* **63**, 1145 (1970).

837. Osmani, Z., and Naidu, M. B. Evidence of sex attractant in female *Dysdercus cingulatus* Fabr. *Indian J. Exp. Biol.* **5**, 51 (1967).

838. Otto, D. Attraktivstoff-Reaktionen bei Forstinsekten und ihre Bedeutung für den Forstschutz. *Biol. Rundsch.* **6**, 22 (1968).

839. Outram, I. Aspects of mating in the spruce budworm, *Choristoneura fumiferana* (Lepidoptera: Tortricidae). *Can. Entomol.* **103**, 1121 (1971).

840. Ouye, M. T., and Butt, B. A. A natural sex lure extracted from female pink bollworms. *J. Econ. Entomol.* **55**, 419 (1962).

841. Ouye, M. T., Graham, H. M., Richmond, C. A., and Martin, D. F. Mating studies of the pink bollworm. *J. Econ. Entomol.* **57**, 222 (1964).

842. Pain, J., Barbier, M., Bogdanovsky, D., and Lederer, E. Chemistry and biological activity of the secretions of queen and worker honeybees (*Apis mellifica* L.). *Comp. Biochem. Physiol* **6**, 233 (1962).

843. Pain, J., and Ruttner, F. Les extraits des glandes mandibulaires des reines d'abeilles attirent les mâles, lors due vol nuptial. *C. R. Acad. Sci.* **256**, 512 (1963).

844. Park, R. J., and Sutherland, M. D. Volatile constituents of the bronze orange bug, *Rhoecocoris sulciventris*. *Aust. J. Chem.* **15**, 172 (1962).

845. Parker, H. L. *Macrocentrus gifuensis* Ashmead, a polyembryonic braconid parasite in the European corn borer. *U.S., Dep. Agr., Tech. Bull.* **230**, 1–62 (1931).

846. Parkes, A. S. Olfactory effects in mammalian reproduction. *Endocrinol. Jap.* **9**, 247 (1962).

847. Patrick, J. C., and Hensley, S. D. Recapture of males released at different distances from a trap baited with virgin female sugarcane borers. *J. Econ. Entomol.* **63**, 1341 (1970).

848. Pattenden, G. A synthesis of propylure, sex attractant of the pink bollworm moth. *J. Chem. Soc., C* p. 2385 (1968).

849. Pattenden, G., and Staddon, B. W. Observations on the metasternal scent glands of *Lethocerus* spp. (Heteroptera: Belostomatidae). *Ann. Entomol. Soc. Amer.* **63**, 900 (1970).

850. Pavan, M., and Quilico, A. Nuove prospettive nella lotta contro gli insetti nocivi. *Atti. Accad. Naz. Lincei* **364,** 37 (1969).

851. Payne, T. L. Electrophysiological investigations on response to pheromones in bark beetles. *Contrib. Boyce Thompson Inst.* **24,** 275 (1970).

852. Payne, T. L. Bark beetle olfaction. I. Electroantennogram responses of the southern pine beetle (Coleoptera: Scolytidae) to its aggregation pheromone frontalin. *Ann. Entomol. Soc. Amer.* **64,** 266 (1971).

853. Payne, T. L., Shorey, H. H., and Gaston, L. K. Sex pheromones of noctuid moths. XXIII. Factors influencing antennal responsiveness in males of *Trichoplusia ni. J. Insect Physiol.* **16,** 1043 (1970).

854. Payne, T. L., and Stewart, J. W. Evidence for a sex pheromone in the cottonwood twig borer. *J. Econ. Entomol.* **64,** 987 (1971).

855. Peacock, J. W., Lincoln, A. C., Simeone, J. B., and Silverstein, R. M. Field attraction of *Scolytus multistriatus* (Coleoptera: Scolytidae) to a virgin-female-produced pheromone. *Abstr. Nat. Meet., Entomol. Soc. Amer., Miami Beach, Fla.* p. 51 (1970).

855a. Peacock, J. W., Lincoln, A. C., Simeone, J. B., and Silverstein, R. M. Attraction of *Scolytus multistriatus* (Coleoptera: Scolytidae) to a virgin-female-produced pheromone in the field. *Ann. Entomol. Soc. Amer.* **64,** 1143 (1971).

856. Percy, J. E., Gardiner, E. J., and Weatherston, J. Studies of physiologically active arthropod secretions. VI. Evidence for a sex pheromone in female *Orgyia leucostigma* (Lepidoptera: Lymantridae). *Can. Entomol.* **103,** 706 (1971).

856a. Percy, J. E., and Weatherston, J. Studies of physiologically active arthropod secretions. IX. Morphology and histology of the pheromone-producing glands of some female Lepidoptera. *Can. Entomol.* **103,** 1733 (1971).

857. Pérez, R. P. Sex-attractant and mating behavior of the sugarcane moth borer. *Bull. P. R. Agr. Exp. Sta.* **188,** 1–28 (1964).

858. Pérez, R. P., and Long, W. H. Sex attractant and mating behavior in the sugarcane borer. *J. Econ. Entomol.* **57,** 688 (1964).

859. Perttunen, V., Oksanen, H., and Kangas, E. Aspects of the external and internal factors affecting the olfactory orientation of *Blastophagus piniperda* (Coleoptera: Scolytidae). *Contrib. Boyce Thompson Inst.* **24,** 293 (1970).

860. Petersen, W. Die Morphologie der Generationsorgane der Schmetterlinge und ihre Bedeutung für die Artbildung. *Mem. Acad. Sci. St. Petersburg* [8] **16,** (8), 1–84 (1904).

861. Pettersson, J. An aphid sex attractant. I. Biological studies. *Entomol. Scand.* **1,** 63 (1970).

861a. Pettersson, J. An aphid sex attractant. II. Histological, ethological and comparative studies. *Entomol. Scand.* **2,** 81 (1971).

862. Pfeiffer, W. Odoriferous substances in the life of insects. *Dragoco Rep. Ger. Ed.* **13,** 161 and 183 (1966).

863. Pinder A. R., and Staddon, B. W. The odoriferous secretion of the water bug, *Sigara falleni* (Fieb.). *J. Chem. Soc. London* p. 2955 (1965).

864. Pitman, G. B. Studies on the pheromone of *Ips confusus.* III. Influence of host material on pheromone production. *Contrib. Boyce Thompson Inst.* **23,** 147 (1966).

865. Pitman, G. B. Bark beetle manipulation with natural and synthetic attractants. *Insect-Plant Interactions. Rep. Conf. Univ. Calif. Santa Barbara,* p. 56 (1968).

866. Pitman, G. B. Pheromone response in pine bark beetles; influence of host volatiles. *Science* **166,** 905 (1969).

867. Pitman, G. B. Field manipulation of the mountain pine beetle with *trans*-verbenol and alpha pinene. *Abstr. Nat. Meet., Entomol. Soc. Amer., Miami Beach, Fla.* p. 57 (1970).

868. Pitman, G. B. *trans*-Verbenol and alpha-pinene: Their utility in manipulation of the mountain pine beetle. *J. Econ. Entomol.* **64**, 426 (1971).

869. Pitman, G. B., Kliefoth, R. A., and Vité, J. P. Studies on the pheromone of *Ips confusus* (LeConte). II. Further observations on the site of production. *Contrib. Boyce Thompson Inst.* **23**, 13 (1965).

870. Pitman, G. B., Renwick, J. A. A., and Vité, J. P. Studies on the pheromone of *Ips confusus* (LeConte). IV. Isolation of the attractive substance by gas-liquid chromatography. *Contrib. Boyce Thompson Inst.* **23**, 243 (1966).

871. Pitman, G. B., and Vité, J. P. Studies on the pheromone of *Ips confusus* (Lec.). I. Secondary sexual dimorphism in the hindgut epithelium. *Contrib. Boyce Thompson. Inst.* **22**, 221 (1963).

872. Pitman, G. B., and Vité, J. P. Aggregation behavior of *Dendroctonus ponderosae* (Coleoptera: Scolytidae) in response to chemical messengers. *Can. Entomol.* **101**, 143 (1969).

873. Pitman, G. B., and Vité, J. P. Field response of *Dendroctonus pseudotsugae* (Coleoptera: Scolytidae) to synthetic frontalin. *Ann. Entomol. Soc. Amer.* **63**, 661 (1970).

874. Pitman, G. B., and Vité, J. P. Predator-prey response to western pine beetle attractants. *J. Econ. Entomol.* **64**, 402 (1971).

875. Pitman, G. B., Vité, J. P., Kinzer, G. W., and Fentiman, A. F., Jr. Bark beetle attractants: *trans-* Verbenol isolated from *Dendroctonus*. *Nature (London)* **218**, 168 (1968).

876. Pitman, G. B., Vité, J. P., Kinzer, G. W., and Fentiman, A. F., Jr. Specificity of population-aggregating pheromones in *Dendroctonus*. *J. Insect Physiol.* **15**, 363 (1969).

877. Pitman, G. B., Vité, J. P., and Renwick, J. A. A. Variation in olfactory behavior of *Ips confusus* (LeC.) (Coleoptera: Scolytidae) between laboratory and field bioassays. *Naturwissenschaften* **53**, 46 (1966).

878. Plattner, J. J., Bhalerao, U. T., and Rapoport, H. Synthesis of *dl*-sirenin. *J. Amer. Chem. Soc.* **91**, 4933 (1969).

879. Pliske, T. E., and Eisner, T. Sex pheromone of the queen butterfly: Biology. *Science* **164**, 1170 (1969).

880. Pointing, P. J. The biology and behavior of the European pine shoot moth, *Rhyacionia buoliana* (Schiff.) in Southern Ontario. I. Adult. *Can. Entomol.* **93**, 1098 (1961).

881. Pointing, P. J., and Green, G. W. A review of the history and biology of the European pine shoot moth, *Rhyacionia buoliana* (Schiff.) (Lepidoptera: Olethreutidae) in Ontario. *Proc. Entomol. Soc. Ont.* **92, 1961**, 58 (1962).

882. Polimanti, O. On the thele-perception of sex in silkworm moths. *J. Anim. Behav.* **4**, 289 (1914).

883. Poulton, E. B. An adaptation which tends to prevent inbreeding in certain Lepidoptera. *Proc. Entomol. Soc. London* **2**, 75 (1927).

884. Poulton, E. B. Notes on *Nemoria viridata* L., by Dr. R. C. L. Perkins: The distance at which the male is attracted by the female. *Proc. Entomol. Soc. London* **3**, 20 (1928).

885. Poulton, E. B. The suggestion that "assembling" moths are attracted by wireless. *Proc. Entomol. Soc. London* **3**, 20 (1928).

886. Priesner, E. Die interspezifischen Wirkungen der Sexuallockstoffe der Saturniidae (Lepidoptera). *Z. Vergl. Physiol.* **61**, 263 (1968).
887. Priesner, E. A new approach to insect pheromone specificity. *In* "Olfaction and Taste" (C. Pfaffmann, ed.), p. 235. Rockefeller Univ. Press, New York, 1969.
888. Pritchard, G. Laboratory observations on the mating behaviour of the Island fruit fly *Rioxa pornia* (Diptera: Tephritidae). *J. Aust. Entomol. Soc.* **6**, 127 (1967).
889. Proujan, C. Trapping insects with scent. *Sci. World* **12** (13), 6 (1966).
890. Proverbs, M. D. The sterile male technique for codling moth control. *West. Fruit Grower* **19** (4), 19 (1965).
891. Proverbs, M. D., Newton, J. R., and Logan, D. M. Orchard assessment of the sterile male technique for control of the codling moth, *Carpocapsa pomonella* (L.) (Lepidoptera: Olethreutidae). *Can. Entomol.* **98**, 90 (1966).
892. Provost, M. W., and Haeger, J. S. Mating and pupal attendance in *Deinocerites cancer* and comparisons with *Opifex fuscus* (Diptera: Culicidae). *Ann. Entomol. Soc. Amer.* **60**, 565 (1967).
893. Prüffer, J. Observations et expériences sur les phénomènes de la vie sexuelle de *Lymantria dispar* L. *Bull. Int. Acad. Pol. Sci. Lett., Cl. Sci. Math. Natur., Ser. B* No. 1, p. 1 (1924).
894. Prüffer, J. Untersuchungen über die Innervierung der Fühler bei *Saturnia pyri* L. *Zool. Jahrb., Abt. Anat. Ontog. Tiere* **51**, 1 (1929).
895. Prüffer, J. Weitere Untersuchungen über die Männchenanlockung bei *Lymantria dispar* L. (Lep.). *Zool. Pol.* **2**, 43 (1937).
896. Pukowski, E. Ökologische Untersuchungen an *Necrophorus* F. *Z. Morphol. Oekol. Tiere* **27**, 518 (1933).
897. Puls, W., Haberland, G. L., and Schlossmann. K. Über Wirkungen des Hexade-cadien-(10,12)-ol-(1) auf den Stoffwechsel der Ratte. *Naturwissenschaften* **50**, 671 (1963).
898. Putman, W. L. The codling moth, *Carpocapsa pomenella* (L.) (Lepidoptera: Tortricidae): A review with special reference to Ontario. *Proc. Entomol. Soc. Ont.* **93 1962**, 22 (1963).
899. Putman, W. L. Some aspects of sex in the European red mite, *Panonychus ulmi. Can. Entomol.* **102**, 612 (1970).
900. Pyatnova, Yu. B., Ivanov, L. L., and Kyskina, A. S. Sex attractants of insects. *Usp. Khim.* **38**, 248 (1969).
901. Rahn, R. La teigne du poireau *Acrolepia assectella* Zeller, éléments de biologie et mise au point d'avertissements agricoles fondés sur le piégeage sexuel des mâles. *Proc. Acad. Agr. Fr.* p. 997 (1966).
902. Rahn, R. Rôle de la plante-hôte sur l'attractivité sexuelle chez *Acrolepia assectella* Zeller (Lep. Plutellidae). *C. R. Acad. Sci., Ser. D* **266**, 2004 (1968).
903. Rahn, R. Etude expérimentale préliminaire de l'attractivité sexuelle de femelles d'*Acrolepia assectella* (Zeller) en conditions naturelles. *Ann. Epiphyt.* **19**, 367 (1968).
904. Rahn, R., and Labeyrie, V. Importance de l'hétérogénéité de l'attraction sexuelle exercée par les femelles d'*Acrolepia assectella* Zeller. *C. R. Acad. Sci., Ser. D* **265**, 427 (1967).
905. Rainwater, C. F. Prospects for the eradication of the boll weevil. *J. Wash. Acad. Sci.* **60**, 48 (1970).
906. Ralls, K. Mammalian scent marking. *Science* **171**, 443 (1971).
907. Rathmayer, W. Sexuallockstoffe bei Hymenopteren. *Naturwiss. Rundsch.* **14**, 153 (1961).

908. Rau, P. Sexual selection experiments in the cecropia moth. *Trans. Acad. Sci. St. Louis* **20**, 275 (1911).
909. Rau, P. Further observations on copulation and oviposition in *Samia cecropia* Linn. *Trans. Acad. Sci. St. Louis* **20**, 309 (1911).
910. Rau, P. A note on the courtship of *Telea polyphemus. Can. Entomol.* **56**, 271 (1924).
911. Rau, P., and Rau, N. L. The sex attraction and rhythmic periodicity in the giant saturniid moths. *Trans. Acad. Sci. St. Louis* **26**, 83 (1929).
912. Read, J. S., Warren, F. L., and Hewitt, P. H. Identification of the sex pheromone of the false codling moth. *Chem. Commun.* p. 792 (1968).
913. Redfern, R. E., Butt, B. A., and Cantu, E. Bioassay of the sex pheromone of the southern armyworm. *J. Econ. Entomol.* **63**, 658 (1970).
914. Redfern, R. E., Cantu, E., Jones, W. A., and Jacobson, M. Response of the male southern armyworm in a field cage to prodenialure A and prodenialure B. *J. Econ. Entomol.* **64**, 1570 (1971).
915. Reece, C. A., Rodin, O. Brownlee, R. G., Duncan, W. G., and Silverstein, R. M. Synthesis of the principal components of the sex attractant from male *Ips confusus* frass: 2-methyl-6-methylene-7-octen-4-ol, 2-methyl-6-methylene-2,7-octadien-4-ol, and (+)-*cis*-verbenol. *Tetrahedron* **24**, 4249 (1968).
916. Regnier, F. E., and Law, J. H. Insect pheromones. *J. Lipid Res.* **9**, 541 (1968).
917. Reichenau, J. (1880). As cited by Dickins (*349*).
918. Rejesus, R. S., and Reynolds, H. T. Demonstration of the presence of a female sex pheromone in the cotton leaf perforator. *J. Econ. Entomol.* **63**, 961 (1970).
919. Renner, M. Das Duftorgan der Honigbiene und die physiologische Bedeutung ihres Lockduftes. *Z. Vergl. Physiol.* **43**, 411 (1960).
920. Renner, M., and Baumann, M. Über Komplexe von subepidermalen Drüsenzellen (Duftdrüsen ?) der Bienenkönigin. *Naturwissenschaften* **51**, 68 (1964).
921. Renwick, J. A. A. Identification of two oxygenated terpenes from the bark beetles *Dendroctonus frontalis* and *Dendroctonus brevicomis. Contrib. Boyce Thompson Inst.* **23**, 355 (1967).
922. Renwick, J. A. A. Chemical aspects of bark beetle aggregation. *Contrib. Boyce Thompson Inst.* **24**, 337 (1970).
923. Renwick, J. A. A., Pitman, G. B., and Vité, J. P. Detection of a volatile compound in hindguts of male *Ips confusus* (LeConte) (Coleoptera: Scolytidae). *Naturwissenschaften* **53**, 83 (1966).
924. Renwick, J. A. A., and Vité, J. P. Isolation of the population aggregating pheromone of the southern pine beetle. *Contrib. Boyce Thompson Inst.* **24**, 65 (1968).
925. Renwick, J. A. A., and Vité, J. P. Bark beetle attractants: Mechanism of colonization by *Dendroctonus frontalis. Nature (London)* **224**, 1222 (1969).
926. Renwick, J. A. A., and Vité, J. P. Systems of chemical communication in *Dendroctonus. Contrib. Boyce Thompson Inst.* **24**, 283 (1970).
927. Ribbands, C. R. "The Behavior and Social Life of Honeybees." Bee Res. Ass., Ltd., London, 1953.
928. Rice, R. E. Response of some predators and parasites of *Ips confusus* (Lec.) (Coleoptera: Scolytidae) to olfactory attractants. *Contrib. Boyce Thompson Inst.* **24**, 189 (1969).
929. Rice, R. E., and Moreno, D. S. Marking and recapture of California red scale for field studies. *Ann. Entomol. Soc. Amer.* **62**, 558 (1969).
930. Rice, R. E., and Moreno, D. S. Comparative production of pheromone by the California red scale reared on lemons and potatoes. *J. Econ. Entomol.* **62**, 958 (1969).

931. Rice, R. E., and Moreno, D. S. Flight of male California red scale. *Ann. Entomol. Soc. Amer.* **63,** 91 (1970).
932. Richards, O. W. Sexual selection and allied problems in the insects. *Biol. Rev.* **2,** 298 (1927).
933. Richards, O. W., and Thomson, W. S. A contribution to the study of the genera *Ephestia*, GN. (including *Strymax*, Dyar), and *Plodia*, GN. (Lepidoptera, Phycitidae), with notes on parasites of the larvae. *Trans. Entomol. Soc. London* **80,** 169 (1932).
934. Riddiford, L. M. *trans*-2-Hexenal: Mating stimulant for polyphemus moths. *Science* **158,** 139 (1967).
935. Riddiford, L. M. Oak leaves and the mating and ovipositional behavior of the polyphemus moth. *Insect-Plant Interactions., Rep. Conf. Univ. Calif., Santa Barbara, 1968* p. 60 (1968).
936. Riddiford, L. M. Antennal proteins of saturniid moths—their possible role in olfaction. *J. Insect Physiol.* **16,** 653 (1970).
937. Riddiford, L. M., and Williams, C. M. Volatile principle from oak leaves: Role in sex life of the polyphemus moth. *Science* **155,** 589 (1967).
938. Riddiford, L. M., and Williams, C. M. Chemical signaling between polyphemus moths and between moths and host plant. *Science* **156,** 541 (1967).
939. Riddiford, L. M., and Williams, C. M. Role of the corpora cardiaca in the behavior of Saturniid moths. I. Release of sex pheromone. *Biol. Bull.* **140,** 1 (1971).
940. Riemschneider, R., and Kasang, G. Stereoisomere des Heptadecadien-(10.12)-ols-(1). *Z. Naturforsch. B* **18,** 646 (1963).
941. Riemschneider, R., Kasang, G., and Böhme, C. 13-Acetoxy-heptadecadien-(8,10)-ol-(1)-, Octadien-(3,5)-ol-(1)-Isomere und verwandte Verbindungen. *Monatsh. Chem.* **96,** 1766 (1965).
942. Riley, C. V. The senses of insects. *Insect Life* **7,** 33 (1895).
943. Ritcher, P. O., and Beer, F. M. Notes on the biology of *Pleocoma dubitalis dubitalis* Davis. *Pan-Pac. Entomol.* **32,** 181 (1956).
944. Ritter, F. J. Some recent developments in the field of insect pheromones. *Meded. Fakult. Landbouw. Wetenschup. Gent* **36,** 874 (1971).
945. Roach, S. H., Ray, L., Hopkins, A. R., and Taft, H. M. Comparison of attraction of wing traps and cotton trap plots baited with male boll weevils for overwintered weevils. *Ann. Entomol. Soc. Amer.* **64,** 530 (1971).
946. Roach, S. H., Ray, L., Taft, H. M., and Hopkins, A. R. Wing traps baited with male boll weevils for determining spring emergence of overwintered weevils and subsequent infestations in cotton. *J. Econ. Entomol.* **64,** 107 (1971).
947. Roach, S. H., Taft, H. M., Ray, L., and Hopkins, A. R. Population dynamics of the boll weevil in an isolated cotton field in South Carolina. *Ann. Entomol. Soc. Amer.* **64,** 394 (1971).
948. Roderick, W. R. Current ideas on the chemical basis of olfaction. *J. Chem. Educ.* **43,** 510 (1966).
949. Rodin, J. O., Silverstein, R. M., Burkholder, W. E., and Gorman, J. E. Sex attractant of female dermestid beetle *Trogoderma inclusum* LeConte. *Science* **165,** 904 (1969).
950. Roelofs, W. L. Sex attractants used to combat insects. *Farm Res.* **32,** (2), 2 (1966).
951. Roelofs, W. L., and Arn, H. Red-banded leaf roller sex attractant characterized. *Food Life Sci. (New York)* **1** (1), 12 (1968).
952. Roelofs, W. L., and Arn, H. Sex attractant of the red-banded leaf roller moth. *Nature (London)* **219,** 513 (1968).

953. Roelofs, W. L., and Cardé, R. T. Hydrocarbon sex pheromone in tiger moths (Arctiidae). *Science* 171, 684 (1971).

953a. Roelofs, W. L., Cardé, R., Benz, G., and von Salis, G. Sex attractant of the larch bud moth found by electroantennogram method. *Experientia* 27, 1438 (1971).

954. Roelofs, W. L., and Comeau, A. Sex pheromone perception. *Nature (London)* 220, 600 (1968).

955. Roelofs, W. L., and Comeau, A. Sex pheromones as a taxonomic principle. *Abstr. East. Br. Meet. Entomol. Soc. Amer.* (1968).

956. Roelofs, W. L., and Comeau, A. Sex pheromone specificity: Taxonomic and evolutionary aspects in lepidoptera. *Science* 165, 398 (1969).

957. Roelofs, W. L., and Comeau, A. Lepidopterous sex attractants discovered by field screening tests. *J. Econ. Entomol.* 63, 969 (1970).

958. Roelofs, W. L., and Comeau, A. Sex pheromone perception: Synergists and inhibitors for the red-banded leaf roller attractant. *J. Insect Physiol.* 17, 435 (1971).

959. Roelofs, W., Comeau, A., Hill, A., and Milicevic, G. Sex attractant of the codling moth: Characterization with electroantennogram technique. *Science* 174, 297 (1971).

960. Roelofs, W. L., Comeau, A., and Selle, R. Sex pheromone of the oriental fruit moth. *Nature (London)* 224, 723 (1969).

961. Roelofs, W. L., and Feng, K. C. Isolation and bioassay of the sex pheromone of the red-banded leaf roller, *Argyrotaenia velutinana* (Lepidoptera: Tortricidae). *Ann. Entomol. Soc. Amer.* 60, 1199 (1967).

962. Roelofs, W. L., and Feng, K. C. Sex pheromone specificity tests in the Tortricidae—an introductory report. *Ann. Entomol. Soc. Amer.* 61, 312 (1968).

963. Roelofs, W. L., Glass, E. H., Arn, H., and Comeau, A. Modulating sex attractant reception for insect control. *Food Life Sci. (New York)* (2), 19 (1970).

964. Roelofs, W. L., Glass, E. H., Tette, J., and Comeau, A. Sex pheromone trapping for red-banded leaf roller control: Theoretical and actual. *J. Econ. Entomol.* 63, 1162 (1970).

965. Roelofs, W. L., Pulver, D. W., Feng, K. C., and Gambrell, F. L. Attractancy tests with chromatographed European chafer extracts. *J. Econ. Entomol.* 60, 869 (1967).

966. Roelofs, W. L., and Tette, J. P. Sex pheromone of the oblique-banded leaf roller moth. *Nature (London)* 226, 1152 (1970).

967. Roever, K. Bionomics of *Agathymus* (Megathymidae). *J. Res. Lepidoptera* 3, 103 (1964).

968. Rogoff, W. M., Biological aspects of the house fly sex pheromone. *Abstr. Pac. Br. Meet., Entomol. Soc. Amer.* p. 7 (1964).

969. Rogoff, W. M., Beltz, A. D., Johnsen, J. O., and Plapp, F. W. A sex pheromone in the housefly, *Musca domestica* L. *J. Insect Physiol.* 10, 239 (1964).

970. Rogoff, W. M., and Gretz, G. H. Female house fly sex pheromone: Attractance demonstrated to optically deficient male responders. *Abstr. Nat. Meet., Entomol. Soc. Amer., Miami Beach, Fla.* p. 55 (1970).

971. Röller, H. A. Research on insect hormones and pheromones. *Insect-Plant Interactions, Rep. Conf. Univ. Calif. Santa Barbara, 1968* p. 62 (1968).

972. Röller, H., Biemann, K., Bjerke, J. S., Norgard, D. W., and McShan, W. H. Isolation and identification of the sex-attractant of *Galleria mellonella* L. (greater waxmoth). *Int. Symp. Insect Endocrinol., Brno, Czech., 1966* (1960).

973. Röller, H., Biemann, K., Bjerke, J. S., Norgard, D. W., and McShan, W. H. Sex

pheromones of pyralid moths. I. Isolation and identification of the sex attractant of *Galleria mellonella* (greater waxmoth). *Acta Entomol. Bohemoslov.* **65,** 208 (1968).

974. Röller, H., Piepho, H., and Holz, I. Zum Problem der Hormonabhängigkeit des Paarungsverhaltens bei Insekten. Untersuchungen an *Galleria mellonella* (L.). *J. Insect Physiol.* **9,** 187 (1963).

975. Roth, L. M. Hypersexual activity induced in females of the cockroach *Nauphoeta cinerea. Science* **138,** 1267 (1962).

976. Roth, L. M. Control of reproduction in female cockroaches with special reference to *Nauphoeta cinerea.* II. Gestation and postparturition. *Psyche* **71,** 198 (1964).

977. Roth, L. M. Loci of sensory end-organs used by cockroaches in mating behavior. *Proc. Int. Congr. Entomol., 12th, 1964* p. 305 (1965).

978. Roth, L. M. Male pheromneos. *In* "McGraw-Hill Yearbook of Science and Technology," p. 293. McGraw-Hill, New York, 1967.

979 Roth, L. M. The evolution of male tergal glands in the Blattaria. *Ann. Entomol. Soc. Amer.* **62,** 176 (1969).

980. Roth, L. M. Interspecific mating in Blattaria. *Ann. Entomol. Soc. Amer.* **63,** 1282 (1970).

981. Roth, L. M., and Barth, R. H., Jr. The control of sexual receptivity in female cockroaches. *J. Insect Physiol.* **10,** 965 (1964).

982. Roth, L. M., and Barth, R. H., Jr. The sense organs employed by cockroaches in mating behavior. *Behaviour* **28,** 58 (1967).

983. Roth, L. M., and Dateo, G. P. A sex pheromone produced by males of the cockroach *Nauphoeta cinerea. J. Insect Physiol.* **12,** 255 (1966).

984. Roth, L. M., and Hartman, H. B. Sound production and its evolutionary significance in the Blattaria. *Ann. Entomol. Soc. Amer.* **60,** 740 (1967).

985. Roth, L. M., and Willis, E. R. A study of cockroach behavior. *Amer. Midl. Natur.* **47,** 66 (1952).

986. Roth, L. M., and Willis, E. R. Observations on the behavior of the webbing clothes moth. *J. Econ. Entomol.* **45,** 20 (1952).

987. Roth, L. M., and Willis, E. R. The reproduction of cockroaches. *Smithson. Misc. Collect.* **122,** 1 (1954).

988. Roth, L. M., and Willis, E. R. Intra-uterine nutrition of the "beetle-roach" *Diploptera dytiscoides* (Serv.) during embryogenesis, with notes on its biology in the laboratory (Blattaria: Diplopteridae). *Psyche* **62,** 55 (1955).

989. Rothschild, G. H. L. Sex pheromone of potato moth. *CSIRO Div. Entomol., Canberra, Aust., Annu. Rep.* p. 33 (1969–1970).

990. Rothschild, G. H. L. Sex pheromone of oriental fruit moth. *CSIRO Div. Entomol., Canberra, Aust., Annu. Rep.* p. 33 (1969–1970).

991. Rothschild, G. H. L. Personal communication (1971).

991a. Rothschild, G. H. L., and Minks, A. K. Field studies with sex pheromone of the oriental fruit moth. *CSIRO Div. Entomol., Canberra, Aust., Annu. Rep.* p. 34 (1970–1971).

991b. Rothschild, G. H. L., and Minks, A. K. Sex pheromone of oriental fruit moth in population survey. *CSIRO Div. Entomol., Canberra, Aust., Annu. Rep.* p. 63 (1970–1971).

992. Rudinsky, J. A. Response of *Dendroctonus pseudotsugae* Hopkins to volatile attractants. *Contrib. Boyce Thompson Inst.* **22,** 23 (1963).

993. Rudinsky, J. A. Host selection and invasion by the Douglas-fir beetle, *Dendroc-*

*tonus pseudotsugae* Hopkins, in coastal Douglas-fir forests *Can. Entomol.* **98**, 98 (1966).

994. Rudinsky, J. A. Scolytid beetles associated with Douglas fir: Response to terpenes. *Science* **152**, 218 (1966).

995. Rudinsky, J. A. Observations on olfactory behavior of scolytid beetles (Coleoptera: Scolytidae) associated with Douglas-fir forests. *Z. Angew. Entomol.* **58**, 356 (1966).

996. Rudinsky, J. A. Pheromone-mask by the female *Dendroctonus pseudotsugae* Hopk., an attraction regulator. *Pan. Pac. Entomol.* **44**, 248 (1968).

997. Rudinsky, J. A. Studies on timber beetles. *Insect-Plant Interactions, Rep. Conf. Univ. Calif., Santa Barbara, 1968* p. 64 (1968).

998. Rudinsky, J. A. Masking of the aggregation pheromone in *Dendroctonus pseudotsugae* Hopk. *Science* **166**, 884 (1969).

999. Rudinsky, J. A. Sequence of Douglas-fir beetle attraction and its ecological significance. *Contrib. Boyce Thompson Inst.* **24**, 311 (1970).

1000. Rudinsky, J. A., and Daterman, G. E. Response of the ambrosia beetle *Trypodendron lineatum* (Oliv.) to a female-produced pheromone. *Z. Angew. Entomol.* **54**, 300 (1964).

1001. Rudinsky, J. A., and Daterman, G. E. Field studies on flight patterns and olfactory responses of ambrosia beetles in Douglas-fir forests of western Oregon. *Can. Entomol.* **96**, 1339 (1964).

1002. Rudinsky, J. A., Novak, V., and Svihra, P. Pheromone and terpene attraction in the bark beetle *Ips typographus* L. *Experientia* **27**, 161 (1971).

1003. Rudinsky, J. A., Novak, V., and Svihra, P. Attraction of the bark beetle *Ips typographus* L. to terpenes and a male-produced pheromone. *Z. Angew. Zool.* **67**, 179 (1971).

1004. Rudinsky, J. A., and Schneider, I. Olfactory responses of the ambrosia beetle, *Trypodendron lineatum. Abstr. Int. Congr. Entomol., 13th, 1968* p. 223 (1971).

1005. Russell, G. F., and Hills, J. I. Odor differences between enantiomeric isomers. *Science* **172**, 1043 (1971).

1006. Ruttner, F. Die Sexualfunktionen der Honigbienen im Dienste ihrer sozialen Gemeinschaft. *Z. Vergl. Physiol.* **39**, 577 (1957).

1007. Ruttner, F., and Kaissling, K. E. Über die interspezifischer Wirkung des Sexuallockstoffes von *Apis mellifica* und *Apis cerana. Z. Vergl. Physiol.* **59**, 362 (1968).

1008. Ruttner, F., and Ruttner, H. Die Flugaktivität und das Paarungsverhalten der Drohnen. *Bienevater (Wien)* **84**, 297 (1963).

1009. Ruttner, F., and Ruttner, H. Beobachtungen an Drohnensammelplatzen. 2. Untersuchungen über die Flugaktivität und das Paarungsverhalten der Drohnen. *Z. Bienenforsch.* **8**, 1 (1965).

1010. Ruttner, F., and Ruttner, H. Untersuchungen über die Flugaktivität und das Paarungsverhalten der Drohnen. *Z. Bienenforsch.* **8**, 332 (1966).

1011. Ryan, E. P. Pheromone: Evidence in a decapod crustacean. *Science* **151**, 340 (1966).

1012. Saario, C. A., Shorey, H. H., and Gaston, L. K. Sex pheromones of noctuid moths. XIX. Effect of environmental and seasonal factors on captures of males of *Trichoplusia ni* in pheromone-baited traps. *Ann. Entomol. Soc. Amer.* **63**, 667 (1970).

1013. Sadler, E. A. Assembling *Pachythelia villosella. Entomol. Rec.* **81**, 322 (1969).

1014. Sanders, C. J. Extrusion of the female sex pheromone gland in the Eastern spruce budworm, *Choristoneura fumiferana* (Lepidoptera: Tortricidae). *Can. Entomol.* **101**, 760 (1969).

1015. Sanders, C. J. Daily activity patterns and sex pheromone specificity as sexual isolating mechanisms in two species of *Choristoneura* (Lepidoptera: Tortricidae). *Can. Entomol.* **103,** 498 (1971).

1015a. Sanders, C. J. Laboratory bioassay of the sex pheromone of the female eastern spruce budworm, *Choristoneura fumiferana* (Lepidoptera: Tortricidae). *Can. Entomol.* **103,** 631 (1971).

1016. Sanders, C. J. Sex pheromone specificity and taxonomy of budworm moths (*Choristoneura*). *Science* **171,** 911 (1971).

1017. Sannasi, A., and Rajulu, G. S. 9-Oxodec-*trans*-2-enoic acid in the Indian honeybees. *Life Sci.* **10,** Part II, 195 (1971).

1018. Sáringer, G., Wégh, G., and Rada, K. Sexual attractiveness of virgin plum fruit moth, *Grapholitha funebrana* Tr. (Lepidoptera: Tortricidae) females examined by ³²P labelled males. *Acta Phytopathol.* **3,** 373 (1968).

1019. Saunders, H. C. The stereochemical theory of olfaction. 5. Some odor observations in terms of the Amoore theory. *Proc. Toilet Goods Ass., Sci. Sect.* **37,** Suppl., 46 (1962).

1020. Scales, A. L. Female tarnished plant bugs attract males. *J. Econ. Entomol.* **61,** 1466 (1968).

1021. Schedl, K. E. Der Schwammspinner (*Porthetria dispar* L.) in Euroasien, Afrika und Neuengland. *Monogr. Angew. Entomol.* **12,** 1–242 (1936).

1022. Schenk, O. Die antennale Hautsinnesorgane einiger Lepidopteren und Hymenopteren mit besonderer Berücksichtigung der sexuellen Unterschiede. *Zool. Jahrb., Abt. Anat. Ontog. Tiere* **17,** 573 (1903).

1023. Schlinger, E. I., and Hall, J. C. The biology, behavior and morphology of *Praon pallitans* Muesebeck, an internal parasite of the spotted alfalfa aphid, *Therioaphis maculata* (Buckton) (Hymenoptera: Braconidae, Aphidiinae). *Ann. Entomol. Soc. Amer.* **53,** 144 (1960).

1024. Schneider, D. Elektrophysiologische Untersuchungen von Chemo- und Mechanorezeptoren der Antenne des Seidenspinners *Bombyx mori* L. *Z. Vergl. Physiol.* **40,** 8 (1957).

1025. Schneider, D. Electrophysiological investigation on the antennal receptors of the silk moth during chemical and mechanical stimulation. *Experientia* **13,** 89 (1957).

1026. Schneider, D. Untersuchungen zum Bau und zur Funktion der Riechorgane von Schmetterlingen und Käfern. *Bern. Phys.-Med. Ges. Wurzburg* [N.S.] **70,** 158 (1961).

1027. Schneider, D. Der Geruchssinn bei den Insekten. *Dragoco Ber.* **2,** 27 (1961).

1028. Schneider, D. Electrophysiological investigation on the olfactory specificity of sexual attracting substances in different species of moths. *J. Insect Physiol.* **8,** 15 (1962).

1029. Schneider, D. Electrophysiological investigation of insect olfaction. *Proc. Int. Symp. Olfaction Taste, 1st, 1963* p. 85 (1963).

1030. Schneider, D. Function of insect olfactory sensilla. *Proc. Int. Congr. Zool., 16th, 1963* Vol. 3, p. 84 (1963).

1031. Schneider, D. Vergleichende Rezeptorphysiologie am Beispiel der Riechorgane von Insekten. *Jahrb. Max-Planck-Ges. Ford. Wiss., 1963* p. 150 (1964).

1032. Schneider, D. Insect antennae. *Annu. Rev. Entomol.* **9,** 103 (1964).

1033. Schneider, D. Chemical sense communication in insects. *Symp. Soc. Exp. Biol.* **20,** 273 (1966).

1034. Schneider, D. Wie arbeitet der Geruchssinn bei Mensch und Tier? *Naturwiss. Rundsch.* **20,** 319 (1967).

1035. Schneider, D. Basic problems of olfactory research. *In* "Theories of Odors and Odor Measurement," Proc. NATO Summer School, p. 201 Robert College, Istanbul, Turkey, 1968.

1036. Schneider, D. Insect olfaction: Deciphering system for chemical messages. *Science* **163,** 1031 (1969).

1037. Schneider, D. Olfactory receptors for the sexual attractant (bombykol) of the silk moth. *In* "The Neurosciences: Second Study Program" (F. O. Schmitt, ed.), p. 511. Rockefeller Univ. Press, New York, 1970.

1038. Schneider, D. Molekulare Grundlagen der chemischen Sinne bei Insekten. *Naturwissenschaften* **58,** 194 (1971).

1039. Schneider, D., Block, B. C., Boeckh, J., and Priesner, E. Die Reaktion der männlichen Seidenspinner auf Bombykol und seine Isomeren: Elektroantennogramm und Verhalten. *Z. Vergl. Physiol.* **54,** 192 (1967).

1040. Schneider, D., and Hecker, E. Zur Elektrophysiologie der Antenne des Seidenspinners *Bombyx mori* bei Reizung mit angereicherten Extrakten des Sexuallockstoffes. *Z. Naturforsch.* B **11,** 121 (1956).

1041. Schneider, D., and Kaissling, K. E. Der Bau der Antenne des Seidenspinners *Bombyx mori* L. I. Architektur und Bewegungsapparat der Antenne sowie Struktur der Cuticula. *Zool. Jahrb., Abt. Anat. Ontog. Tiere* **75,** 287 (1956).

1042. Schneider, D., and Kaissling, K. E. Der Bau der Antenne des Seidespinners *Bombyx mori* L. II. Sensillen, cuticulare Bildung und innerer Bau. *Zool. Jahrb., Abt. Anat. Ontog. Tiere* **76,** 223 (1957).

1043. Schneider, D., and Kaissling, K. E. Der Bau der Antenne des Seidenspinners *Bombyx mori* L. III. Das Bindegewebe und das Blutgefäss. *Zool. Jahrb., Abt. Anat. Ontog. Tiere* **77,** 111 (1959).

1044. Schneider, D., Kasang, G., and Kaissling, K. E. Bestimmung der Riechschwelle von *Bombyx mori* mit Tritium-markiertem Bombykol. *Naturwissenschaften* **55,** 395 (1968).

1045. Schneider, D., Lacher, V., and Kaissling, K. E. Die Reaktiosweise und das Reaktionsspektrum von Riechzellen bei *Antheraea pernyi* (Lepidoptera, Saturniidae). *Z. Vergl. Physiol.* **48,** 632 (1964).

1046. Schneider, D., and Seibt, U. Sex pheromone of the queen butterfly: Electroantennogram responses. *Science* **164,** 1173 (1969).

1047. Schneider, D., and Steinbrecht, R. A. Checklist of insect olfactory sensilla. *Symp. Zool. Soc. London* **23,** 279 (1968).

1048. Schneider, I., and Rudinsky, J. A. The site of pheromone production in *Trypodendron lineatum* (Coleoptera: Scolytidae): Bioassay and histological studies of the hindgut. *Can. Entomol.* **101,** 1181 (1969).

1049. Schoenleber, L. G., Butt, B. A., and Hathaway, D. O. A trap with sex attractant for monitoring time of codling moth flights. *U.S. Dep. Agr., ARS* **ARS 42–177,** 1–7 (1970).

1050. Schönherr, J. Evidence of an aggregating pheromone in the ash-bark beetle, *Leperisinus fraxini* (Coleoptera: Scolytidae). *Contrib. Boyce Thompson Inst.* **24,** 305 (1970).

1051. Schultz, G. A., and Boush, G. M. Suspected sex pheromone glands in three economically important species of *Dacus. J. Econ. Entomol.* **64,** 347 (1971).

1052. Schwarz, M., Jacobson, M., and Cuthbert, F. P., Jr. Chemical studies of the sex attractant of the banded cucumber beetle. *J. Econ. Entomol.* **64,** 769 (1971).

1053. Schwinck, I. Über den Sexualduftstoff der Pyraliden. *Z. Vergl. Physiol.* **35,** 167 (1953).

1054. Schwinck, I. Experimentelle Untersuchungen über Geruchssinn und Strömungswahrnehmung in der Orientierung bei Nachtschmetterlingen. *Z. Vergl. Physiol.* **37**, 19 (1954).

1055. Schwinck, I. Freilandversuche zur Frage der Artspezifität des weiblichen Sexualduftstoffes der Nonne (*Lymantria monacha* L.) und des Schwammspinners (*Lymantria dispar* L.). *Z. Angew. Entomol.* **37**, 349 (1955).

1056. Schwinck, I. Weitere Untersuchungen zur Frage des Geruchsorientierung der Nachtschmetterlinge: Partielle Fühleramputation bei Spinnermännchen, insbesondere am Seidenspinner *Bombyx mori* L. *Z. Vergl. Physiol.* **37**, 439 (1955).

1057. Schwinck, I. A. study of olfactory stimuli in the orientation of moths. *Proc. Int. Congr. Entomol., 10th, 1956* Vol. 2, p. 577 (1958).

1058. Seiler, J. Resultate aus der Artkreuzung zwischen *Solenobia triquetrella* F. R. und *Solenobia fumosella* H. mit Intersexualität in $F_1$. *Arch. Julius Klaus-Stift. Vererbungsforsch., Sozialanthropol. Rassenhyg.* **24**, 124 (1949).

1059. Seiler, J., and Puchta, O. Die Fortpflanzungsbiologie der Solenobien (Lepid. Psychidae). Verhalten bei Artkreuzungen und $F_1$-Resultate. *Wilhelm Roux' Arch. Entwicklungsmech. Organismen* **149**, 115 (1956).

1060. Seitz, A. (1894). As cited by Urbahn (*1212*).

1061. Sekul, A. A., and Cox, H. C. Mating stimulant in the fall armyworm, *Laphygma frugiperda* (J. E. Smith). *Bull. Entomol. Soc. Amer.* **10**, 167 (1964).

1062. Sekul, A. A., and Cox, H. C. Sex pheromone in the fall armyworm, *Spodoptera frugiperda* (J. E. Smith). *BioScience* **15**, 670 (1965).

1063. Sekul, A. A., and Cox, H. C. Response of males to the female sex pheromone of the fall armyworm, *Spodoptera frugiperda* (Lepidoptera: Noctuidae): A laboratory evaluation. *Ann. Entomol. Soc. Amer.* **60**, 691 (1967).

1064. Sekul, A. A., and Sparks, A. N. Sex pheromone of the fall armyworm moth: Isolation, identification, and synthesis. *J. Econ. Entomol.* **60**, 1270 (1967).

1065. Sengün, A. Über die biologische Bedeutung des Duftstoffes von *Bombyx mori* L. *Istanbul Univ. Fen Fak. Mecm., Seri B* **19**, 281 (1954).

1066. Sexton, O. J., and Hess, E. H. A pheromone-like dispersant affecting the local distribution of the European house cricket, *Acheta domestica. Biol. Bull.* **134**, 490 (1968).

1067. Shamshurin, A. A., Kovalev, B. G., and Donya, A. P. A new synthesis of *trans*-1-acetoxy-10-propyltrideca-5, 9-diene ("propylure"), a field attractant for the cotton moth. *Dokl. Akad. Nauk SSSR* **190**, 1362 (1970).

1068. Sharma, D. Pest control by sex attractants. *In* "Pesticide Symposia" (S. K. Majumdar, ed.), p. 210. Acad. Pest Contr. Sci., Mysore, India, 1968.

1069. Sharma, R. K., Rice, R. E., Reynolds, H. T., and Shorey, H. H. Seasonal influence and effect of trap location on catches of pink bollworm males in sticky traps baited with hexalure. *Ann. Entomol. Soc. Amer.* **64**, 102 (1971).

1070. Sharma, R. K., Shorey, H. H., and Gaston, L. K. Sex pheromones of noctuid moths. XXIV. Evaluation of pheromone traps for males of *Trichoplusia ni. J. Econ. Entomol.* **64**, 361 (1971).

1071. Sharma, S. L., and Hussein, M. Z. A pest of wild Saccharums in Bihar: *Mahasena graminivorum* Hampson (family Psychidae: Lepidoptera). *Indian J. Entomol.* **17**, 89 (1955).

1072. Shaw, J. G., Moreno, D. S., and Fargerlund, J. Virgin female California red scales used to detect infestations *J. Econ. Entomol.* **64**, 1305 (1971).

1073. Shearer, D. A., and Boch, R. Citral in the Nassanoff pheromone of the honey bee. *J. Insect Physiol.* **12**, 1513 (1966).

1074. Shearer, D. A., Boch, R., Morse, R. A., and Laigo, F. M. Occurrence of 4-oxodec-trans-2-enoic acid in queens of *Apis dorsata, Apis cerana,* and *Apis mellifera. J. Insect Physiol.* **16,** 1437 (1970).

1075. Shirck, F. H. The flight of sugar-beet wireworm adults in southwestern Idaho. *J. Econ. Entomol.* **35,** 423 (1942).

1076. Shorey, H. H. Nature of the sound produced by *Drosophila melanogaster* during courtship. *Science* **137,** 677 (1962).

1077. Shorey, H. H. Sex pheromones of noctuid moths. II. Mating behavior of *Tricho-plusia ni* (Lepidoptera: Noctuidae) with special reference to the role of the sex pheromone. *Ann. Entomol. Soc. Amer.* **57,** 371 (1964).

1078. Shorey, H. H. The biology of *Trichoplusia ni* (Lepidoptera: Noctuidae). IV. Environmental control of mating. *Ann. Entomol. Soc. Amer.* **59,** 502 (1966).

1079. Shorey, H. H. Sex pheromones of Lepidoptera. *In* "Control of Insect Behavior by Natural Products" (D. L. Wood, R. M. Silverstein, and M. Nakajima, eds.), p. 249. Academic Press, New York, 1970.

1080. Shorey, H. H., Andres, L. A., and Hales, R. L., Jr. The biology of *Trichoplusia ni* (Lepidoptera: Noctuidae). I. Life history and behavior. *Ann. Entomol. Soc. Amer.* **55,** 591 (1962).

1081. Shorey, H. H., and Bartell, R. J. Role of a volatile female sex pheromone in stimulating male courtship behaviour in *Drosophila melanogaster. Anim. Behav.* **18,** 159 (1970).

1082. Shorey, H. H., Bartell, R. J., and Barton-Browne, L. B. Sexual stimulation of males of *Lucilia cuprina* (Calliphoridae) and *Drosophila melanogaster* (Droso-philidae) by the odors of aggregation sites. *Ann. Entomol. Soc. Amer.* **62,** 1419 (1969).

1083. Shorey, H. H., and Gaston, L. K. Sex pheromones of noctuid moths. III. Inhibi-tion of male responses to the sex pheromone in *Trichoplusia ni* (Lepidoptera: Noctuidae). *Ann. Entomol. Soc. Amer.* **57,** 775 (1964).

1084. Shorey, H. H., and Gaston, L. K. Sex pheromones of noctuid moths. V. Circadian rhythm of pheromone-responsiveness in males of *Autographa californica, Heliothis virescens, Spodoptera exigua,* and *Trichoplusia ni* (Lepidoptera: Noctuidae). *Ann. Entomol. Soc. Amer.* **58,** 597 (1965).

1085. Shorey, H. H., and Gaston, L. K. Sex pheromones of noctuid moths. VII. Quanti-tative aspects of the production and release of pheromone by females of *Tri-choplusia ni* (Lepidoptera: Noctuidae). *Ann. Entomol. Soc. Amer.* **59,** 604 (1965).

1086. Shorey, H. H., and Gaston, L. K. Sex pheromones of noctuid moths. VIII. Orien-tation to light by pheromone-stimulated males of *Trichoplusia ni* (Lepidoptera: Noctuidae). *Ann. Entomol. Soc. Amer.* **58,** 833 (1965).

1087. Shorey, H. H., and Gaston, L. K. Pheromones. *In* "Pest Control: Biological, Physical and Selected Chemical Methods" (W. W. Kilgore and R. Doutt, eds.), p. 241. Academic Press, New York, 1967.

1088. Shorey, H. H., and Gaston, L. K. Sex pheromones of noctuid moths. XII. Lack of evidence for masking of sex pheromone activity in extracts from females of *Heliothis zea* and *H. virescens* (Lepidoptera: Noctuidae). *Ann. Entomol. Soc. Amer.* **60,** 847 (1967).

1089. Shorey, H. H., and Gaston, L. K. Sex pheromones of noctuid moths. XX. Short-range visual orientation by pheromone-stimulated males of *Trichoplusia ni. Ann. Entomol. Soc. Amer.* **63,** 829 (1970).

1090. Shorey, H. H., and Gaston, L. K. Use of sex pheromones for behavioral control of the cabbage looper. *Abstr. Nat. Meet., Entomol. Soc. Amer., Miami Beach, Fla.* p. 55 (1970).

1091. Shorey, H. H., Gaston, L. K., and Fukuto, T. R. Sex pheromones of noctuid moths. I. A quantitative bioassay for the sex pheromone of *Trichoplusia ni* (Lepidoptera:Noctuidae). *J. Econ. Entomol.* **57**, 252 (1964).

1092. Shorey, H. H., Gaston, L. K., and Jefferson, R. N. Insect sex pheromones. *Advan. Pest Contr. Res.* **8**, 57 (1968).

1093. Shorey, H. H., Gaston, L. K., and Roberts, J. S. Sex pheromones of noctuid moths. VI. Absence of behavioral specificity for the female sex pheromones of *Trichoplusia ni* versus *Autographa californica*, and *Heliothis zea* versus *H. virescens* (Lepidoptera: Noctuidae). *Ann. Entomol. Soc. Amer.* **58**, 600 (1965).

1094. Shorey, H. H., Gaston, L. K., and Saario, C. A. Sex pheromones of noctuid moths. XIV. Feasibility of behavioral control by disrupting pheromone communication in cabbage loopers. *J. Econ. Entomol.* **60**, 1541 (1967).

1095. Shorey, H. H. McFarland, S. U., and Gaston, L. K. Sex pheromones of noctuid moths. XIII. Changes in pheromone quantity, as related to reproductive age and mating history, in females of seven species of Noctuidae (Lepidoptera). *Ann. Entomol. Soc. Amer.* **61**, 372 (1968).

1096. Shorey, H. H., Morin, K. L., and Gaston, L. K. Sex pheromones of noctuid moths. XV. Timing of development of pheromone-responsiveness and other indicators of reproductive age in males of eight species. *Ann. Entomol. Soc. Amer.* **61**, 857 (1968).

1097. Siebold, C. T. von. Die Spermatozoën wirbellosen Tiere. IV. Die Spermatozoën in dem befruchteten Insektenweibchen. *Arch. Anat. Physiol.* p. 381 (1837).

1097a. Silhacek, D. L., Carlson, D. A., Mayer, M. S., and James, J. D. Composition and sex attractancy of cuticular hydrocarbons from houseflies: Effects of age, sex, and mating. *J. Insect Physiol.* **18**, 347 (1972).

1097b. Silhacek, D. L., Mayer, M. S., Carlson, D. A., and James, J. D. Chemical classification of a male housefly attractant. *J. Insect Physiol.* **18**, 43 (1972).

1098. Silverstein, R. M. Terpenes and insect behavior. *Insect-Plant Interactions, Rep. Conf. Univ. Calif., Santa Barbara, 1968* p. 73 (1968).

1099. Silverstein, R. M. Methodology for isolation and identification of insect pheromones—examples from Coleoptera. *In* "Control of Insect Behavior by Natural Products" (D. L. Wood, R. M. Silverstein, and M. Nakajima, eds.), p. 285. Academic Press, New York, 1970.

1100. Silverstein, R. M. Attractant pheromones of Coleoptera. *In* "Chemicals Controlling Insect Behavior" (M. Beroza, ed.), p. 21. Academic Press, New York, 1970.

1101. Silverstein, R. M., Brownlee, R. G., Bellas, T. E., Wood, D. L., and Browne, L. E. Brevicomin: Principal sex attractant in the frass of the female western pine beetle. *Science* **159**, 889 (1968).

1102. Silverstein, R. M., and Rodin, J. O. Spectrometric identification of organic compounds on a milligram scale. The use of complementary information. *Microchem. J.* **9**, 301 (1965).

1103. Silverstein, R. M., and Rodin, J. O. Insect pheromone collection with absorption columns. I. Studies on model organic compounds. *J. Econ. Entomol.* **59**, 1152 (1966).

1104. Silverstein, R. M., Rodin, J. O., Burkholder, W. E., and Gorman, J. E. Sex attractant of the black carpet beetle. *Science* **157**, 85 (1967).

1105. Silverstein, R. M., Rodin, J. O., and Wood, D. L. Sex attractants in frass produced by male *Ips confusus* in ponderosa pine. *Science* **154**, 509 (1966).

1106. Silverstein, R. M., Rodin, J. O., and Wood, D. L. Methodology for isolation and identification of insect pheromones with reference to studies on California five-spined *Ips*. *J. Econ. Entomol.* **60**, 944 (1967).

1107. Silverstein, R. M., Rodin, J. O., Wood, D. L., and Browne, L. E. Identification of two new terpene alcohols from frass produced by *Ips confusus* in ponderosa pine. *Tetrahedron* **22**, 1929 (1966).

1108. Simpson, J. Queen perception by honey bee swarms. *Nature (London)* **199**, 94 (1963).

1109. Singh, B. Studies toward the synthesis of the proposed structure for the cockroach sex attractant. *J. Org. Chem.* **31**, 181 (1966).

1110. Sladen, F. W. L. "The Humble-bee," pp. 12–14. Macmillan, New York. 1912.

1111. Slifer, E. H. The structure of arthropod receptors. *Annu. Rev. Entomol.* **15**, 121 (1970).

1112. Smissman, E. E. Some aspects of insect chemistry. *J. Pharm. Sci.* **54**, 1395 (1965).

1113. Smith, E. H. Laboratory rearing of the peach tree borer and notes on its biology. *J. Econ. Entomol.* **58**, 228 (1965).

1114. Smith, J. B. Notes on some structural peculiarities of *Sanninoidea exitiosa* Say. *Entomol. News* **9**, 114 (1898).

1115. Smith, L. B., and Hadley, C. H. The Japanese beetle. *U.S., Dep. Agr., Circ.* **363**, 1–66 (1926).

1116. Smith, S. G. A partial breakdown of temporal and ecological isolation between *Choristoneura* species (Lepidoptera: Tortricidae). *Evolution* **8**, 206 (1954).

1117. Smyth, T. Mating behavior of the Madeira cockroach. *Abstr. East. Br. Meet., Entomol. Soc. Amer.* p. 22 (1963).

1118. Snow, J. W., and Callahan, P. S. Laboratory mating studies of the corn earworm, *Heliothis zea* (Lepidoptera: Noctuidae). *Ann. Entomol. Soc. Amer.* **60**, 1066 (1967).

1119. Snow, J. W., and Copeland, W. W. Fall armyworm: Use of virgin female traps to detect males and to determine seasonal distribution. *U.S., Dep. Agr., Prod. Res. Rep.* **110**, 1–9 (1969).

1120. Solomon, J. D., and Morris, R. C. Sex attraction of the carpenterworm moth. *J. Econ. Entomol.* **59**, 1534 (1966).

1121. Soo Hoo, C. F., and Roberts, R. J. Sex attraction in *Rhopaea* (Coleoptera: Scarabaeidae). *Nature (London)* **205**, 724 (1965).

1122. Soule, C. G. Notes on the mating of *Attacus cecropia* and others. *Psyche* **9**, 224 (1901).

1123. Soule, C. G. Notes on hybrids of *Samia cynthia* and *Attacus promethea*. *Psyche* **9**, 411, (1902).

1124. Soule, C. G., and Elliott, P. L. As cited by Rau and Rau (*911*).

1125. Sower, L. L., Gaston, L. K., and Shorey, H. H. Parameters of sex pheromone communication distance for cabbage looper moths. *Abstr. Nat. Meet., Entomol. Soc. Amer., Miami Beach, Fla.* p. 55 (1970).

1125a. Sower, L. L., Gaston, L. K., and Shorey, H. H. Sex pheromones of noctuid moths. XXVI. Female release rate, male response threshold, and communication distance for *Trichoplusia ni*. *Ann. Entomol. Soc. Amer.* **64**, 1148 (1971).

1126. Sower, L. L., Shorey, H. H., and Gaston, L. K. Sex pheromones of noctuid moths. XXI. Light:dark cycle regulation and light inhibition of sex pheromone release by females of *Trichoplusia ni*. *Ann. Entomol. Soc. Amer.* **63**, 1090 (1970).

1127. Sower, L. L., Shorey, H. H., and Gaston, L. K. Sex pheromones of noctuid moths. XXV. Effects of temperature and photoperiod on circadian rhythms of sex pheromone release by females of *Trichoplusia ni*. *Ann. Entomol. Soc. Amer.* **64**, 488 (1971).

1128. Sparks, A. N. The use of infra-red photography to study mating of the European corn borer. *Proc. N. Cent. Br. Entomol. Soc. Amer.* **18**, 96 (1963).

1129. Spener, F., Mangold, H. K., Sansone, G., and Hamilton, J. G. Long-chain alkyl acetates in the preputial gland of the mouse. *Biochim. Biophys. Acta* **192**, 516 (1969).

1130. Spieth, H. T. Mating behavior within the genus *Drosophila* (Diptera). *Bull. Amer. Mus. Natur. Hist.* **99**, 397 (1952).

1131. Ställberg-Stenhagen, S., and Stenhagen, E. Chemical communication in the insect world. *Sv. Kem. Tidskr.* **80**, 178 (1968).

1132. Standfuss, M., "Handbuch der Paläarktischen Gross-Schmetterlinge für Forscher und Sammler," 2nd ed. Fischer, Jena, 1896.

1133. Stanić, V., Zlotkin, E., and Shulov, A. Localization of pheromone excretion in the female of *Trogoderma granarium* (Dermestidae). *Entomol. Exp. Appl.* **13**, 342 (1970).

1133a. Stark, R. W. Substances attractives chez les Scolytides. *Mitt. Schweiz. Entomol. Ges.* **41**, 245 (1968).

1134. Starks, K. J., Callahan, P. S., McMillian, W. W., and Cox, H. C. A photoelectric counter to monitor olfactory response of moths. *J. Econ. Entomol.* **59**, 1015 (1966).

1135. Statler, M. W. Effects of gamma radiation on the ability of the adult female gypsy moth to attract males. *J. Econ. Entomol.* **63**, 163 (1970).

1136. Stefanović, D. K., Grujić-Injac, B., and Mićić, D. Effect of optical activity on the synthetic of the sexual odor of the gypsy moth female (gyptol and the higher homolog of gyplure). *Zast. Bilja* **73**, 235 (1963).

1137. Stefanović, G., and Grujić, B. Chemical investigation of the active substance from the female gypsy moth. II. *Zast. Bilja* **56**, 94 (1959).

1138. Stefanović, G., Grujić, B., and Prekajski, P. Short communication from the Institute of Chemistry of the Faculty of Science on the chemical investigation of the active substance from the female gypsy moth. *Zast. Bilja* **52–53**, 176 (1959).

1139. Stefanović, G., Grujić-Injac, B., and Mićić, D. Preparation of chromatographically pure (+)-12-acetoxy-*cis*-9-octadecene-1-ol (gyplure). *Glas* **257**, 175 (1965).

1140. Stefanović, G., Grujić-Injac, D. B., Petrović, D., Milovanović, J., and Lajsić, S. Gyplure and gyptol, synthetic attractive substances. *Bull. Acad. Serbe Sci. Arts, Cl. Sci. Math. Natur., Sci. Natur.* **12**, 73 (1969).

1141. Stefanović, G., Grujić-Injac, B., Petrović, D., Milovanović, J., and Lajsić, S. On the synthetic sex attractants gyplure and gyptol. *Glas* **276**, 113 (1969).

1142. Stein, G. Über den Sexuallockstoff von Hummelmännchen. *Naturwissenschaften* **50**, 305 (1963).

1143. Stein, G. Untersuchungen über den Sexuallockstoff der Hummelmännchen. *Biol. Zentralbl.* **82**, 343 (1963).

1144. Steinbrecht, R. A. Feinstruktur und Histochemie der Sexualduftdrüse des Seidenspinners *Bombyx mori* L. *Z. Zellforsch. Mikrosk. Anat.* **64**, 227 (1964).

1145. Steinbrecht, R. A. Die Abhängigkeit der Lockwirkung des Sexualduftorgans weiblicher Seidenspinner (*Bombyx mori*) von Alter und Kopulation. *Z. Vergl. Physiol.* **48**, 341 (1964).

1146. Steinbrecht, R. A. Zur Morphometrie der Antenne des Seidenspinners, *Bombyx mori* L.: Zahl und Verteilung der Riechsensillen (Insecta, Lepidoptera). *Z. Morphol. Oekal. Tiere* **68**, 93 (1970).

1147. Steinbrecht, R. A., and Schneider, D. Die Faltung der äusseren Zellmembran in den Sexuallockstoff-Drüsenzellen des Seidenspinners. *Naturwissenschaften* **51**, 41 (1964).

1148. Stockel, J. Utilisation du piégeage sexuel pour l'étude du déplacement de l'alucite *Sitotroga cerealella* (Lepidoptère: Gelechiidae) vers les cultures de mais. *Entomol. Exp. Appl.* **14**, 39 (1971).

1149. Stoll, M., and Flament, I. Synthèse du "Propylure," phéromone sexuelle de *Pectinophora gossypiella* Saunders. *Helv. Chim. Acta* **52**, 1996 (1969).

1150. Stowell, J. C. A short synthesis of the sex pheromone of the pink bollworm moth. *J. Org. Chem.* **35**, 244 (1970).

1151. Strang, G. E. A study of honey bee drone attraction in the mating response. *J. Econ. Entomol.* **63**, 641 (1970).

1152. Strong, F. E. A computer-generated model to simulate mating behavior of lygus bugs. *J. Econ. Entomol.* **64**, 46 (1971).

1153. Strong, F. E., Sheldahl, J. A., Hughes, P. R., and Hussein, E. M. K. Reproductive biology of *Lygus hesperus* Knight. *Hilgardia* **40**, 122 (1970).

1154. Struble, D. L. A sex pheromone in the forest tent caterpillar. *J. Econ. Entomol.* **63**, 295 (1970).

1155. Struble, D. L., and Jacobson, L. A. A sex pheromone in the red-backed cutworm. *J. Econ. Entomol.* **63**, 841 (1970).

1156. Stuart, A. M. The role of chemicals in termite communication. *In* "Communication by Chemical Signals" (J. W. Johnston, Jr., D. G. Moulton, and A. Turk, eds.), p. 88. Appleton, New York, 1970.

1157. Stürckow, B. The electroantennogram (EAG) as an assay for the reception of odours by the gypsy moth. *J. Insect Physiol.* **11**, 1573 (1965).

1158. Stürckow, B., and Bodenstein, W. G. Location of the sex pheromone in the American cockroach, *Periplaneta americana* (L.). *Experienta* **22**, 851 (1966).

1159. Sturtevant, A. H. Experiments on sex recognition and the problem of sexual selection in *Drosophila*. *J. Anim. Behav.* **5**, 351 (1915).

1160. Suter, P. R. Biologie von *Echidnophaga gallinacea* (Westw.) und Vergleich mit andern Verhaltenstypen bei Flöhen. *Acta Trop.* **21**, 193 (1964).

1161. Sutton, G. P. Zygaenidae attracted by *Lasiocampa quercus* female. *Entomologist* **55**, 280 (1922).

1162. Syrjämäki, J., and Ulmanen, I. Further experiments on male sexual behaviour in *Stictochironomus crassiforceps* (Kieff.) (Diptera: Chironomidae). *Ann. Zool. Fenn.* **7**, 216 (1970).

1163. Takahashi, S., and Kitamura, C. An apparatus for bioassaying the pheromones of moths. *Bochu-Kagaku* **35**, 130 (1970).

1164. Takahashi, S., Kitamura, C., and Kuwahara, Y. The comparative pheromone activity of acetates of unsaturated alcohol to males of the almond moth. *Bochu-Kagaku* **36**, 24 (1971).

1165. Takahashi, S., Kitamura, C., Kuwahara, Y., and Fukami, H. Studies on sex pheromones of Pyralididae. II. Mass rearing of virgin females of the almond moth, *Cadra cautella* Walker (Phycitinae). *Bochu-Kagaku* **33**, 163 (1968).

1166. Tamaki, G., Butt, B. A., and Landis, B. J. Arrest and aggregation of male *Myzus persicae* (Hemiptera: Aphididae). *Ann. Entomol. Soc. Amer.* **63**, 955 (1970).

1167. Tamaki, Y., Noguchi, H., and Yushima, T. Mating behavior of the smaller tea tortrix, *Adoxophyes orana* Fischer von Röslerstamm and evidence of sex pheromone production. *Bochu-Kagaku* **34**, 97 (1969).

1168. Tamaki, Y., Noguchi, H., and Yushima, T. Attractiveness of black light, virgin female and sex pheromone extract for the smaller tea tortrix. *Adoxophyes orana* Fischer von Röslerstamm. *Bochu-Kagaku* **34**, 102 (1969).

1169. Tamaki, Y., Noguchi, H., and Yushima, T. Artificial control of mating activity of the smaller tea tortrix, *Adoxophyes orana* Fischer von Röslerstamm, and a quantitative bioassay for the sex pheromone. *Bochu-Kagaku* **34**, 107 (1969).

1169a. Tamaki, Y., Noguchi, H., Yushima, T., and Hirano, C. Two sex pheromones of the smaller tea tortrix: Isolation, identification, and synthesis. *Appl. Entomol. Zool.* **6**, 139 (1971).

1169b. Tamaki, Y., Noguchi, H., Yushima, T., Hirano, C., Honma, K., and Sugawara, H. Sex pheromone of the summerfruit tortrix: Isolation and identification. *Konchu* **39**, 338 (1971).

1170. Taschenberg, E. P. Die Insekten, Tausendfüssler und Spinnen. *In* "Tierleben" (J. A. Brehm), 2nd ed., Vol. 9, p. 75. Verlag Bibliographischen Instituts, Leipzig. 1877.

1171. Tashiro, H., and Beavers, J. B. Growth and development of the California red scale, *Aonidiella aurantii*. *Ann. Entomol. Soc. Amer.* **61**, 1009 (1968).

1172. Tashiro, H., Beavers, J. B., and Moreno, D. Comparative response of two strains of California red scale, *Aonidiella aurantii*, males to pheromone extract and to females of the reciprocal strain. *Ann. Entomol. Soc. Amer.* **62**, 279 (1969).

1173. Tashiro, H., and Chambers, D. L. Reproduction in the California red scale, *Aonidiella aurantii* (Homoptera: Diaspididae). I. Discovery and extraction of a female sex pheromone. *Ann. Entomol. Soc. Amer.* **60**, 1166 (1967).

1174. Tashiro, H., Chambers, D. L., Moreno, D., and Beavers, J. Reproduction in the California red scale, *Aonidiella aurantii*. III. Development of an olfactometer for bioassay of the female sex pheromone. *Ann. Entomol. Soc. Amer.* **62**, 935 (1969).

1175. Tashiro, H., and Moffitt, C. Reproduction in the California red scale, *Aonidiella aurantii*. II. Mating behavior and postinsemination changes. *Ann. Entomol. Soc. Amer.* **61**, 1014 (1968).

1176. Teetes, G. L., and Randolph, N. M. Color preference and sex attraction among sunflower moths. *J. Econ. Entomol.* **63**, 1358 (1970).

1177. Teudt, H. Eine Erklärung der Geruchserscheinungen. *Biol. Zentralbl.* **33**, 716 (1913).

1177a. Thomas, L. Notes of a biology watcher: A fear of pheromones. *Sciences* **11** (10), 4 (1971).

1178. Tinbergen, N. An objectivistic study of the innate behaviour of animals. *Bibl. Biotheor. (Leiden), ser. D* **1**, 1–98 (1942).

1179. Tinbergen, N. "Social Behaviour in Animals," p. 150. Methuen, London, 1953.

1180. Tinbergen, N. "Animal Behavior," pp. 62–65 and 74–75. Time, Inc., New York, 1965.

1181. Tinbergen, N., Meeuse, B. J. D., Boerema, L. K., and Varossieau, W. W. Die Balz des Samtfalters, *Eumenis (Satyrus) semele* (L.). *Z. Tierpsychol.* **5**, 182 (1943).

1182. Titschack, E. Beiträge zu einer Monographie der Kleidermotte, *Tineola biselliella* Hum. *Z. Tech. Biol.* **10**, 1 (1922).

1183. Toba, H. H., Green, N., Kishaba, A. N., Jacobson, M., and Debolt, J. W. Response of male cabbage loopers to 15 isomers and congeners of the looper pheromone. *J. Econ. Entomol.* **63**, 1048 (1970).

1184. Toba, H. H., Kishaba, A. N., and Wolf, W. W. Bioassay of the synthetic female sex pheromone of the cabbage looper. *J. Econ. Entomol.* **61**, 812 (1968).

1185. Toba, H. H., Kishaba, A. N., and Wolf, W. W. A polyethylene bag for dispensing the synthetic sex pheromone of the cabbage looper. *J. Econ. Entomol.* **62**, 517 (1969).

1186. Toba, H. H., Kishaba, A. N., Wolf, W. W., and Gibson, T. Spacing of screen traps baited with synthetic sex pheromone of the cabbage looper. *J. Econ. Entomol.* **63**, 197 (1970).

1186a. Tobin, E. N., and Smith, L. W., Jr. Note on the mating behavior of the cigarette beetle [*Lasioderma serricorne* (F.): Anobiidae]. *Entomol. News* **82** (1), 23 (1971).

1187. Todd, J. H. The chemical languages of fishes. *Sci. Amer.* **224** (5), 98 and 108 (1971).

1188. Tomida, I. Sex attractant of insects. *J. Soc. Org. Syn. Chem., Tokyo* **23**, 577 (1965).

1189. Tomida, I. Pheromone and its application to agricultural chemicals. *Jap. Agr. Res. Quart.* **3** (3), 14 (1968).

1190. Tomida, I., and Ishii, S. Sex pheromone of the eri-silkworm moth, *Philosamia cynthia ricini* Donovan (Lepidoptera: Saturniidae). *Appl. Entomol. Zool.* **3**, 103 (1968).

1191. Torgersen, T. R. The bionomics of the European pine shoot moth *Rhyacionia buoliana* (Schiffermüller) (Lepidoptera: Tortricidae) in Wisconsin. Ph.D. Thesis, University of Wisconsin, Madison (1964).

1192. Travis, B. V. Habits of the June beetle, *Phyllophaga lanceolata* (Say) in Iowa. *J. Econ. Entomol.* **32**, 690 (1939).

1193. Traynier, R. M. M. Sex attraction in the Mediterranean flour moth, *Anagasta kühniella:* Location of the female by the male. *Can. Entomol.* **100**, 5 (1968).

1194. Traynier, R. M. M. Habituation of the response to sex pheromone in two species of Lepidoptera, with reference to a method of control. *Entomol. Exp. Appl.* **13**, 179 (1970).

1195. Traynier, R. M. M. Sexual behaviour of the Mediterranean flour moth, *Anagasta kühniella:* Some influences of age, photoperiod, and light intensity. *Can. Entomol.* **102**, 534 (1970).

1196. Traynier, R. M. M., and Burton, D. J. Male response to females in the marsh crane fly, *Tipula paludosa* Mg. (Diptera: Tipulidae). *J. Entomol. Soc. Brit. Columbia* **67**, 21 (1970).

1197. Tribble, M. T. Insect sex atractants. Part I. Synthesis of the sex attractant of the queen honey bee (*Apis mellifera*) and related substances. Part II. Investigation of the sex attractant of the sugarcane borer moth (*Diatraea saccharalis* (F.)). Master's Thesis, Dept. Chem., Louisiana State University, Baton Rouge (1966).

1198. Truscheit, E., and Eiter, K. Synthese der vier isomeren Hexadecadien- (10.12)-ole-(1). *Justus Liebig's Ann. Chem.* **658**, 65 (1962).

1199. Truscheit, E., and Eiter, K. Ungesättigte aliphatische Alkoholen. Ger. Pat. 1,163,313 (1964).

1200. Truscheit, E., Eiter, K., Butenandt, A., and Hecker, E. Verfahren zur Herstellung sterisch einheitlicher Dien-(10,12)-Ole-(1). Aust. Pat. 223,182 (1962).

1201. Truscheit, E., Eiter, K., Butenandt, A., and Hecker, E., Verfahren zur Herstellung von 10-*trans*, 12-*cis*- oder 10-*cis*,12-*cis*-Dien-(10,12)-olen- (1). Ger. Pat. 1,138,037 (1962).

1202. Truscheit, E., Haberland, G., Puls, W., and Schlossmann, K. 10-Hydroxy-7-hexadecyn-1-oic acid. Fr. Pat. 1,357,951 (1964).

1203. Tschinkel, W. R. Chemical studies on the sex pheromone of *Tenebrio molitor* (Coleoptera: Tenebrionidae). *Ann. Entomol. Soc. Amer.* **63,** 626 (1970).

1204. Tschinkel, W. R., Willson, C., and Bern, H. A. Sex pheromone of the mealworm beetle (*Tenebrio molitor*). *J. Exp. Zool.* **164,** 81 (1967).

1205. Tumlinson, J. H. III. Isolation, identification and partial synthesis of the boll weevil sex attractant. Ph.D. Thesis, Chem. Dept., Mississippi State University, State College, Mississippi (1969).

1206. Tumlinson, J. H., Gueldner, R. C., Hardee, D. D., Thompson, A. C., Hedin, P. A., and Minyard, J. P. The boll weevil sex attractant. *In* "Chemicals Controlling Insect Behavior" (M. Beroza, ed.), p. 41. Academic Press, New York, 1970.

1207. Tumlinson, J. H., Gueldner, R. C., Hardee, D. D., Thompson, A. C., Hedin, P. A., and Minyard, J. P. Identification and synthesis of the four compounds comprising the boll weevil sex attractant. *J. Org. Chem.* **36,** 2616 (1971).

1208. Tumlinson, J. H., Hardee, D. D., Gueldner, R. C., Thompson, A. C., Hedin, P. A., and Minyard, J. P. Sex pheromones produced by male boll weevil: Isolation, identification, and synthesis. *Science* **166,** 1010 (1969).

1209. Tumlinson, J. H., Hardee, D. D., Minyard, J. P., Thompson, A. C., Gast, R. T., and Hedin, P. A. Boll weevil sex attractant: Isolation studies. *J. Econ. Entomol.* **61,** 470 (1968).

1210. Tunstall, J. P. Sex attractant studies in *Diparopsis. Pest Art. News Summ., Ser. A* **11,** 212 (1965).

1211. Tvermyr, S. Sex pheromone in females of *Erannis aurantiaria* Hb. and *Erannis defoliaria* Cl. (Lep., Geometridae). *Nor. Entomol. Tidsskr.* **16,** 25 (1969).

1212. Urbahn, E. Abdominale Duftorgane bei weiblichen Schmetterlingen. *Jena. Z. Naturwiss.* **59,** 277 (1913).

1213. Valentine, J. M. The olfactory sense of the adult mealworm beetle *Tenebrio molitor* (Linn.) *J. Exp. Zool.* **58,** 165 (1931).

1214. Vareschi, E., and Kaissling, K. E. Dressur von Bienenarbeiterinnen und Drohnen auf Pheromone und andere Duftstoffe. *Z. Vergl. Physiol.* **66,** 22 (1970).

1215. Velasquez Antich, A. Atraccion por olor en ninfas y adultos de *Rhodnius prolixus* (Stal.). *Rev. Inst. Med. Trop. Sao Paulo* **10,** 242 (1968).

1216. Velthuis, H. H. W. Queen substances from the abdomen of the queen honey bee. *Z. Vergl. Physiol.* **70,** 210 (1970).

1217. Vick, K. W., Burkholder, W. E., and Gorman, J. E. Interspecific response to sex pheromones of *Trogoderma* species (Coleoptera: Dermestidae). *Ann. Entomol. Soc. Amer.* **63,** 379 (1970).

1218. Vité, J. P. Die Wirkung pflanzen- und insekteneigener Lockstoffe auf *Pityophthorus* und *Pityogenes* (Coleoptera: Scolytidae). *Naturwissenschaften* **52,** 267 (1965).

1219. Vité, J. P. Sex attractants in frass from bark beetles. *Science* **156,** 105 (1967).

1220. Vité, J. P. Pest management systems using synthetic pheromones. *Contrib. Boyce Thompson Inst.* **24,** 343 (1970).

1221. Vité, J. P., and Crozier, R. G. Studies on the attack behavior of the southern pine beetle. IV. Influence of host condition on aggregation pattern. *Contrib. Boyce Thompson Inst.* **24,** 87 (1968).

1222. Vité, J. P., and Gara, R. I. A field method for observation on olfactory responses of bark beetles (Scolytidae) to volatile materials. *Contrib. Boyce Thompson Inst.* **21,** 175 (1961).

1223. Vité, J. P., and Gara, R. I. Volatile attractants from ponderosa pine attacked by bark beetles (Coleoptera, Scolytidae). *Contrib. Boyce Thompson Inst.* **21,** 251 (1962).

1224. Vité, J. P., Gara, R. I., and Kliefoth, R. A. Collection and bioassay of a volatile fraction attractive to *Ips confusus* (Lec.) (Coleoptera: Scolytidae). *Contrib. Boyce Thompson Inst.* **22,** 39 (1963).

1225. Vité, J. P., Gara, R. I., and von Scheller, H. D. Field observations on the response to attractants of bark beetles infesting southern pines. *Contrib. Boyce Thompson Inst.* **22,** 461 (1964).

1226. Vité, J. P., and Pitman, G. B. Bark beetle aggregation: Effects of feeding on the release of pheromones in *Dendroctonus* and *Ips. Nature (London)* **218,** 169 (1968).

1227. Vité, J. P., and Pitman, G. B. Insect and host odors in the aggregation of the western pine beetle. *Can. Entomol.* **101,** 113 (1969).

1228. Vité, J. P., and Pitman, G. B. Aggregation behavior of *Dendroctonus brevicomis* in response to synthetic pheromones. *J. Insect Physiol.* **15,** 1617 (1969).

1229. Vité, J. P., and Pitman, G. B. Management of western pine beetle populations: Use of chemical messengers. *J. Econ. Entomol.* **63,** 1132 (1970).

1230. Vité, J. P., Pitman, G. B., and Renwick, J. A. A. Bark beetle attractants: Insect and host compounds in mass aggregation. *Abstr. Div. Pestic. Chem., Amer. Chem. Soc. Nat. Meet., 1970* Pap. no 22.

1231. Vité, J. P., and Renwick, J. A. A. Insect and host factors in the aggregation of the southern pine beetle. *Contrib. Boyce Thompson Inst.* **24,** 61 (1968).

1232. Vité, J. P., and Renwick, J. A. A. Differential diagnosis and isolation of population attractants. *Contrib. Boyce Thompson Inst.* **24,** 323 (1970).

1233. Vité, J. P., and Williamson, D. L. *Thanasimus dubius:* Prey perception. *J. Insect Physiol.* **16,** 233 (1970).

1234. Voelkel, H. Zur Biologie und Bekämpfung des Khrapräkäfers, *Trogoderma granarium* Everts. *Arch. Biol. Reichsanst.* **13,** 129 (1924).

1235. Vogel, R. Zur Kenntnis der Geruchsorgane der Wespen und Bienen. *Zool. Anz.* **53,** 20 (1921).

1236. Vöhringer, K. Zur Biologie der grossen Wachsmotte (*Galleria mellonella* L.), III. Morphologische und biologische Untersuchungen am Falter der grossen Wachsmotte (*Galleria mellonella* L.). *Zool. Jahrb., Abt. Anat. Ontog. Tiere* **58,** 275 (1934).

1237. Volkov, Yu P., Poleschchuk, V. D., Zharov, V. G., and Vashkov, V. I. Study of the sex attractant substances of the female rust-colored cockroach (*Blattella germanica* L.). *Med. Parazitol. Parazit. Bolez.* **36,** 45 (1967).

1238. Wagner, W. Anlockung der Schlupfwespen-Männchen durch Weibchen, die noch im Cocon sassen. *Z. Wiss. Insektenbiol.* **5,** 245 (1909).

1239. Wakabayashi, N. A new synthesis of 2,2-dimethyl-3-isopropylidenecyclopropyl propionate. *J. Org. Chem.* **32,** 489 (1967).

1240. Waku, Y., and Sumimoto, K. Ultrastructure and secretory mechanism of the alluring gland cell in the silkworm, *Bombyx mori* L. (Lepidoptera: Bombycidae). *Appl. Entomol. Zool.* **4,** 135 (1969).

1241. Warren, F. L. The sex pheromones of insects. *J. S. Afr. Chem. Inst.* **22,** Spec. Issue, S150 (1969).

1242. Warthen, D. Synthesis of *cis*-9-tetradecen-1-ol acetate, the sex pheromone of the fall armyworm. *J. Med. Chem.* **11,** 371 (1968).

1243. Warthen, D., and Green, N. Insect sex attractants. XI. Analysis of cis and trans isomers of fatty alcohol acetates by gas chromatography. *J. Amer. Oil Chem. Soc.* **46,** 191 (1969).

1244. Warthen, D., and Jacobson, M. Insect sex attractants. VII. 5,9- Tridecadien-1-ol acetates. *J. Med. Chem.* **10**, 1190 (1967).

1245. Warthen, D., and Jacobson, M. Insect sex attractants. X. 5-Dodecen-1-ol acetates, analogs of the cabbage looper sex attractant. *J. Med. Chem.* **11**, 373 (1968).

1246. Warthen, J. D., Jr., Rudrum, M., Moreno, D. S., and Jacobson, M. *Aonidiella aurantii* sex pheromone isolation. *J. Insect Physiol.* **16**, 2207 (1970).

1247. Wasserman, H. H., and Barber, E. H. Carbonyl epoxide rearrangements. Synthesis of brevicomin and related [3.2.1] bicyclic systems. *J. Amer. Chem. Soc.* **91**, 3674 (1969).

1248. Waters, R. M., and Jacobson, M. Attractiveness of gyplure masked by impurities. *J. Econ. Entomol.* **58**, 370 (1965).

1249. Waters, R. M., and Jacobson, M. A rack for holding male gypsy moths during laboratory bioassays of airborne attractants. *J. Econ. Entomol.* **61**, 873 (1968).

1249a. Weatherston, J. Aspects of the chemical defenses of millipedes. *Proc. Int. Congr. Pestic. Chem., 2nd, 1971* p. 183 (1971).

1250. Weatherston, J., and Percy, J. E. Studies of physiologically active arthropod secretions. I. Evidence for a sex pheromone in female *Vitula edmandsae* (Lepidoptera: Phycitidae). *Can. Entomol.* **100**, 1065 (1968).

1251. Weatherston, J., and Percy, J. E. Studies of physiologically active arthropod secretions. IV. Topography of the sex pheromone-producing gland of the eastern spruce budworm, *Choristoneura fumiferana* (Clem.) (Lepidoptera: Tortricidae). *Can. J. Zool.* **48**, 569 (1970).

1251a. Weatherston, J., Roelofs, W., Comeau, A., and Sanders, C. J. Studies of physiologically active arthropod secretions. X. Sex pheromone of the eastern spruce budworm, *Choristoneura fumiferana* (Lepidoptera: Tortricidae). *Can. Entomol.* **103**, 1741 (1971).

1251b. Weatherston, J., Tyrrell, D., and Percy, J. E. Long-chain alcohol acetates in the defensive secretion of the millipede *Blaniulus guttulatus. Chem. Phys. Lipids* **7**, 98 (1971).

1252. Weismann, A. Über Duftschuppen. *Zool. Anz.* **1**, 98 (1878).

1253. Wellso, S. G. Sexual attraction and biology of *Xenorhipis brendeli* (Coleoptera: Buprestidae). *J. Kans. Entomol. Soc.* **39**, 242 (1966).

1254. Wendt, H. "The Sex Life of the Animals," pp. 131–133 and 146. Simon & Schuster, New York, 1965.

1255. Werner, R. A. Factors influencing the olfactory response of *Ips grandicollis* to frass and frass extracts from males attacking loblolly pine. *Abstr. Nat. Meet., Entomol. Soc. Amer., Miami Beach, Fla.* p. 57 (1970).

1256. Wharton, D. R. A., Black, E. D., and Merritt, C., Jr. Sex attractant of the American cockroach. *Science* **142**, 1257 (1963).

1257. Wharton, D. R. A., Black, E. D., Merritt, C., Jr., Wharton, M. L., Bazinet, M., and Walsh, J. T. Isolation of the sex attractant of the American cockroach. *Science* **137**, 1062 (1962).

1258. Wharton, D. R. A., Miller, G. L., and Wharton, M. L. The odorous attractant of the American cockroach, *Periplaneta americana* (L.). I. Quantitative aspects of the response to the attractant. *J. Gen. Physiol.* **37**, 461 (1954).

1259. Wharton, D. R. A., Miller, G. L., and Wharton, M. L. The odorous attractant of the American cockroach, *Periplaneta americana* (L.). II. A bioassay method for the attractant. *J. Gen. Physiol.* **37**, 471 (1954).

1260. Wharton, M. L., and Wharton, D. R. A. The production of sex attractant substance and of oothecae by the normal and irradiated American cockroach, *Periplaneta americana* L. *J. Insect Physiol.* **1**, 229 (1957).

1261. White, L. D., and Hutt, R. B. Codling moth catches in sex and light traps after exposure to 0, 25, or 40 krad of gamma irradiation. *J. Econ. Entomol.* **64**, 1249 (1971).

1262. Whiting, P. W. Reproductive reactions of sex mosaics of a parasitic wasp, *Habrobracon juglandis. J. Comp. Psychol.* **14**, 345 (1932).

1263. Whitten, W. K. Pheromones and mammalian reproduction. *Advan. Reprod. Physiol.* **1**, 155 (1966).

1264. Wilde, J. de, Lambers-Suverkropp, K. H. R., and van Tol, A. Responses to air flow and airborne plant odour in the Colorado beetle. *Neth. J. Plant Pathol.* **75**, 53 (1969).

1265. Wilkinson, R. C. Attraction and development of *Ips* bark beetle populations in artificially infested pine bolts exposed on firetowers and turntables in Florida. *Fla. Entomol.* **47**, 58 (1964).

1266. Willcocks, F. C. "The Insects and Related Pests of Egypt." Vol. 1, Part 2, p. 546. Cairo, Egypt, 1937.

1267. Williamson, D. L. Olfactory discernment of prey by *Medetera bistriata* (Diptera: Dolichopodidae). *Ann. Entomol. Soc. Amer.* **64**, 586 (1971).

1268. Wilson, E. O. Pheromones. *Sci. Amer.* **208** (5), 100 (1963).

1269. Wilson, E. O. Chemical communication in the social insects. *Science* **149**, 1064 (1965).

1270. Wilson, E. O. Chemical systems. *In* "Animal Communication, Techniques of Study and Results of Research" (T. A. Sebeok, ed.), p. 75. Univ. of Indiana Press, Bloomington, 1968.

1271. Wilson, E. O. Chemical communication within animal species. *In* "Chemical Ecology" (E. Sondheimer and J. B. Simeone, eds.), p. 133. Academic Press, New York, 1970.

1272. Wilson, E. O., and Bossert, W. H. Chemical communication among animals. *Recent Progr. Horm. Res.* **19**, 673 (1963).

1273. Wilson, E. O., Bossert, W. H., and Regnier, F. E. A general method for estimating threshold concentrations of odorant molecules. *J. Insect Physiol.* **15**, 597 (1969).

1274. Wilson, K. H. Sex attractants in moths. *Proc. Int. Congr. Entomol., 10th, 1956* Vol. 2, p. 4 (1958).

1275. Withycombe, C. L. Notes on the biology of some British neuroptera (Planipennia). *Trans. Roy. Entomol. Soc. London* **74**, 501 (1922).

1276. Wolbarsht, M. L. Receptor sites in insect chemoreceptors. *Cold Spring Harbor Symp. Quant. Biol.* **30**, 281 (1965).

1277. Wolf, W. W., Hartsock, J. G., Ford, J. H., Henneberry, T. J., Hills, O. A., and Debolt, J. W. Combined use of sex pheromone and electric traps for cabbage looper control. *Trans. Amer. Soc. Agr. Eng.* **12**, 329 (1969).

1278. Wolf, W. W., Kishaba, A. N., Howland, A. F., and Henneberry, T. J. Sand as a carrier for synthetic sex pheromone of cabbage loopers used to bait blacklight and carton traps. *J. Econ. Entomol.* **60**, 1182 (1967).

1279. Wolf, W. W., Kishaba, A. N., and Toba, H. H. Proposed method for determining density of traps required to reduce an insect population. *J. Econ. Entomol.* **64**, 872 (1971).

1280. Wong, T. T. Y., Cleveland, M. L., and Davis, D. G. Sex attraction and mating of lesser peach tree borer moths. *J. Econ. Entomol.* **62**, 789 (1969).

1281. Wong, T. T. Y., Cleveland, M. L., Ralston, D. F., and Davis, D. G. Virgin female traps to determine activity and populations of red-banded leaf rollers. *J. Econ. Entomol.* **64,** 132 (1971).

1282. Wong, T. T. Y., Cleveland, M. L., Ralston, D. F., and Davis, D. G. Time of sexual activity of codling moths in the field. *J. Econ. Entomol.* **64,** 553 (1971).

1283. Wong, T. T. Y., Kamasaki, H., Dolphin, R. E., Cleveland, M. L., Ralston, D. F., Davis, D. G., and Mouzin, T. E. Distribution and abundance of the lesser peachtree borer on Washington Island, Wisconsin. *J. Econ. Entomol.* **64,** 879 (1971).

1283a. Wong, T. T. Y., Mouzin, T. E., Ralston, D. F., Davis, D. G., Burnside, J. A., and Dolphin, R. E. Populations of codling moths on Washington Island, Wisconsin, in 1970. *J. Econ. Entomol.* **64,** 1411 (1971).

1284. Wood, D. L. The attraction created by males of a bark beetle *Ips confusus* (LeConte) attacking ponderosa pine (Coleoptera: Scolytidae). *Pan-Pac. Entomol.* **38,** 141 (1962).

1285. Wood, D. L. Bark beetle pheromones. *Insect-Plant Interactions, Rep. Conf. Univ. Calif., Santa Barbara, 1968* p. 88 (1962).

1286. Wood, D. L. Pheromones of bark beetles. *In* "Control of Insect Behavior by Natural Products" (D. L. Wood, R. M. Silverstein, and M. Nakajima, eds.), p. 301. Academic Press, New York, 1970.

1287. Wood, D. L., Bedard, W. D., and Browne, L. E. Synergism and blocking of the *Dendroctonus brevicomis* pheromones. *Abstr. Nat. Meet., Entomol. Soc. Amer., Miami Beach, Fla.* p. 57 (1970).

1288. Wood, D. L., Browne, L. E., Bedard, W. D., Tilden, P. E., Silverstein, R. M., and Rodin, J. O. Response of *Ips confusus* to synthetic sex pheromones in nature. *Science* **159,** 1373 (1968).

1289. Wood, D. L., Browne, L. E., Silverstein, R. M., and Rodin, J. O. Sex pheromones of bark beetles. I. Mass production, bioassay, source, and isolation of the sex pheromone of *Ips confusus* (LeC.) *J. Insect Physiol.* **12,** 523 (1966).

1290. Wood, D. L., and Bushing, R. W. The olfactory response of *Ips confusus* (LeConte) (Coleoptera: Scolytidae) to the secondary attraction in the laboratory. *Can. Entomol.* **95,** 1066 (1963).

1291. Wood, D. L., and Silverstein, R. M. Bark beetle pheromones. *Nature (London)* **225,** 557 (1970).

1292. Wood, D. L., Silverstein, R. M., and Nakajima, M. Pest control. *Science* **164,** 203 (1969).

1293. Wood, D. L., Silverstein, R. M., and Rodin, J. O. Sex attractants in frass from bark beetles. *Science* **156,** 105 (1967).

1294. Wood, D. L., Stark, R. W., Silverstein, R. M., and Rodin, J. O. Unique synergistic effects produced by the principal sex attractant compounds of *Ips confusus* (LeConte) (Coleoptera: Scolytidae). *Nature (London)* **215,** 206 (1967).

1295. Wood, D. L., and Vité, J. P. Studies on the host selection behavior of *Ips confusus* (LeConte) (Coleoptera, Scolytidae) attacking *Pinus ponderosa*. *Contrib. Boyce Thompson Inst.* **21,** 79 (1961).

1296. Wray, C., and Farrier, M. H. Response of the Nantucket pine tip moth to attractants. *J. Econ. Entomol.* **56,** 714 (1963).

1297. Wright, D. Female *Spilosoma lutea* Hufn. attracting *Arctia caia* L. *Entomol. Rec.* **64,** 24 (1952).

1298. Wright, R. H. Odour and chemical constitution. *Nature (London)* **173,** 831 (1954).

1299. Wright, R. H. A theory of olfaction and of the action of mosquito repellent. *Can. Entomol.* **89,** 518 (1957).
1300. Wright, R. H. The olfactory guidance of flying insects. *Can. Entomol.* **90,** 81 (1958).
1301. Wright, R. H. How insects follow a scent. *New Sci.* **14,** 339 (1962).
1302. Wright, R. H. Molecular vibration and insect sex attractants. *Nature (London)* **198,** 455 (1963).
1303. Wright, R. H. Chemical control of chosen insects. *New Sci.* **20,** 598 (1963).
1304. Wright, R. H. Insect control by nontoxic means. *Science* **144,** 487 (1964).
1305. Wright, R. H. Olfactory guidance of insects. *Can. Entomol.* **96,** 146 (1964).
1306. Wright, R. H. Odor and molecular vibration: The far infrared spectra of some perfume chemicals. *Ann. N. Y. Acad. Sci.* **116,** 552 (1964).
1307. Wright, R. H. "The Science of Smell." Basic Books, New York, 1964.
1308. Wright, R. H. After pesticides—what? *Nature (London)* **204,** 121 (1964).
1309. Wright, R. H. "Metarchon," a new term for a class of nontoxic pest control agents. *Nature (London)* **204,** 603 (1964).
1310. Wright, R. H. Finding metarchons for pest control. *Nature (London)* **207,** 103 (1965).
1311. Wright, R. H. An insect olfactometer. *Can. Entomol.* **98,** 282 (1966).
1312. Wright, R. H. Primary odors and insect attraction. *Can. Entomol.* **98,** 1083 (1966).
1313. Wright, R. H. How animals distinguish odours. *Science J.* **4** (7), 57 (1968).
1314. Wright, R. H. Some alternatives to insecticides. *Pesticide Sci.* **1,** 24 (1970).
1315. Wright, R. H. Correlation of far infrared spectra and Mediterranean fruit fly (Diptera: Tephritidae) attraction. *Can. Entomol.* **103,** 284 (1971).
1316. Wright, R. H., Chambers, D. L., and Keiser, I. Insect attractants, antiattractants, and repellents. *Can. Entomol.* **103,** 627 (1971).
1317. Wurzell, B. Unnatural selection. *Sunday Times (London)*, *Mag. Sect.* pp. 72 and 75 (1970).
1318. Yamada, M. Extracellular recording from single neurones in the olfactory centre of the cockroach. *Nature (London)* **217,** 778 (1968).
1319. Yamada, M. Electrophysiological investigation of insect olfaction. *In* "Control of Insect Behavior by Natural Products" (D. L. Wood, R. M. Silverstein, and M. Nakajima, eds.), p. 317. Academic Press, New York, 1970.
1320. Yamada, M., Ishii, S., and Kuwahara, Y. Preliminary report on olfactory neurons specific to the sex pheromone of the American cockroach. *Bochu-Kagaku* **33,** 37 (1968).
1321. Yamada, M., Ishii, S., and Kuwahara, Y. Odour discrimination: "Sex pheromone specialists" in the olfactory lobe of the cockroach. *Nature (London)* **227,** 855 (1970).
1322. Yamamoto, R. Collection of the sex attractant from female American cockroaches. *J. Econ. Entomol.* **56,** 119 (1963).
1323. Yinon, U., and Shulov, A. A bioassay of the pheromone of *Trogoderma granarium* males as an attractant for both sexes of the species. *Entomol. Exp. Appl.* **10,** 453 (1967).
1324. Yinon, U., and Shulov, A. New findings concerning pheromones produced by *Trogoderma granarium* (Everts), (Coleoptera, Dermestidae). *J. Stored Prod. Res.* **3,** 251 (1967).
1325. Yinon, U., and Shulov, A. Response of some stored-product insects to *Trogoderma granarium* pheromones. *Ann. Entomol. Soc. Amer.* **62,** 172 (1969).

1326. Yinon, U., Shulov, A., and Ikan, R. The olfactory responses of granary beetles towards natural and synthetic fatty acid esters. *J. Insect Physiol.* **17**, 1037 (1971).

1327. Yoshio, T., Noguchi, H., and Yashima, T. Artificial control of mating activity of the smaller tea tortrix, *Adoxophyes orana*, and a quantitative bioassay for the sex pheromone. *Bochu-Kagaku* **34**, 107 (1969).

1328. Zajic, J. E., and Kuehn, H. H. Influence of insect lures on hyphal growth and sporulation of *Choanephora trispora*. *Can J. Microbiol.* **7**, 807 (1961).

1329. Zayed, S. M. A. D., and Hussein, T. M. Über die Lipide aus den Hinterenden der weiblichen Schmetterlinge *Prodenia litura* F. *HoppeSeyler's Z. Physiol. Chem.* **341**, 91 (1965).

1330. Zayed, S. M. A. D., Hussein, T. M., and Emran, A. The attraction of male moths of the cotton leaf worm by traps baited with living females. *Z. Angew. Entomol.* **66**, 52 (1970).

1331. Zayed, S. M. A. D., Hussein, T. M., and Fakr, I. M. Über den Sexuallockstoff des Baumwollwurms (*Prodenia litura* F.) *Z. Naturforsch. B* **18**, 265 (1963).

1332. Zehmen, H. von. Ein Beitrag zur Frage der Anlockstoffe weiblicher Falterschädlinge. *Zentrbl. Ges. Forstw.* **68**, 57 (1942).

1333. Zethner-Moller, O., and Rudinsky, J. A. Studies on the site of sex pheromone production in *Dendroctonus pseudotsugae* (Coleoptera: Scolytidae). *Ann. Entomol. Soc. Amer.* **60**, 575 (1967).

1334. Zmarlicki, E., and Morse, R. A. Drone congregation areas. *J. Apicult. Res.* **2**, 64 (1963).

1335. Zmarlicki, E., and Morse, R. A. Queen mating. Drones apparently congregate in certain areas to which queens fly to mate. *Amer. Bee J.* **103**, 414 (1963).

1336. Zurflüh, R., Dunham, L. L., Spain, V. L., and Siddall, J. B. Synthetic studies on insect hormones. IX. Stereoselective total synthesis of a racemic boll weevil pheromone. *J. Amer. Chem. Soc.* **92**, 425 (1970).

1337. Zwick, R. W., and Peifer, F. W. Observations on the emergence and trapping of male *Pleocoma minor* Linsley with black light and female-baited traps. *PanPac. Entomol.* **41**, 118 (1965).

# SUBJECT INDEX

## A

*Abelia grandiflora*, attraction for *Tricho-*
*plusia ni*, 145
*Acanthomyops claviger*, 60
  scent glands, 60
*Acanthoscelides obtectus*, 57–58
  pheromone, collection of, 58, 193
    identification of, 194
    isolation of, 58, 193–194
Acarina, 5, 55, 122, 165
*d*-10-Acetoxy-*cis*-7-hexadecen-1-ol, 188, 226
*d*-12-Acetoxy-*cis*-9-octadecen-1-ol, 226
*Acherontia atropos*, 63
  scent glands, 63
*Acheta domesticus*, 71
*Achroia grisella*, 27, 38, 63–64
  pheromone, bioassay of, 133
    collection of, 170–171
    in control, 278
    identification of, 171
    production of, 161
  scent glands, 97
*Achroia* sp., 27
*Acleris glomerana*, 27
*Acrolepia assectella*, 27

pheromone, production of, 82
  in survey, 243–244
  trapping, 243–244
*Acronicta psi*, 27
  scent glands, 81
*Actias selene*, 27
*Actias villica*, 28, 43
*Adoxophyes fasciata*, 28
  pheromone, collection of, 171
    identification of, 171
    production of, 150
    in survey, 244
*Adoxophyes orana*, 28, 210
  pheromone, bioassay of, 28, 133–134
    collection of, 171
    identification of, 171
    production of, 158
    in survey, 244
    synthesis of, 210, 229
African migratory locust, *see Locusta*
  *migratoria migratorioides*
*Agathymus baueri*, 28
*Agathymus polingi*, 28
*Aglia tau*, 28
  pheromone, bioassay of, 28
  scent glands, 28, 81